Molecular Genetics of Inflammatory Bowel Disease

Mauro D'Amato • John D. Rioux
Editors

Molecular Genetics of Inflammatory Bowel Disease

Springer

Editors
Mauro D'Amato
Department of Biosciences
and Nutrition
Karolinska Institutet
Stockholm, Sweden

John D. Rioux
Department of Medicine & MHI
Research Center
Université de Montréal and the Montreal
Heart Institute
Montreal, QC, Canada

ISBN 978-1-4939-4256-5 ISBN 978-1-4614-8256-7 (eBook)
DOI 10.1007/978-1-4614-8256-7
Springer New York Heidelberg Dordrecht London

Printed on acid-free paper

Springer is part of Springer Science+Business Media (www.springer.com)

Preface

If you are picking up this book for the very first time you might be asking yourself why it is entirely dedicated to the topic of the *Molecular Genetics of IBD*—or more specifically, why IBD and why now? The simplest answer is that research in Crohn's disease (CD) and ulcerative colitis (UC), together known as the inflammatory bowel diseases (IBD), has truly seen a revolution in the last 5–10 years. For example, in 2005 there were less than a handful of validated genetic risk factors known for IBD, while in 2013 that number increased to 163!

In the following chapters you will discover how these genetic discoveries have led to the identification of biological functions not previously associated with IBD pathophysiology (e.g., autophagy), how multiple genetic risk factors for IBD converge on given biological functions and that together the identified variants in these genes have predisposing and protective roles (e.g., the multiple variants in the receptor for the IL23 cytokine and its signaling cascade), and how having such a large number of known genetic risk factors has changed our understanding not only about the genetic and molecular overlap between CD and UC but also between these diseases and other chronic inflammatory diseases (e.g., psoriasis, multiple sclerosis, type 1 diabetes, and many others).

While it is clear that the genetic makeup of an individual is tremendously important in determining whether they will develop CD or UC, nongenetic risk factors play an equally important role. Of these environmental risk factors, two have predominated in the field of IBD: smoking and the gut microbiome. Although much work needs still to be done, great advances have been made recently in our ability to assess and understand the impact of environmental risk factors, be it at the level of how these environmental risk factors lead to structural changes to the DNA or their influence on the local gut milieu which can be considered not only as a meeting point for gut flora, epithelial cells, and immune cells but also as an important interface between genetic and nongenetic risk factors.

Ultimately, we all hope, these advances in our knowledge will lead to better diagnosis, prognosis and treatment of these debilitating, lifelong diseases whose prevalence can be as high as one in every 150 individuals[1]. At a conceptual level, it is easy to understand that different patients have developed IBD due to different combinations of genetic and nongenetic risk factors. It then follows that a better understanding of these individual differences will move the field forward in its goals for better clinical management of IBD patients. Part of the answer lies in the identification and understanding of the key biological mechanisms/pathogenic pathways that lead to IBD, as this will help in the stratification of this complex mix of IBD patients and to tailor the development and application of therapies accordingly. As you will read, another part of the equation is that in IBD there is a multitude of informative cellular and animal models, and access to relevant clinical tissues (e.g., immune cells, gut biopsies), that give a certain advantage over multiple other diseases in terms of resources to better understand disease mechanisms as well as to develop, screen and evaluate novel therapies for IBD. Finally, as you progress through this book, each chapter written by experts in the field, we hope that you will share our optimism that IBD is well positioned to deliver on the promise of translating basic research into a better clinical management of IBD, and hopefully may even plant the seeds that will lead to achieving a more long-term and audacious goal of preventing and curing these chronic diseases once and for all.

Stockholm, Sweden — Mauro D'Amato
Montreal, QC, Canada — John D. Rioux

[1]Rocchi et al. (2012) Can J Gastroenterol 26(11):811–817.

Acknowledgments

We wish to express our gratitude to the colleagues and friends who provided their expert contribution to our project, this book could only take shape thanks to their scientific input and advice. Evangelia Papadaki's assistance in proofreading the book is also greatly acknowledged. Mauro D'Amato's work in IBD is supported by funds from the Swedish Research Council (Vetenskapsrådet), the AstraZeneca—Karolinska Institutet Translational Research Program, the EU IBD-Character Consortium, Karolinska Institutet's faculty and KID-funding, and the Olle Engkvist Foundation. John D. Rioux's work in IBD is supported by funds from the Canadian Institutes of Health Research (CIHR), the U.S. National Institutes of Health (NIDDK and NHGRI), the Canadian Foundation for Innovation, and the Crohn's and Colitis Foundation of Canada (CCFC). Dr. Rioux is the holder of the Canada Research Chair in Genetics and Genomic Medicine.

Contents

Contributors

Jeffrey C. Barrett, B.S., D.Phil. Wellcome Trust Sanger Institute, Wellcome Trust Genome Campus, Hinxton, Cambridge, CB, UK

Stephan Beck, M.Sc., Ph.D. UCL Cancer Institute, London, UK

Atul K. Bhan, M.B.B.S., M.D. Department of Pathology, Massachusetts General Hospital, Boston, MA, USA

Adam B. Castoreno, Ph.D. Gastrointestinal Unit, Center for the Study of Inflammatory Bowel Disease, and Center for Computational and Integrative Biology, Massachusetts General Hospital, Boston, MA, USA

Broad Institute of Harvard University and Massachusetts Institute of Technology, Cambridge, MA, USA

Isabelle Cleynen, Ph.D. Department of Clinical and Experimental Medicine, TARGID, KU Leuven, Leveun, Belgium

Division of Gastroenterology, University Hospitals Leuven, Leuven, Belgium

Mauro D'Amato, Ph.D. Department of Biosciences and Nutrition, NOVUM, Karolinska Institutet, Stockholm, Sweden

Mark J. Daly, Ph.D. Simches Research Center, Massachusetts General Hospital, Boston, MA, USA

Analytic and Translational Genetics Unit, Department of Medicine, Massachusetts General Hospital, Boston, MA, USA

Matthias A. Engel, M.D. Division of Gastroenterology, First Department of Medicine, Universitätsklinikum Erlangen, Friedrich-Alexander-Universität Erlangen-Nurnberg, Erlangen, Germany

Andre Franke, Ph.D. Institute of Clinical Molecular Biology, Christian—Albrechts—University of Kiel, Kiel, Schleswig-Holstein, Germany

Robert Häsler, Ph.D. Institute of Clinical Molecular Biology, Christian-Albrechts-University of Kiel, Kiel, Germany

Jonas Halfvarson, M.D., Ph.D. Department of Internal Medicine, Division of Gastroenterology, Örebro University Hospital, Örebro, Sweden

Faculty of Health and Medical Sciences, Örebro University, Örebro, Sweden

Maria Kaparakis-Liaskos, Ph.D. Centre for Innate Immunity and Infectious Diseases, Monash Institute of Medical Research, Monash University, Clayton, VIC, Australia

Stefan Koch, Ph.D. Department of Pathology and Laboratory Medicine, Epithelial Pathobiology Unit, Emory University, Atlanta, GA, USA

Division of Molecular Embryology, German Cancer Research Center (DKFZ), Heidelberg, Germany

Petric Kuballa, Ph.D. Gastrointestinal Unit, Center for the Study of Inflammatory Bowel Disease, and Center for Computational and Integrative Biology, Massachusetts General Hospital, Boston, MA, USA

Broad Institute of Harvard University and Massachusetts Institute of Technology, Cambridge, MA, USA

Catherine Labbé, M.Sc., Ph.D. Montreal Heart Institute, Montréal, Québec, Canada

Université de Montréal, Montréal, Québec, Canada

Charlie W. Lees, M.B.B.S., M.R.C.P., Ph.D. Gastrointestinal Unit, Western General Hospital, Edinburgh, UK

Nouara Lhocine, Ph.D. Institut Pasteur Paris, Unité de Pathogénie Microbienne Moléculaire, Bâtiment Duclau, Paris, France

Richard P. MacDermott, M.D. Division of Gastroenterology, Department of Medicine, Albany Medical College, Albany, NY, USA

Kevin J. Maloy, Ph.D. Sir William Dunn School of Pathology, University of Oxford, Oxford, UK

Dermot P.B. McGovern, M.D., Ph.D., F.R.C.P. Cedars-Sinai Medical Center, Los Angeles, CA, USA

Emiko Mizoguchi, M.D., Ph.D. Gastroinestinal Unit, Massachusetts General Hospital, Boston, MA, USA

Atsushi Mizoguchi, M.D., Ph.D. Gastroinestinal Unit, Massachusetts General Hospital, Boston, MA, USA

Natalia B. Nedelsky, Ph.D. Gastrointestinal Unit, Center for the Study of Inflammatory Bowel Disease, and Center for Computational and Integrative Biology, Massachusetts General Hospital, Boston, MA, USA

Markus F. Neurath, M.D., Ph.D. Division of Gastroenterology, First Department of Medicine, Universitätsklinikum Erlangen, Friedrich-Alexander-Universität Erlangen-Nürnberg, Erlangen, Germany

Asma Nusrat, M.D. Department of Pathology and Laboratory Medicine, Epithelial Pathobiology and Mucosal Inflammation Research Unit, Emory University, Atlanta, GA, USA

Miles Parkes, M.A., M.B.B.S., F.R.C.P., D.M. IBD Research Unit, Department of Medicine, University of Cambridge, Cambridge, UK

Charles A. Parkos, M.D., Ph.D. Department of Pathology and Laboratory Medicine, Epithelial Pathobiology and Mucosal Inflammation Research Unit, Emory University, Atlanta, GA, USA

Dana J. Philpott, Ph.D. Department of Immunology, University of Toronto, Toronto, ON, Canada

John D. Rioux Department of Medicine & MHI Research Center, Université de Montréal and the Montreal Heart Institute, Montréal, QC, Canada

Philip Rosenstiel, M.D. Institute of Clinical Molecular Biology, Christian-Albrechts-University of Kiel, Kiel, Germany

Philippe J. Sansonetti, M.D. Institut Pasteur Paris, Unité de Pathogénie Microbienne Moléculaire, Paris, France

Stefan Schreiber, M.D. Department of General Internal Medicine, University Hospital Schleswig-Holstein, Campus Kiel, Kiel, Germany

Institute of Clinical Molecular Biology, Christian-Albrechts-University of Kiel, Kiel, Germany

Séverine Vermeire, M.D., Ph.D. Department of Clinical and Experimental Medicine, TARGID, KU Leuven, Leveun, Belgium

Division of Gastroenterology, University Hospitals Leuven, Leuven, Belgium

Ramnik J. Xavier, M.D., Ph.D. Gastrointestinal Unit, Center for the Study of Inflammatory Bowel Disease, and Center for Computational and Integrative Biology, Massachusetts General Hospital, Harvard Medical School, Boston, USA

Broad Institute of Harvard University and Massachusetts Institute of Technology, Cambridge, MA, USA

Alexandra Zhernakova, M.D., Ph.D. Department of Genetics, University Medical Centre Groningen and University of Groningen, Groningen, The Netherlands

Xinjun Cindy Zhu, M.D. Division of Gastroenterology, Department of Medicine, Albany Medical College, Albany, NY, USA

Part I
The Foundation of IBD Genetics, Human and Animal Studies

Chapter 1
A Primer on IBD: Phenotypes, Diagnosis, Treatment, and Clinical Challenges

Xinjun Cindy Zhu and Richard P. MacDermott

Abstract The inflammatory bowel diseases are chronic, relapsing disorders characterized by inflammation and ulceration in part of or the entire gastrointestinal tract. IBD affects people worldwide but is most prevalent in northern Europe and North America. Etiologically, the current consensus is that the intestinal inflammation is largely caused by an aberrant and excessive immune response to environmental triggers (intestinal bacterial infection, medications, or other agents) in genetically susceptible individuals. IBD exerts a heavy toll on patients' quality of life and imposes a considerable economic burden on the healthcare system. Some forms of IBD lead to severe complications such as formation of fistulae and intestinal strictures, for which management options are very limited. Moreover, many IBD patients develop drug tolerance and toxic responses, which severely compromise the disease control. Over the past decade, modern genetics has led to a new era of IBD research. Up to 163 genes have been identified to be associated with IBD. A growing number of new diagnostic tools and therapeutic agents resulted from the IBD genetics have already been implemented in the management of IBD; however, unmet needs persist. By utilizing new experimental tools such as powerful next-generation DNA sequencing machines along with accessing large cohorts of IBD patients, genetic studies will lead to a better understanding of the causes of individual IBD, more diagnostic tools to aid in evaluating the disease course and responsiveness to treatment, more novel targets to be identified for new drug development, and ultimately the most desirable strategies for IBD prevention.

X.C. Zhu, M.D. (✉) • R.P. MacDermott, M.D.
Division of Gastroenterology, Department of Medicine, Albany Medical College, Albany, NY, USA
e-mail: zhux@mail.amc.edu

M. D'Amato and J.D. Rioux (eds.), *Molecular Genetics of Inflammatory Bowel Disease*,
DOI 10.1007/978-1-4614-8256-7_1, © Springer Science+Business Media New York 2013

Introduction

The inflammatory bowel diseases (IBD) are idiopathic, chronic, relapsing disorders of the small and/or large bowel, characterized pathologically by inflammation and ulceration of the mucosal and submucosal layers or mucosal layer only. The principal categories of IBD are ulcerative colitis (UC), Crohn's disease (CD), and indeterminate colitis (IC) [1].

The incidence of both CD and UC has particular geographic patterns, with the highest prevalence in Western developed countries, including Europe and North America, and intermediate prevalence in countries such as Japan, Korea, Hong Kong, South Africa, and Israel. In less developed countries, the prevalence is low. The overall prevalences of UC and CD in North America are similar. It is estimated that the prevalence of UC is approximately 37–246 cases per 100,000 persons and for CD it is 26–199 cases per 100,000 persons [2]. In general, there are no gender differences in terms of the frequency of UC and CD. The onset of UC and CD occurs at any age, but the peak incidence is more likely to be around late adolescence and early adulthood. In female CD patients there is also a second peak during the sixth and seventh decades of life. Multiple studies suggest there is a tendency in early onset CD for upper GI involvement, including gastric, duodenal, and the proximal small bowel, whereas in late onset CD, colonic inflammation is more common [2].

The etiology of IBD remains unknown. The current consensus is that intestinal inflammation is largely caused by an aberrant and excessive immune response to environmental triggers (intestinal bacterial infection, medications, or other agents) in genetically susceptible individuals [3, 4]. IBD, particularly CD and to a lesser extent UC, tends to cluster in families, suggesting there is a strong genetic component to the pathology [5, 6]. Interestingly, environmental factors appear also to play an important role, as demonstrated by the fact that both CD and UC appeared to be more frequent in northern parts of the United States than in southern and in urban more than rural parts [7, 8]. Intestinal microflora may also be a contributing factor. In geographic areas of low IBD prevalence such as Asia, people tend to have the highest frequency of indigenous intestinal infections, including helminthic infestations. Recent studies have found that intestinal bacteria are altered in IBD [9]. The role of the microbiota in IBD is an interesting area worthy of intensive investigation. Other environmental factors have also been found to have an impact on IBD. For example, smoking increases the risk of CD, but intriguingly it lowers the risk of UC [10, 11]. Clinical and endoscopic recurrence in CD and disease activity after surgery were affected by active smoking, and transdermal nicotine patches have been shown to improve symptoms in UC patients [12]. The use of nonsteroidal anti-inflammatory drugs has been shown to increase disease flares in IBD patients [13]. In general, the interface between gut mucosa, luminal bacteria, environmental factors, and immune cells may determine the onset and course of disease in susceptible individuals.

Clinical Manifestations and Phenotypes of IBD

In UC, the symptoms tend to begin gradually, typically presenting with rectal bleeding, diarrhea, and abdominal cramping pain. The Mayo score system (ranging from 0 to 12) is commonly used to assess UC activity in clinical trials and is based on stool pattern, rectal bleeding, endoscopic findings, and overall assessment by a physician, with higher scores correlating with greater severity [14]. In contrast to UC, the clinical manifestations of CD are variable and sometimes insidious, featuring diarrhea and abdominal cramping pain associated with iron deficiency anemia, fatigue, weight loss, and fever [15]. Rectal bleeding is less severe in CD, except in CD affecting the colon only, also referred to as Crohn's colitis. Fistula and abscess formation preferentially occur in CD due to transmural bowel inflammation. Patients with CD affecting the upper GI tract are younger at onset, with clinical presentations of abdominal pain and cramps, nausea and vomiting, and general malaise and fatigue [16]. It can present with aphthous ulcers in the mouth and difficulty or pain in swallowing. CD is also characterized by malabsorption of bile acids, iron, calcium, water-soluble vitamins such as folic acid and vitamin B12, fat-soluble vitamins—such as vitamins A, D, E, and K—and trace minerals such as zinc, resulting in diarrhea, gallstones, iron deficiency anemia, vitamin B12 deficiency, vitamin D deficiency, hypocalcemia, and vague abdominal pain. In Crohn's ileitis, excessive oxalate is absorbed in the colon, leading to calcium oxalate kidney stone formation. Perianal complications include perirectal abscesses, anorectal fistulas, and anal fissures, all of which are accompanied by perianal pain and discomfort. Both CD and UC can have extraintestinal manifestations, and about 50–60 % of patients suffer from joint pain due to arthropathies and sacroiliitis [17], osteopenia or osteoporosis [18], skin lesions (such as erythema nodosum and pyoderma gangrenosum) [19], uveitis and iritis [20], which are often correlated with disease activity [21]. Mild abnormality in liver function tests may be present, and primary sclerosing cholangitis (PSC) occurs more frequently in UC than CD. PSC is known to be associated with an increased risk for colorectal cancer [1, 22, 23].

UC is characterized by diffuse mucosal inflammation restricted to the colon but varying in extent from the rectum to the cecum. Based on the affected anatomy, UC can be clinically divided into distal and more extensive forms. Distal UC is defined as colitis with features of inflammation confined to the rectum (proctitis) or rectum and sigmoid colon (proctosigmoiditis). The more extensive forms of UC include left-sided UC with inflammation extending up to the splenic flexure and pancolitis, with inflammation that extends proximal to the splenic flexure, affecting the entire colon. The Montreal consensus classified UC based on anatomic extent into three categories, ulcerative proctitis (E1), distal or left-sided UC (E2), and extensive UC involving the colon proximal to splenic flexure (E3) [1]. The severity of UC was also classified into mild, moderate, and severe based on daily frequency of bowel movements including bloody stool and the presence of systemic toxicity. In addition, age of onset is considered to be one of the important factors for classification of UC subtypes [1]. In contrast, CD can affect any part of the gastrointestinal (GI) tract from mouth to anus, with a characteristic feature of patchy and transmural inflammation. It may be further defined based on the location of inflammation (terminal ileal (L1), colonic (L2), ileocolic (L3), isolated upper GI (L4)) or the pattern of clinical complications (non-stricturing and non-penetrating (B1), structuring (B2), penetrating (B3), or perianal disease (p)) [1, 24–27]. Both L4 and p can coexist with L1–L3 and B1–B3 disease. Additionally, by considering age at diagnosis as a risk factor for developing severe complications, CD is also classified into A1 (<16 years old), A2 (17–40 years old), and A3 (>40 years old) [1]. Although the majority of CD patients (approximately 80 %) have small bowel involvement, 20 % have inflammation limited to the colon only [24, 28]. In contrast to the invariable involvement of the rectum in patients with UC, the rectum is spared in 50 % of CD patients [24]. Further subdivision of CD is as follows: about 50 % of patients have ileocolitis, which refers to involvement of both the ileum and colon, and 14–30 % have disease only involving the distal ileum (ileitis). About 37 % of patients with Crohn's colitis have perianal inflammation, such as perianal CD and perianal fistula formation [29]. Less than 5 % of patients have involvement of the upper GI tract, including the mouth, esophagus, gastroduodenal area, and the proximal small bowel [24, 28]. It is worth noting that about 5–10 % of IBD patients are unclassifiable, referred to as IC, and they have pathological features of both UC and CD. IC occurs more often in children (12.7 %) than in adults (6.0 %) [30–33].

Other chronic, nonspecific colitides include microscopic colitis (either collagenous or lymphocytic), in which the colon appears normal, but lymphocyte infiltrations are present on biopsy [34]; diversion colitis, which occurs in the part of the colon that is excluded from the fecal stream [35]; diverticular colitis, which is limited to portions of the colon with diverticula present [36]; and pouchitis, which occurs in nearly half of patients with UC who have undergone ileal pouch-anal anastomosis [32].

Diagnosis and Management of UC and CD

The diagnosis of IBD can be made by clinical history and physical examination in combination with radiologic, endoscopic, and histologic findings [21]. Endoscopy and histology are considered the gold standard for diagnosing IBD, monitoring the effectiveness of treatment and relapses of disease, and IBD-related cancer surveillance [37]. The severity of UC can be determined by colonoscopy as mild (duller, redder mucosa with a "granular" or fine sandpaper-like texture, and decreased vascular pattern), moderate (gross pitting mucosa with friability), and severe (diffuse ulceration with mucopurulent exudate and spontaneous hemorrhage) (see Fig. 1.1a–c) [1]. The histology of UC includes crypt distortion, crypt atrophy, distorted or branched glands, and neutrophilic microabscesses inside the lumen of crypts. UC also displays diffuse lamina propria inflammation, caused by increased acute and chronic inflammatory cells. Basal plasmacytosis and basal lymphoid hyperplasia are apparently unique to UC [37–39].

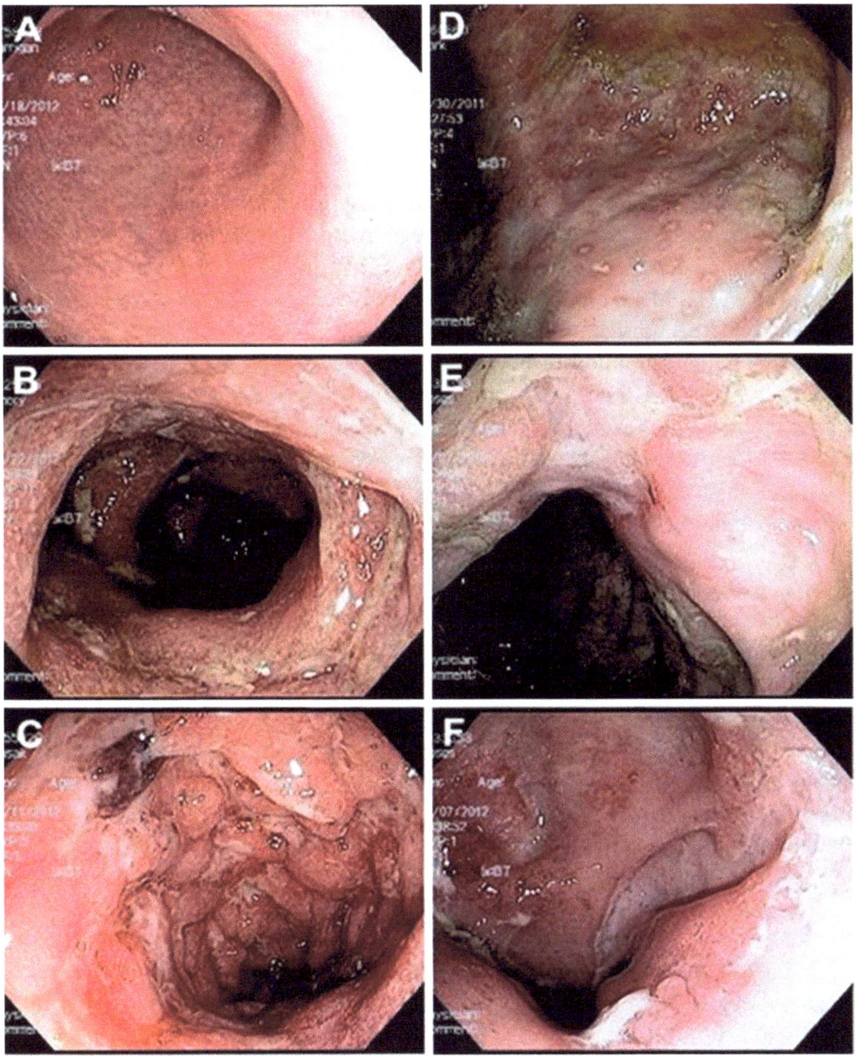

Fig. 1.1 Endoscopic features of UC and CD. Colonoscopic examination determines the severity of UC and CD. (**a**–**c**) UC can be graded as mild (duller, redder mucosa with granular texture) (**a**), moderate (mucosa with ulceration, friability, and loss of normal vascular pattern) (**b**), and severe (diffuse ulceration with mucopurulent exudates and spontaneous hemorrhage) (**c**). (**d**–**f**) CD can be graded as mild (discrete pouched-out aphthous ulcers) (**d**), moderate (stellate ulcers and longitudinal ulcers) (**e**), and severe (macroulcerations and pseudopolyps) (**f**)

Pathological changes in the inflammatory intestine of CD depend on disease severity, ranging from discrete pouched-out aphthae, irregular stellate ulcers, and longitudinal ulcers to macroulcerations and pseudopolyps (see Fig. 1.1d–f). Endoscopic indices, such as Rutgeerts' score, have been used for grading disease severity following ileocolonic resection. Histologically, CD is characterized by patchy, segmental, and transmural inflammation, consisting of small collections of polymorphonuclear cells and chronic inflammatory cells. The hallmark of CD is the presence of epithelioid granuloma. In addition, serum markers of inflammation, such as ESR, CRP, and platelet count, are also used for monitoring disease activity [40].

Therapy for IBD is a fast-evolving field and new agents are continuously emerging. The primary aims of medical treatment for UC and CD are to control inflammation and reduce symptoms, achieve steroid-free clinical remission, and if possible achieve mucosal healing. The choices of therapy for UC and the mode of delivery of medications depend largely on clinical severity and anatomic extent of the disease (Table 1.1). Mildly to moderately active UC can be treated with 5-aminosalicylic acid (5-ASA) derivatives. Oral and topical mesalamine are effective in inducing and maintaining remission in distal UC [41]. An additive benefit is achieved in patients with distal UC who received the combination of topical rectal mesalamine (4 g rectal enema once nightly) and oral mesalamine (2.4 g/day), which produced results similar to those achieved with a higher dose of oral mesalamine (4.8 g/day) [42]. Corticosteroids are only used for controlling flare of UC in patients with more severe symptoms, but are not recommended for long-term use [43]. Moderate to severe UC can be treated with steroid-free regimens such as immunomodulatory agents (AZA or 6-MP) or biologics (such as anti-TNF monoclonal antibodies) or both. AZA and 6-MP have slow onset of action (3–6 months) and are associated with severe adverse events which limit their use, including bone marrow suppression, infection, hepatotoxicity, pancreatitis, and malignancies, particularly hepatosplenic T-cell lymphoma. Two FDA-approved anti-TNF monoclonal antibodies, infliximab and adalimumab, are currently used for the treatment of moderately to severely active UC in adults [44, 45]. Cyclosporine can be used as a salvage therapy for severe and refractory UC for less than 3–6 months and as a bridge for thiopurine therapy [46]. Additional new drugs, such as Tofacitinib, an anti-JAK antibody [47], are currently under development (Table 1.1).

The choice of CD therapy also depends on the anatomic location and severity of the disease [48]. The Crohn's Disease Activity Index (CDAI) is mostly used to evaluate disease severity in clinical trials as follows: asymptomatic remission (CDAI <150), mild to moderate (CDAI 150–220), moderate to severe (CDAI 220–450), and severe-fulminant disease (CDAI >450). A drop in the CDAI of 70 points is considered to be responsiveness. Current medications include oral 5-ASA derivatives, antibiotics, glucocorticoids, nonsystemic glucocorticoids, immunomodulators, and biologic therapies.

The use of 5-ASA medications for CD is not as effective as in UC patients. Generally 5-ASA drugs are used to treat mild ileitis, ileocolitis, or Crohn's colitis. Antibiotics such as metronidazole, ciprofloxacin, and rifaximin are recommended as an adjuvant therapy for active luminal Crohn's colitis, perianal fistulizing CD (especially when requiring draining of abscess), postoperative recurrence of ileocolitis, UC, and pouchitis [49, 50]. Short-course use of oral or intravenous glucocorticoids with tapering is often used to treat patients with moderate to severe disease at the initial presentation or during flares [51]. A steroid-sparing regimen should be used concomitantly with the steroid and continued as maintenance therapy once

Table 1.1 Drugs used in the treatment of IBD

Medications	Clinical applications	References
5-ASA	First line therapy for remission and maintenance in mild UC	[41]
Mesalazine	High dose 4.8 g/day, the faster resolution of symptoms of UC	[41]
Sulfasalazine	May reduce the risk of colorectal cancer by up 75 %	[42]
Topical 5-ASA	Combination with oral 5-ASA has been shown to be more effective in UC	
Metronidazole	Following ileocecal resection, 20 mg/kg/day for 3 months reduces risk of recurrence of CD	[49]
	Treat perianal disease and pouchitis	[50]
Ciprofloxacin	Greater benefit than metronidazole in perianal disease and pouchitis	
Corticosteroids	Potent reagent for inducing remission in moderate to severe UC and CD	[51]
	Rectal steroids are effective adjuvants but less effective than topical 5-ASA	[52]
Budesonide	Therapeutic benefit in ileocecal CD or UC	[53]
Thiopurines	Moderate to severe UC with clinical and endoscopic remission in 53 % patients in UC	[55]
	Effective for both induction and maintenance of remission in moderate and severe CD	[56]
	Modestly prevent postoperative recurrence of CD (at 1 year 8–13 % and 15 % for clinical and endoscopic remission, respectively)	[46]
Methotrexate	Effective for the induction and maintenance of remission in CD in RCT	
Cyclosporine	Rapid effective as a salvage therapy for severe and refractory UC for less than 3–6 months and as a bridge for thiopurine therapy	
Infliximab	Moderate to severe refractory CD with 81 % response rate at 4 weeks and 48 % at week 12	[44]
	Efficacy for fistula closure is 36 %	[45]
	Effective in severe UC, corticosteroid-refractory UC	[59]
Adalimumab	Moderate to severe refractory CD in TNF naïve patients and those who failed infliximab	[57]
	Efficacy for fistula closure is 33 % at week 56 compared with 13 % given placebo	
	Effective to moderate to severe UC and intolerance to infliximab	
Certolizumab	Effective in induction and maintenance of response and remission in complicated CD	[57]
Natalizumab	Moderate to severe refractory CD who failed IFX therapy	[62]

remission is achieved. The nonsystemic glucocorticoid budesonide may be effective for short-term (up to 6 months to 1 year) maintenance of remission in mild to moderate ileitis or ileocolitis [52, 53]. Patients with severe forms of CD at initial presentation, who relapse or fail to respond, or who exhibit steroid dependency usually require a top-down therapy starting with immunomodulators (6MP, AZA, or methotrexate) or biologics (infliximab, adalimumab, certolizumab). 6MP and AZA are effective in inducing remission and maintenance [54, 55] but require 4–6 weeks due to their long half-life. Methotrexate (25 mg/week intramuscularly) is effective for induction and maintenance of remission in CD [57]. All anti-TNF agents are efficacious in induction and maintenance of remission in patients with active luminal CD and penetrating diseases. For perianal diseases, including abscess and fistula formation, combination therapy should be recommended including metronidazole and ciprofloxacin [49, 50] and infliximab, adalimumab, and certolizumab [57, 58, 59], in addition to surgical intervention. To confirm loss of response, patients should undergo endoscopic or radiologic imaging to confirm the presence of active inflammation and to rule out other causes of symptoms such as infection. There are increasing percentage of *Clostridium difficile* infections and hospitalizations among IBD patients [60].

Severe forms of both CD and UC are managed medically using similar combinatorial therapies, including immunomodulators, biologics, or both. Disease relapse in both conditions is fairly common. It is recommended to measure infliximab and human anti-chimeric antibody (HACA) concentrations in patients with disease relapse for therapy stratification. When detectable HACA is present, patients should be switched to another anti-TNF medication. When a subtherapeutic anti-TNF level is detected, the drug dose should be escalated to achieve clinical response. For active CD patients with nonresponse, loss of response, or intolerance to infliximab, switching to adalimumab has shown some efficacy [61]. Natalizumab, an anti-α4 integrin antibody, has shown to be effective in treating moderate to severe refractory CD who failed anti-TNF therapy [62].

If patients develop toxic megacolon with uncontrolled active inflammation, demyelinating disease, congestive heart failure, drug-induced lupus, drug-induced psoriasis, vasculitis, or various infections, including hepatitis, viral infection, and granulomatous infections (tuberculosis, histoplasmosis), it is recommended to change to a drug of a different category or discontinue medical management.

Total colectomy can be considered particularly in patients with moderately to severely active disease who are refractory or intolerant to available medical therapies. However, disease recurrence postoperatively is common in CD and in UC with pouchitis. Anti-TNF medications have been shown to prevent postoperative recurrence of CD, as evidenced by less clinical and endoscopic relapse [63].

Progression of IBD to complications is common. Such complications are stricture, fistula, abscess, malnutrition, surgical resection-related complications, infection, and depression. Therefore, IBD management requires a coordinated effort involving specialists from multiple disciplines, including, but not limited to, surgeon, internist, nutritionist, and psychiatrist.

Management of IBD remains a formidable challenge. Although mild CD can be controlled by conventional medical therapy, most cases of IBD inevitably progress to more severe forms with complications or become tolerant to current treatment regimens and require aggressive therapy such as biologics. However, about 50–60 % of patients respond to aggressive therapies at the beginning and only 30–40 % of patients remain in remission.

Understanding Genetics in IBD Facilitates Coping with Clinical Challenges

Despite much progress in IBD over the past decade, various challenges are still encountered in managing IBD patients to achieve accurate diagnosis, effective control of inflammation, and prevention of severe complications. The goals of understanding the genetics of IBD are to identify those at risk for IBD, to provide tools for diagnosis, to evaluate disease course and responsiveness of treatment, to develop new therapeutics, and to prevent disease from developing.

First of all, there is a need for additional diagnostic tools to assist separation of IBD subtypes. Classification within IBD is required for aiding patient counseling, prediction of disease progression, and ultimately delivery of optimal therapy to the individual patient. Though UC and CD have relatively distinct pathological and clinical characteristics and perhaps unique etiologies, making an accurate diagnosis is sometimes still difficult in IC cases. Moreover, IBD can be further categorized into subtypes based on the location of inflammation, onset of age, and the pattern of clinical complications. Therefore, understanding the genetic determinants underlying the predisposition to clinical phenotypes would also be extremely desirable for guiding treatment and prognosis.

Advances in medical genetics and human genetics, particularly over the past decade, have been truly phenomenal. Genetic epidemiological studies have revealed significantly higher risk among relatives in both UC and CD, suggesting that genetic factors play an important role in the pathogenesis of IBD. This is further exemplified by the fact that Ashkenazi Jews tend to have an overall increased risk for IBD, and there is concordance for IBD between monozygotic twins. In addition, more than 100 loci have been found to be significantly associated with IBD from genome-wide association [64]. These genes are involved in a diverse array of functions, including microbial recognition, lymphocyte activation, survival and proliferation, T-cell activation, IL-7 receptor, cytokine signaling, autophagy, and intestinal epithelial defense [65, 66].

Many IBD susceptibility genes discovered from GWAS are associated with both UC and CD. Interestingly, genetic studies have also uncovered that some of these genes are selectively associated with subtypes of IBD [67, 68]. It is conceivable that these genes could be used as diagnostic markers to aid clinicians in differentiating IBD subtypes from each other.

About 50 % of patients with IC will be eventually diagnosed with either CD or UC. The remaining cases are still undetermined. Therefore it is of great importance to develop a diagnostic test using genetic tools to be able to categorize phenotypic IC as it occurs more frequently among children who tend to develop a much more severe course with a greater chance of requiring colectomy and with pouch failure. In a recent study, several variants in genes, including *ICAM1*, *BTNL2*, and *SH2B1*, were found to be closely associated with IC at a young age [69]. Further study is needed to determine if this finding can be extended to a larger population of IC patients.

Second, it is of importance to stratify risk factors with a goal of appropriate care of patients with IBD to prevent severe complications and delay the time period for requiring surgery. These risk factors include age of onset, involvement of ileum, and

Table 1.2 Gene variants and clinical phenotypes of early-onset IBD

Genes variants	Characteristics	Mechanisms of action	References
OCTN SLC22A4 1672T, SLC22A5-207	Pediatric onset CD; mean 12 years old	Transport organic cations, e.g., carnitine	[71]
IBD5 risk allele	Pediatric onset CD	A region containing immunoregulatory genes	[72]
(*IRF1, OCTN1, OCTN2, PDLIM 4, P4HA2*)	Extensive inflammation	(*IL4, Il 13, IL5* and *IRF1*)	[73]
IL 10 RA rs2228054 and rs2228055	Infantile UC and severe arthritis	Cytokine signaling	[74]
IL 10 RB SNPs	Infantile CD	Cytokine signaling	[74]
IL 10 and IL 10R deficiency	Infantile CD	Cytokine signaling	[75]
NCF2 c.113 G/A R38Q	Infantile CD (L2, L3 and p)	A component of NADPH oxidase complex	[76]
XIAP p.C203Y	Infantile CD; fistulizing CD	Activation of NFκB	[77]
IRGM rs1000113 and rs4958847	Childhood CD	Autophagy pathway	[78]
NOD2/CARD15	Colonic CD, a higher male/female ratio	Microbial recognition	[70]
DLG5 rs2165047	Pediatric CD (<19 years old)	Intestinal epithelial permeability	[79]

disease behaviors. Up to a quarter of patients develop IBD prior to 18 years of age. Despite many similarities of IBD features between adult onset and early onset, early-onset IBD tends to be much more severe with more rapid progression and a higher risk of complications. Pediatric IBD has less involvement of the rectum but more frequent inflammation in the upper GI tract such as the stomach and duodenum. Very-early-onset CD occurs at ages younger than 8 years, featuring less perianal disease, a higher male-to-female ratio, higher anti-*Saccharomyces cerevisiae* antibodies, and seropositivity rates, and preferentially in Jewish individuals [70]. The exact genetic factors that contribute to early susceptibility to IBD remain unknown. Several variants of genes appear to be uniquely associated with early-onset IBD by GWAS and candidate-gene analysis (Table 1.2) [70–79]. These genes govern various biological functions including cytokine signaling, neutrophil and macrophage phagocytosis, involvement of the proinflammatory response and activation of NFκB, autophagy, microbial recognition, and intestinal epithelial permeability. Among these gene candidates, several research groups have repeatedly demonstrated that IL-10 and IL-10 receptor deficiency are associated with early onset of severe forms of IBD [74, 75]. Thus, IL-10 and IL-10R can potentially serve as important genetic indicators for initiation of aggressive treatment.

CD affecting the ileum typically displays a unique clinical phenotype associated with developing severe complications such as formation of fistulae and intestinal stricture. There are very limited options available for effective management of these pathologies. In most cases, surgery becomes inevitable to attenuate disease progression after an average of 7–15 years from diagnosis. It is therefore beneficial to initiate aggressive treatment during the early course of disease or postoperatively, if severe complications are anticipated. Genetics has been shown to play a role in the susceptibility to ileal CD and development of severe complications. Several gene variants have been discovered to be closely associated with ileal CD (Table 1.3), including *NOD2* 1007fs, *ATG16L1* rs2241879 and rs2241880, *IRGM* rs4958847, calcium-activated potassium channel 4 (*KCNN4*) rs2306801, *AK097548* gene rs1363670 G, *IL-10* promoter 627 CA, and *TCF-4* rs3814570 [67, 80–85]. Other gene variants have been linked to structuring disease, such as *NOD2* 1007fs, *ATG16L1* rs2241879 and rs2241880, *AK097548* gene rs1363670 G, *TCF-4* rs3814570, *IL-10* promoter 627 CA, *CXCL16* p.Ala181Val, and *TGFB1* codon 25 [80, 81, 83–90]. Perianal diseases including fistulae formation have distinct gene variants such as *NCF4*, *XIAP*, *IRGM* rs4958847, *NOD2* 1007fs, *DLG5* rs2165047, the carnitine/organic cation transporter (*OCTN*) on 5q31 (*IBD5*), and *CDKAL1* rs6908425 [76–79, 83, 91–94]. Interestingly, some of these gene variants associated with ileal CD were found to overlap with those linked to either stricturing or perianal diseases, suggesting that these genes perhaps reflect severe clinical phenotypes, a connection which might be exploited for clinical application. These gene variants associated with structuring or penetrating disease are fairly distinctive. However, it requires verification in a larger population of IBD patients.

Genetic testing has begun to provide a new approach to better determine the subtypes in IBD patients. For example, mutation of the *NOD2* gene is closely associated with CD, and genetic testing for *NOD2* mutations (or variants) is already available. However, the challenge is how to best utilize these tests for the benefit of patient care in general. Apparently, monitoring the *NOD2* gene alone is not sufficient as a diagnostic test because 70 % of CD patients have no *NOD2* major variants. Moreover, around 10 % of the healthy population carries *NOD2* major variants. Therefore use of a panel of genes in genetic testing may be the better molecular methods for IBD diagnosis. In addition to providing tools for assisting in diagnosis, clinicians may eventually utilize genetic information to guide decision-making in IBD therapy, especially to manage IBD and to avoid severe complications. For example, as noted above, *NOD2* and *IL-10*/*IL-10R* mutations tend to be associated with severe early onset of IBD with severe complications. These gene variants could be used to guide early aggressive therapy (using biologics and immunomodulators) so that severe complications such as irreversible fibrostenotic disease could be prevented.

It is worth mentioning that extraintestinal manifestations are common in patients with IBD, suggesting that IBD may represent an intestinal manifestation of syndromes with multiorgan involvement due to immunological disorders. GWAS results revealed that some genes appear to be associated with IBD along with other chronic inflammatory diseases. These gene variants are *STAT41* rs11889341, *HLA-B**27, *B**58, *HLA-DRB1*0103*, *1031 TNFA*, and *GPR35* rs3749171(Thr/Met) (Table 1.4) [95–98], which could potentially be used as genetic markers to further refine those IBD subgroups with extraintestinal symptoms and to further assist decision-making in treatment choice.

Table 1.3 Association of gene variants with different phenotypes of CD

NOD2	Ileal CD	Microbe recognition	[86]
(1007fs and other variants)			[80]
ATG16L1	Ileal CD	Autophagy	[81]
(rs2241879 and rs2241880)			[88]
AK097548 (rs1363670G)	Ileal CD	Encoding for hypothetical protein near the *IL12B gene*	[83]
TCF-4 rs3814570	Ileal CD	Wnt signaling pathway transcription factor	[85]
IL10 promoter 627 CA	Ileal CD	Cytokine	[84]
CXCL16 p.Ala181Val	Stricturing behavior	A chemokine for defense against bacteria	[89]
TGF beta1 codon 25	Stricturing behavior	Growth factor	[90]
Penetrating (B3)			
NCF4	Perianal CD	A component of NADPH oxidase complex	[91]
NCF2 c.113 G/A R38Q	Infantile CD with penetrating and perianal Disease	A component of NADPH oxidase complex	[76]
XIAP p.C203Y	Infantile CD with fistulizing CD	Activation of NFκB	[77]
IRGM rs4958847	Ileocolonic CD, frequent surgery and perianal fistula	Autophagy	[92]
OCTN, 5q31 (IBD5)	Perianal and penetrating CD	The carnitine/organic cation transporter	[93]
TNF2 allele	Penetrating disease	Cytokine signaling	[94]
CDKAL1 rs6908425	Development of perianal fistula	A methylthiotransferase modifying tRNA-Lysine	[83]
Perianal (p)			
DLG5 rs2165047	Pediatric-onset CD and perianal disease	Intestinal epithelial permeability	[79]
OCTN on 5q31 (IBD5)	Perianal and penetrating CD	The carnitine/organic cation transporter	[93]
NCF2 c.113 G/A R38Q	Infantile CD with penetrating and perianal disease	A component of NADPH oxidase complex	[76]
NCF4	Fistula formation	A component of NADPH oxidase complex	[91]
IRGM rs4958847	Ileocolonic CD and frequent surgery; perianal fistula	Autophagy	[78]
OCTN, 5q31 (IBD5)	Perianal and penetrating CD	The carnitine/organic cation transporter	[93]
CDKAL1 rs6908425	Perianal fistula	A methylthiotransferase modifying tRNA-Lysine	[83]

Table 1.4 Association of related genes and extrainstestinal manifestations in IBD

Genes involved	Characteristics	Mechanisms of action	References
STAT41 rs11889341	Joint pain	Signal transduction and activation of transcription	[95]
*HLA-B*27, B*58*	Uveitis	Autoimmunity	[96]
*HLA-DRB1*0103*	Uveitis	Autoimmunity	[96]
−1031 TNF-alpha	Erythema nodosum	Cytokine	[96]
GPR35 rs3749171 (Thr/Met)	UC and PSC	G-protein coupled receptor signaling	[97]

Thirdly, as managing IBD requires long-term medical therapy, drug tolerance and toxicity clearly become a major issue that severely compromise inflammation control. Knowledge of the genetics of IBD has been increasingly translated into clinical applications (pharmacogenetics). Recent studies revealed that genetic determinants play a role in unresponsiveness to drugs [79, 99, 100]. For example, variants in genes such as *HLA-DR8, ILRA, and NALP1* (*L155H*) are linked to treatment failure of budesonide and intravenous use of steroid. In contrast, *IL23R* genotype status along with disease activity and antineutrophil cytoplasmic antibody (ANCA) positivity are good indicators closely correlating with infliximab response in UC. It is hoped that genetic studies in the near future will pinpoint the genetic determinants underlying drug responsiveness and toxicity in individuals to enable choice of the most appropriate medical regimen or prompt switching to another class of medications.

Lastly, genetic studies have identified many genes associated with IBD, which potentially provide many new therapeutic targets. Physicians continue to face many challenges in managing severe IBD because of a lack of effective therapeutics and patients developing tolerance to conventional therapeutic regimens. Some severe forms of IBD require aggressive treatment, for example, biologics. However, only a portion of patients display a positive response to currently available medications and remain in remission. There is a pressing need for new effective therapeutics. By characterizing the altered intracellular pathways caused by candidate genes identified from genetic studies, new drugs could be developed to target specific subtypes of IBD.

Discovery of a large array of candidate genes associated with IBD from genetic studies will help to interpret the diverse clinical phenotype of IBD, including severity, and drug responsiveness, which will lead to many utilitarian improvements such as diagnostic tools, prognostic indicators, and more refined therapeutic regimens. However, challenges remain, for example, how to sort out the long list of IBD susceptibility genes and translate the genetic discoveries into real clinical applications. Of particular note, loss-of-function mutations in many IBD susceptibility genes do not seem to be sufficient to cause IBD. This supports the notion that IBD disorders are multifactorial, triggered in susceptible individuals when environmental factors become unfavorable. For example, the gut microbiome is now recognized as an important factor and could contribute to the observed familial clustering or geographic prevalence of IBD. It has become increasingly clear that specific genetic variations are associated with increased susceptibility to IBD and that environmental factors such as intestinal bacteria may serve as cofactors or triggers to the development of IBD. It is therefore of interest to understand how interactions between genetics (mutations in IBD susceptibility genes) and environmental factors create a perfect storm that leads to intestinal inflammation. Studying these interactions may lead to discovery of an effective preventive strategy.

In summary, results from genetic studies could provide enormous benefits in guiding overall patient management. The more we know of the genetics of IBD, the environmental factors and triggers, and the genetics of drug responsiveness, the closer we will be to delivering truly personalized medicine that effectively treats IBD.

References

1. Silverberg MS, Satsangi J, Ahmad T, Arnott ID, Bernstein CN, Brant SR et al (2005) Toward an integrated clinical, molecular and serological classification of inflammatory bowel disease: Report of a Working Party of the 2005 Montreal World Congress of Gastroenterology. Can J Gastroenterol 9(Suppl 1):5–36
2. Molodecky NA, Soon IS, Rabi DM, Ghali WA, Ferris M, Chernoff G et al (2012) Increasing incidence and prevalence of the inflammatory bowel diseases with time, based on systematic review. Gastroenterology 142(1):46–54, e42; quiz e30
3. Podolsky DK (2002) The current future understanding of inflammatory bowel disease. Best Pract Res Clin Gastroenterol 16(6):933–943
4. Fiocchi C (1998) Inflammatory bowel disease: etiology and pathogenesis. Gastroenterology 115(1):182–205
5. Farmer RG, Michener WM, Mortimer EA (1980) Studies of family history among patients with inflammatory bowel disease. Clin Gastroenterol 9(2):271–277
6. Tysk C, Lindberg E, Jarnerot G, Floderus-Myrhed B (1988) Ulcerative colitis and Crohn's disease in an unselected population of monozygotic and dizygotic twins. A study of heritability and the influence of smoking. Gut 29(7):990–996
7. Sonnenberg A, McCarty DJ, Jacobsen SJ (1991) Geographic variation of inflammatory bowel disease within the United States. Gastroenterology 100(1):143–149
8. Loftus EV Jr (2004) Clinical epidemiology of inflammatory bowel disease: Incidence, prevalence, and environmental influences. Gastroenterology 126(6):1504–1517
9. Sokol H, Seksik P, Rigottier-Gois L, Lay C, Lepage P, Podglajen I et al (2006) Specificities of the fecal microbiota in inflammatory bowel disease. Inflamm Bowel Dis 12(2):106–111
10. Logan RF, Edmond M, Somerville KW, Langman MJ (1984) Smoking and ulcerative colitis. Br Med J (Clin Res Ed) 288(6419):751–753
11. Somerville KW, Logan RF, Edmond M, Langman MJ (1984) Smoking and Crohn's disease. Br Med J (Clin Res Ed) 289(6450):954–956
12. Pullan RD, Rhodes J, Ganesh S, Mani V, Morris JS, Williams GT et al (1994) Transdermal nicotine for active ulcerative colitis. N Eng J Med 330(12):811–815
13. Ananthakrishnan AN, Higuchi LM, Huang ES, Khalili H, Richter JM, Fuchs CS et al (2012) Aspirin, nonsteroidal anti-inflammatory drug use, and risk for Crohn disease and ulcerative colitis: a cohort study. Ann Intern Med 156(5):350–359, Epub 2012/03/07
14. Schroeder KW, Tremaine WJ, Ilstrup DM (1987) Coated oral 5-aminosalicylic acid therapy for mildly to moderately active ulcerative colitis. A randomized study. N Eng J Med 317(26):1625–1629
15. Podolsky DK (2002) Inflammatory bowel disease. N Eng J Med 347(6):417–429
16. Wagtmans MJ, Verspaget HW, Lamers CB, van Hogezand RA (1997) Clinical aspects of Crohn's disease of the upper gastrointestinal tract: a comparison with distal Crohn's disease. Am J Gastroenterol 92(9):1467–1471

17. de Vlam K, Mielants H, Cuvelier C, De Keyser F, Veys EM, De Vos M (2000) Spondyloarthropathy is underestimated in inflammatory bowel disease: prevalence and HLA association. J Rheumatol 27(12):2860–2865
18. Ghosh S, Cowen S, Hannan WJ, Ferguson A (1994) Low bone mineral density in Crohn's disease, but not in ulcerative colitis, at diagnosis. Gastroenterology 107(4):1031–1039
19. Lebwohl M, Lebwohl O (1998) Cutaneous manifestations of inflammatory bowel disease. Inflamm Bowel Dis 4(2):142–148
20. Mintz R, Feller ER, Bahr RL, Shah SA (2004) Ocular manifestations of inflammatory bowel disease. Inflamm Bowel Dis 10(2):135–139
21. Nikolaus S, Schreiber S (2007) Diagnostics of inflammatory bowel disease. Gastroenterology 133(5):1670–1689
22. Ahmad J, Slivka A (2002) Hepatobiliary disease in inflammatory bowel disease. Gastroenterol Clin North Am 31(1):329–345
23. Talwalkar JA, Lindor KD (2005) Primary sclerosing cholangitis. Inflamm Bowel Dis 11(1):62–72
24. Farmer RG, Hawk WA, Turnbull RB Jr (1975) Clinical patterns in Crohn's disease: a statistical study of 615 cases. Gastroenterology 68(4 Pt 1):627–635
25. Gasche C, Scholmerich J, Brynskov J, D'Haens G, Hanauer SB, Irvine EJ et al (2000) A simple classification of Crohn's disease: report of the working party for the world congresses of gastroenterology, Vienna 1998. Inflamm Bowel Dis 6(1):8–15
26. Farmer RG, Whelan G, Fazio VW (1985) Long-term follow-up of patients with Crohn's disease. Relationship between the clinical pattern and prognosis. Gastroenterology 88(6): 1818–1825
27. Cosnes J, Cattan S, Blain A, Beaugerie L, Carbonnel F, Parc R et al (2002) Long-term evolution of disease behavior of Crohn's disease. Inflamm Bowel Dis 8(4):244–250
28. Mekhjian HS, Switz DM, Melnyk CS, Rankin GB, Brooks RK (1979) Clinical features and natural history of Crohn's disease. Gastroenterology 77(4 Pt 2):898–906
29. Lapidus A, Bernell O, Hellers G, Lofberg R (1998) Clinical course of colorectal Crohn's disease: a 35-year follow-up study of 507 patients. Gastroenterology 114(6):1151–1160
30. Martland GT, Shepherd NA (2007) Indeterminate colitis: definition, diagnosis, implications and a plea for nosological sanity. Histopathology 50(1):83–96, Epub 2007/01/06
31. Tremaine WJ (2012) Is indeterminate colitis determinable? Curr Gastroenterol Rep 14(2):162–165, Epub 2012/02/09
32. Murrell ZA, Melmed GY, Ippoliti A, Vasiliauskas EA, Dubinsky M, Targan SR et al (2009) A prospective evaluation of the long-term outcome of ileal pouch-anal anastomosis in patients with inflammatory bowel disease-unclassified and indeterminate colitis. Dis Colon Rectum 52(5):872–878, Epub 2009/06/09
33. Prenzel F, Uhlig HH (2009) Frequency of indeterminate colitis in children and adults with IBD—a metaanalysis. J Crohns Colitis 3(4):277–281
34. Chetty R, Govender D (2012) Lymphocytic and collagenous colitis: an overview of so-called microscopic colitis. Nat Rev Gastroenterol Hepatol 9(4):209–218, Epub 2012/02/22
35. Geraghty JM, Talbot IC (1991) Diversion colitis: histological features in the colon and rectum after defunctioning colostomy. Gut 32(9):1020–1023
36. Lamps LW, Knapple WL (2007) Diverticular disease-associated segmental colitis. Clin Gastroenterol Hepatol 5(1):27–31, Epub 2007/01/20
37. Rameshshanker R, Arebi N (2012) Endoscopy in inflammatory bowel disease when and why. World J Gastrointest Endosc 4(6):201–211
38. Kornbluth A, Sachar DB, Practice Parameters Committee of the American College of Gastroenterology (2010) Ulcerative colitis practice guidelines in adults: American College Of Gastroenterology, Practice Parameters Committee. Am J Gastroenterol 105(3):501–523
39. Nostrant TT, Kumar NB, Appelman HD (1987) Histopathology differentiates acute self-limited colitis from ulcerative colitis. Gastroenterology 92(2):318–328
40. Sands BE (2004) From symptom to diagnosis: clinical distinctions among various forms of intestinal inflammation. Gastroenterology 126(6):1518–1532

41. Hanauer SB, Sandborn WJ, Kornbluth A, Katz S, Safdi M, Woogen S et al (2005) Delayed-release oral mesalamine at 4.8 g/day (800 mg tablet) for the treatment of moderately active ulcerative colitis: the ASCEND II trial. Am J Gastroenterol 100(11):2478–2485
42. Marteau P, Probert CS, Lindgren S, Gassul M, Tan TG, Dignass A et al (2005) Combined oral and enema treatment with Pentasa (mesalazine) is superior to oral therapy alone in patients with extensive mild/moderate active ulcerative colitis: a randomised, double blind, placebo controlled study. Gut 54(7):960–965
43. Truelove SC, Watkinson G, Draper G (1962) Comparison of corticosteroid and sulphasalazine therapy in ulcerative colitis. Br Med J 2(5321):1708–1711
44. Rutgeerts P, Sandborn WJ, Feagan BG, Reinisch W, Olson A, Johanns J et al (2005) Infliximab for induction and maintenance therapy for ulcerative colitis. N Eng J Med 353(23):2462–2476
45. Oussalah A, Laclotte C, Chevaux JB, Bensenane M, Babouri A, Serre AA et al (2008) Long-term outcome of adalimumab therapy for ulcerative colitis with intolerance or lost response to infliximab: a single-centre experience. Aliment Pharmacol Ther 28(8):966–972
46. Shibolet O, Regushevskaya E, Brezis M, Soares-Weiser K (2005) Cyclosporine A for induction of remission in severe ulcerative colitis. Cochrane Database Syst Rev 1, CD004277
47. Perrier C, Rutgeerts P (2012) New Drug Therapies on the Horizon for IBD. Dig Dis 30(Suppl 1):100–105
48. Lichtenstein GR, Hanauer SB, Sandborn WJ, Practice Parameters Committee of American College of Gastroenterology (2009) Management of Crohn's disease in adults. Am J Gastroenterol 104(2):465–483; quiz 4, 84
49. Rutgeerts P, Hiele M, Geboes K, Peeters M, Penninckx F, Aerts R et al (1995) Controlled trial of metronidazole treatment for prevention of Crohn's recurrence after ileal resection. Gastroenterology 108(6):1617–1621
50. Thia KT, Mahadevan U, Feagan BG, Wong C, Cockeram A, Bitton A et al (2009) Ciprofloxacin or metronidazole for the treatment of perianal fistulas in patients with Crohn's disease: a randomized, double-blind, placebo-controlled pilot study. Inflamm Bowel Dis 15(1):17–24
51. Benchimol EI, Seow CH, Steinhart AH, Griffiths AM (2008) Traditional corticosteroids for induction of remission in Crohn's disease. Cochrane Database Syst Rev 2, CD006792
52. Seow CH, Benchimol EI, Griffiths AM, Otley AR, Steinhart AH (2008) Budesonide for induction of remission in Crohn's disease. Cochrane Database Syst Rev 3, CD000296
53. Lichtenstein GR, Bengtsson B, Hapten-White L, Rutgeerts P (2009) Oral budesonide for maintenance of remission of Crohn's disease: a pooled safety analysis. Aliment Pharmacol Ther 29(6):643–653
54. Prefontaine E, Macdonald JK, Sutherland LR (2009) Azathioprine or 6-mercaptopurine for induction of remission in Crohn's disease. Cochrane Database Syst Rev 4, CD000545
55. Prefontaine E, Sutherland LR, Macdonald JK, Cepoiu M (2009) Azathioprine or 6-mercaptopurine for maintenance of remission in Crohn's disease. Cochrane Database Syst Rev 1, CD000067
56. Feagan BG, Rochon J, Fedorak RN, Irvine EJ, Wild G, Sutherland L et al (1995) Methotrexate for the treatment of Crohn's disease. The North American Crohn's Study Group Investigators. N Eng J Med 332(5):292–297
57. Vavricka SR, Schoepfer AM, Bansky G, Binek J, Felley C, Geyer M et al (2011) Efficacy and safety of certolizumab pegol in an unselected Crohn's disease population: 26-week data of the FACTS II survey. Inflamm Bowel Dis 17(7):1530–1539
58. Rutgeerts P (2004) Review article: treatment of perianal fistulizing Crohn's disease. Aliment Pharmacol Ther 20(Suppl 4):106–110
59. Colombel JF, Sandborn WJ, Rutgeerts P, Enns R, Hanauer SB, Panaccione R et al (2007) Adalimumab for maintenance of clinical response and remission in patients with Crohn's disease: the CHARM trial. Gastroenterology 132(1):52–65
60. Sinh P, Barrett TA, Yun L (2011) Clostridium difficile Infection and Inflammatory Bowel Disease: A Review. Gastroenterol Res Pract 2011:136064, Epub 2011/09/15

61. Afif W, Loftus EV Jr, Faubion WA, Kane SV, Bruining DH, Hanson KA et al (2010) Clinical utility of measuring infliximab and human anti-chimeric antibody concentrations in patients with inflammatory bowel disease. Am J Gastroenterol 105(5):1133–1139
62. Targan SR, Feagan BG, Fedorak RN, Lashner BA, Panaccione R, Present DH et al (2007) Natalizumab for the treatment of active Crohn's disease: results of the ENCORE Trial. Gastroenterology 132(5):1672–1683
63. Sandborn WJ (2009) The future of inflammatory bowel disease care. Rev Gastroenterol Disord 9(3):E69–E77
64. Jostins L, Ripke S, Weersma RK, Duerr RH, McGovern DP, Hui KY et al (2012) Host-microbe interactions have shaped the genetic architecture of inflammatory bowel disease. Nature 491(7422):119–124, Epub 2012/11/07
65. Lees CW, Barrett JC, Parkes M, Satsangi J (2011) New IBD genetics: common pathways with other diseases. Gut 60(12):1739–1753, Epub 2011/02/09
66. Cho JH, Brant SR (2011) Recent insights into the genetics of inflammatory bowel disease. Gastroenterology 140(6):1704–1712
67. Brand S, Staudinger T, Schnitzler F, Pfennig S, Hofbauer K, Dambacher J et al (2005) The role of Toll-like receptor 4 Asp299Gly and Thr399Ile polymorphisms and CARD15/NOD2 mutations in the susceptibility and phenotype of Crohn's disease. Inflamm Bowel Dis 11(7):645–652
68. Sehgal R, Berg A, Hegarty JP, Kelly AA, Lin Z, Poritz LS et al (2010) NOD2/CARD15 mutations correlate with severe pouchitis after ileal pouch-anal anastomosis. Dis Colon Rectum 53(11):1487–1494
69. Christodoulou K, Wiskin AE, Gibson J, Tapper W, Willis C (2012) Afzal NA, et al. Next generation exome sequencing of paediatric inflammatory bowel disease patients identifies rare and novel variants in candidate genes, Gut
70. Russell RK, Drummond HE, Nimmo EE, Anderson N, Smith L, Wilson DC et al (2005) Genotype-phenotype analysis in childhood-onset Crohn's disease: NOD2/CARD15 variants consistently predict phenotypic characteristics of severe disease. Inflamm Bowel Dis 11(11):955–964
71. Babusukumar U, Wang T, McGuire E, Broeckel U, Kugathasan S (2006) Contribution of OCTN variants within the IBD5 locus to pediatric onset Crohn's disease. Am J Gastroenterol 101(6):1354–1361
72. Reinhard C, Rioux JD (2006) Role of the IBD5 susceptibility locus in the inflammatory bowel diseases. Inflamm Bowel Dis 12(3):227–238
73. Mirza MM, Fisher SA, King K, Cuthbert AP, Hampe J, Sanderson J et al (2003) Genetic evidence for interaction of the 5q31 cytokine locus and the CARD15 gene in Crohn disease. Am J Hum Genet 72(4):1018–1022
74. Moran CJ, Walters TD, Guo CH, Kugathasan S, Klein C, Turner D et al (2012) IL-10R polymorphisms are associated with very-early-onset ulcerative colitis. Inflamm Bowel Dis 19(1):115–123
75. Kotlarz D, Beier R, Murugan D, Diestelhorst J, Jensen O, Boztug K et al (2012) Loss of interleukin-10 signaling and infantile inflammatory bowel disease: implications for diagnosis and therapy. Gastroenterology 143(2):347–355
76. Muise AM, Xu W, Guo CH, Walters TD, Wolters VM, Fattouh R et al (2012) NADPH oxidase complex and IBD candidate gene studies: identification of a rare variant in NCF2 that results in reduced binding to RAC2. Gut 61(7):1028–1035
77. Worthey EA, Mayer AN, Syverson GD, Helbling D, Bonacci BB, Decker B et al (2011) Making a definitive diagnosis: successful clinical application of whole exome sequencing in a child with intractable inflammatory bowel disease. Genet Med 13(3):255–262
78. Latiano A, Palmieri O, Cucchiara S, Castro M, D'Inca R, Guariso G et al (2009) Polymorphism of the IRGM gene might predispose to fistulizing behavior in Crohn's disease. Am J Gastroenterol 104(1):110–116

79. de Ridder L, Weersma RK, Dijkstra G, van der Steege G, Benninga MA, Nolte IM et al (2007) Genetic susceptibility has a more important role in pediatric-onset Crohn's disease than in adult-onset Crohn's disease. Inflamm Bowel Dis 13(9):1083–1092
80. Lichtenstein GR, Targan SR, Dubinsky MC, Rotter JI, Barken DM, Princen F et al (2011) Combination of genetic and quantitative serological immune markers are associated with complicated Crohn's disease behavior. Inflamm Bowel Dis 17(12):2488–2496
81. Glas J, Konrad A, Schmechel S, Dambacher J, Seiderer J, Schroff F et al (2008) The ATG16L1 gene variants rs2241879 and rs2241880 (T300A) are strongly associated with susceptibility to Crohn's disease in the German population. Am J Gastroenterol 103(3):682–691
82. Simms LA, Doecke JD, Roberts RL, Fowler EV, Zhao ZZ, McGuckin MA et al (2010) KCNN4 gene variant is associated with ileal Crohn's Disease in the Australian and New Zealand population. Am J Gastroenterol 105(10):2209–2217
83. Henckaerts L, Van Steen K, Verstreken I, Cleynen I, Franke A, Schreiber S et al (2009) Genetic risk profiling and prediction of disease course in Crohn's disease patients. Clin Gastroenterol Hepatol 7(9):972–980, e2
84. Marrakchi R, Moussa A, Ouerhani S, Bougatef K, Bouhaha R, Messai Y et al (2009) Interleukin 10 promoter region polymorphisms in inflammatory bowel disease in Tunisian population. Inflamm Res 58(3):155–160
85. Koslowski MJ, Kubler I, Chamaillard M, Schaeffeler E, Reinisch W, Wang G et al (2009) Genetic variants of Wnt transcription factor TCF-4 (TCF7L2) putative promoter region are associated with small intestinal Crohn's disease. PLoS One 4(2):e4496
86. Brand S (2012) Homozygosity for the NOD2 p.Leu1007fsX1008 variant is the main genetic predictor for fibrostenotic Crohn's disease. Inflamm Bowel Dis 18(2):393–394
87. Ippoliti A, Devlin S, Mei L, Yang H, Papadakis KA, Vasiliauskas EA et al (2010) Combination of innate and adaptive immune alterations increased the likelihood of fibrostenosis in Crohn's disease. Inflamm Bowel Dis 16(8):1279–1285
88. Weersma RK, Zhernakova A, Nolte IM, Lefebvre C, Rioux JD, Mulder F et al (2008) ATG16L1 and IL23R are associated with inflammatory bowel diseases but not with celiac disease in the Netherlands. Am J Gastroenterol 103(3):621–627
89. Seiderer J, Dambacher J, Leistner D, Tillack C, Glas J, Niess JH et al (2008) Genotype-phenotype analysis of the CXCL16 p.Ala181Val polymorphism in inflammatory bowel disease. Clin Immunol 127(1):49–55
90. Hume GE, Fowler EV, Lincoln D, Eri R, Templeton D, Florin TH et al (2006) Angiotensinogen and transforming growth factor beta1: novel genes in the pathogenesis of Crohn's disease. J Med Genet 43(10):e51
91. Eglinton TW, Barclay ML, Gearry RB, Frizelle FA (2012) The spectrum of perianal Crohn's disease in a population-based cohort. Dis Colon Rectum 55(7):773–777
92. Sehgal R, Berg A, Polinski JI, Hegarty JP, Lin Z, McKenna KJ et al (2012) Mutations in IRGM are associated with more frequent need for surgery in patients with ileocolonic Crohn's disease. Dis Colon Rectum 55(2):115–121
93. Vermeire S, Pierik M, Hlavaty T, Claessens G, van Schuerbeeck N, Joossens S et al (2005) Association of organic cation transporter risk haplotype with perianal penetrating Crohn's disease but not with susceptibility to IBD. Gastroenterology 129(6):1845–1853
94. Santana G, Bendicho MT, Santana TC, Reis LB, Lemaire D, Lyra AC (2011) The TNF-alpha -308 polymorphism may affect the severity of Crohn's disease. Clinics 66(8):1373–1378
95. Moon CM, Cheon JH, Kim SW, Shin DJ, Kim ES, Shin ES et al (2010) Association of signal transducer and activator of transcription 4 genetic variants with extra-intestinal manifestations in inflammatory bowel disease. Life Sci 86(17–18):661–667
96. Orchard TR, Chua CN, Ahmad T, Cheng H, Welsh KI, Jewell DP (2002) Uveitis and erythema nodosum in inflammatory bowel disease: clinical features and the role of HLA genes. Gastroenterology 123(3):714–718, Epub 2002/08/29

97. Ellinghaus E, Stuart PE, Ellinghaus D, Nair RP, Debrus S, Raelson JV et al (2012) Genome-wide meta-analysis of psoriatic arthritis identifies susceptibility locus at REL. J Invest Dermatol 132(4):1133–1140
98. Parkes J (2011) on behalf of the authors. Validating non-invasive markers of fibrosis: the need for a new histological reference standard—The Authors' response. Gut
99. Gelbmann CM, Rogler G, Gierend M, Gross V, Scholmerich J, Andus T (2001) Association of HLA-DR genotypes and IL-1ra gene polymorphism with treatment failure of budesonide and disease patterns in Crohn's disease. Eur J Gastroenterol Hepatol 13(12):1431–1437
100. Toedter G, Li K, Marano C, Ma K, Sague S, Huang CC et al (2011) Gene expression profiling and response signatures associated with differential responses to infliximab treatment in ulcerative colitis. Am J Gastroenterol 106(7):1272–1280

Chapter 2
Genetic Epidemiology of Inflammatory Bowel Disease, Early Twin and Family Studies

Jonas Halfvarson

Abstract The influence of genetics in the etiology of inflammatory bowel disease (IBD) was initially demonstrated by epidemiological data, including differences in prevalence among different ethnic groups, familial aggregation of IBD, concordance in twins, and association with genetic syndromes. These early observations commenced the successful era of molecular genetics in IBD that has illuminated the complexity of the genetic and environmental interaction in IBD. Recent advances in molecular genetics have dramatically improved the resolution of the IBD genome. Yet, some of the key epidemiological observations are still unanswered from a molecular perspective. Based on analyzes of the DNA sequence, the observed high heritability in Crohn's disease is only partly understood. Possible contributions of other molecular mechanisms of heritability, such as epigenetics, are yet to be explored. Similarly, pronounced phenotypic similarities have been observed within multiplex families and especially within concordant monozygotic twin pairs with Crohn's disease, suggesting that genetics also influences the phenotype of the disease. However, it has been difficult to establish any firm genotype–phenotype associations beyond *NOD2*. Thus, it can be questioned if the observed phenotypic concordance within epidemiological studies rather reflects the exposure to shared environmental factors. The possible influence of disease tolerance, that is, differences in susceptibility to tissue damage, in contrast to disease resistance, also needs to be taken into consideration in this perspective. Future longitudinal studies with periodic measurements in subjects at high risk, that is, siblings and offspring below or around the peak age of onset of IBD, will probably become an important tool to elucidate the genetic and environmental interactions underlying these archetypal complex diseases.

J. Halfvarson, M.D., Ph.D. (✉)
Department of Internal Medicine, Division of Gastroenterology, Örebro University Hospital, Örebro 70185, Sweden

Faculty of Health and Medical Sciences, Örebro University, Örebro, Sweden
e-mail: jonas.halfvarsson@orebroll.se

M. D'Amato and J.D. Rioux (eds.), *Molecular Genetics of Inflammatory Bowel Disease*,
DOI 10.1007/978-1-4614-8256-7_2,

There has been great progress within the field of molecular genetics in inflammatory bowel disease (IBD). The influence of genetics in the etiology of IBD was initially supported by epidemiological data, including differences in prevalence among different ethnic groups, familial aggregation of IBD, concordance in twins, and association with genetic syndromes. These early epidemiological findings, and especially the observed concordance within twin pairs, were the major triggers that commenced the successful era of molecular genetics in IBD. Even today, these key observations can bring important pieces to our understanding the pathogenesis of IBD.

Ethnic Differences

IBD has traditionally been associated with considerable geographic and ethnic differences in incidence and prevalence [1, 2]. Generally, the incidence of both Crohn's disease and ulcerative colitis has gradually increased since the Second World War, especially in northern Europe and North America, where the highest incidence rates have been reported [3]. During the 1970s and 1980s, the incidence of Crohn's disease seemed to level off in some areas, like certain parts of Scandinavia and the United States (USA), with a stable high incidence [4–9]. Other areas have reported a continued increase, either during the entire period or after a temporary plateau [10–14]. Especially striking is the increasing age-specific incidence of pediatric Crohn's disease [12, 13, 15]. However, in several areas with traditionally low occurrence of IBD, such as Asia and Africa, increasing figures have been reported in recent years [16–18]. Japan is a striking example, where the reported incidence of Crohn's disease has increased from 2.9 cases per 10^5 persons in 1986 to 13.5 cases per 10^5 persons in 1998 [19]. Similarly, a quadrupled prevalence of ulcerative colitis from 7.57 cases per 10^5 persons in 1997 to 30.87 cases per 10^5 persons in 2005 has been reported from South Korea (see Fig. 2.1) [20]. A recent, comprehensive, systematic review has confirmed a statistically significant increase in the incidence of IBD over time [2]. A continuous increase in the incidence of Crohn's disease beyond the year 1980 was also observed in the majority of studies. For ulcerative colitis the pattern was less consistent, with approximately one-third of the studies showing a continuous increase. On the other hand, a significant decrease in the incidence of ulcerative colitis was only observed in 6 % of the studies and none for Crohn's disease. Although historical differences could be influenced by a number of factors, including different types of biases, these shifts in the risk of developing IBD within limited period of times can barely be explained by changes in the genome but rather provide evidence for the importance of exposure to environmental factors in the disease pathogenesis.

However, different prevalence rates of IBD have also been observed in different ethnic groups within the same geographic region. The prevalence of IBD among Caucasians in the USA has been reported to be higher than among non-Caucasians [21]. Similarly, the prevalence of Crohn's disease per 100,000 was reported to be 43.6 among whites, 29.8 among African-Americans, 4.1 among Hispanics, and 5.6

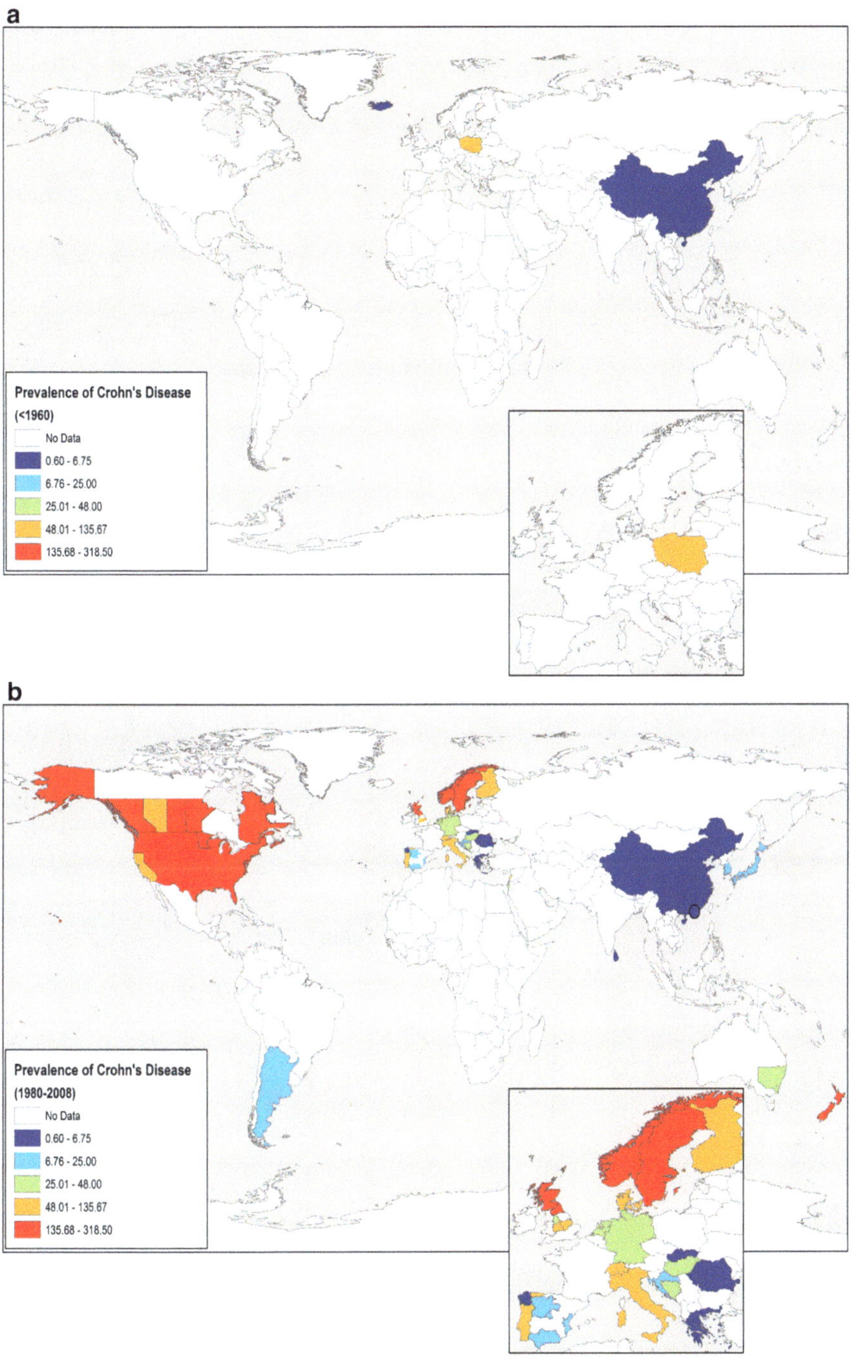

Fig. 2.1 Worldwide Crohn's disease prevalence for countries reporting data before (**a**) 1960 and (**b**) after 1980. Worldwide ulcerative colitis prevalence for countries reporting data, before (**c**) 1960 and (**d**) after 1980. Prevalence rates were ranked into quintiles representing low (*dark* and *light blue*) to intermediate (*green*) to high (*yellow* and *red*) incidence of disease. Reprinted from Gastroenterology, 142/1, Natalie A. Molodecky, Ing Shian Soon, Doreen M. Rabi, William A. Ghali, Mollie Ferris, Greg Chernoff, Eric I. Benchimol, Remo Panaccione, Subrata Ghosh, Herman W. Barkema, Gilaad G. Kaplan, Increasing Incidence and Prevalence of the Inflammatory Bowel Diseases With Time, Based on Systematic Review, 46–54.e42, Copyright (2013), with permission from Elsevier

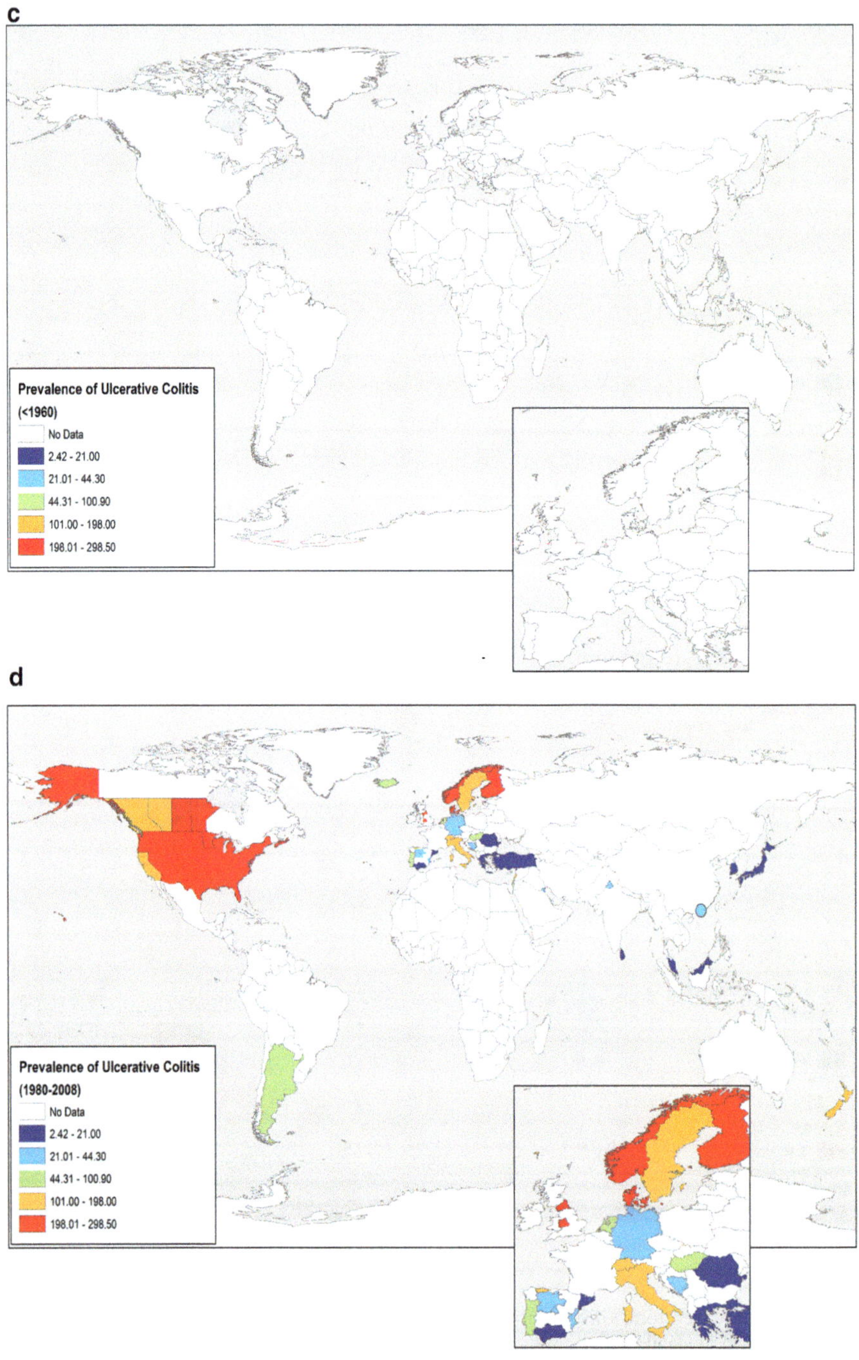

Fig. 2.1 (continued)

among Asians in a mail-based study performed in 1982–1988 [22]. These findings could reflect different genetic susceptibility within different ethnic groups, but could also be explained by differences in socioeconomic circumstances and access to and/or organization of health care as well as environmental exposure. A more recent study reported the incidence rates of Crohn's disease and ulcerative colitis per 100,000 to be 7–12 and 5–7, respectively, in African-Americans in Georgia, suggesting that previous observed interethnic differences within certain geographic areas might be less pronounced today [23]. The observed low occurrence of IBD in first-generation Asian immigrants in the United Kingdom (UK), but a high occurrence in the second generation [24], is further reason to question the role of disparities in genetic susceptibility in different ethnic groups. This observation rather speaks in favor of the importance of exposure to environmental factors early in life, especially since these second-generation immigrants were at higher risk of developing IBD than the indigenous population in the UK.

Although many earlier observed geographic differences in the occurrence of IBD seem to have diminished, some ethnic differences still exist. The prevalence of IBD is 2–4 times higher among Jews than in any other population [25], with the highest numbers in Ashkenazi Jews compared with Sephardic or Oriental Jews [26, 27]. The difference between Jewish and non-Jewish populations, seem to persist, irrespective of geographic location or time period, reflecting a more pronounced genetic predisposition in Jews [28]. Similarly, the higher rate of Crohn's disease in Ashkenazi Jews than in non-Ashkenazi Jews has been reported from different geographic regions, such as California and Israel [29, 30].

Family Studies

Risk for Relatives

The familial nature of IBD was first recognized in 1909 [31] and was soon reported by several others [32–37]. In 1963 Kirsner et al. published the first controlled family study of IBD [38]. Since then, aggregation of cases of IBD in families has been widely confirmed, with 5–23 % of patients with IBD having an affected first-degree relative (Table 2.1) [28, 39–51]. In general, population-based studies have reported figures in the lower interval, that is, 5–10 % of patients with IBD reporting a positive family history [41, 43, 45, 52]. In contrast, the higher figures have been reported in case series that probably have been biased due to the effect of referral. Families with multiple affected individuals, the so-called multiplex families, are most often concordant for disease type, that is, family members having either Crohn's disease or ulcerative colitis [3]. However, approximately one-fourth of the families are mixed, with relatives having Crohn's disease and ulcerative colitis. This argues for a model in which some genetic variants are disease specific and some variants are common to both ulcerative colitis and Crohn's disease, whereby

Table 2.1 Studies of first-degree relatives in a proband with IBD

	Proband with CD		Proband with UC	
Population	First-degree relatives with CD	First-degree relatives with IBD	First-degree relatives with UC	First-degree relatives with IBD
Welsh [42]	5.0 % (*n*=139)	9.3 % (*n*=139)	–	–
USA [41]	15.1 % (*n*=522)	16.7 % (*n*=522)	15.5 % (*n*=316)	15.8 % (*n*=316)
Dutch [46]	8.0 % (*n*=400)	9.5 % (*n*=400)	–	–
Swedish [43]	5.7 % (*n*=963)	5.7 % (*n*=963)	–	–
USA [47]	16.2 % (*n*=80)	22.5 % (*n*=80)	8.9 % (*n*=101)	13.9 % (*n*=101)
Swedish [45]	6.9 % (*n*=1,048)	6.9 % (*n*=1,048)	–	–
Danish [39]	2.2 % (*n*=133)	5.2 % (*n*=133)	7.5 % (504)	8.1 (*n*=504)
USA [28]	7.4 % (*n*=258)	14.0 % (*n*=258)	7.1 % (*n*=269)	8.6 % (*n*=269)
UK [51]	9.4 % (*n*=424)	10.4 % (*n*=424)	6.2 % (*n*=469)	6.8 % (*n*=469)
UK [59]	Not stated	11.5 % (*n*=433)	–	–
USA [44]	12.2 % (*n*=554)	Not stated	–	–
Belgian [48]	13.6 % (*n*=640)	14.5 % (*n*=640)	–	–
French [50]	7.5 % (*n*=1,316)	8.4 % (*n*=1,316)	–	–
Finnish [116]	10.9 % (*n*=257)	15.6 % (*n*=257)	11.3 % (*n*=436)	13.8 % (*n*=436)
Canadian [117]	8.7 % (*n*=1,000)	Not stated	–	–

CD Crohn's disease, *IBD* inflammatory bowel disease, *UC* ulcerative colitis

environmental factors might influence disease phenotype. This theory is supported by the effect of smoking, being the most extensively described environmental factor in IBD. Active smoking increases the risk of Crohn's disease, but protects against ulcerative colitis [53]. Bridger et al. studied the influence of smoking in 23 mixed sibling pairs, with one sibling suffering from Crohn's disease and one from ulcerative colitis, discordant for smoking habits at diagnosis [54]. In 21/23 pairs Crohn's disease occurred in the smoker and ulcerative colitis in the nonsmoker. This supports the hypothesis that environmental factors can act on genetic predisposition in IBD and shift the disease phenotype towards either Crohn's disease or ulcerative colitis. More recently, molecular genetics has confirmed that certain genetic variants are shared by Crohn's disease and ulcerative colitis and that other variants are disease specific [55–57]. An association between methylation of the genome and disease status has also been shown in Crohn's disease [58]. This may indicate that environmental factors influence disease phenotype in IBD by epigenetic modifications of the genome.

The greatest risk for developing IBD is having a relative with the disease. The population relative risk λ_R (lambda) of developing IBD in first-degree relatives of affected individuals, in comparison with the general population, is an estimate of the genetic contribution to disease pathogenesis. This relative risk can be assessed by either cohort studies or case–control studies. Danish cohort data, by Orholm et al., demonstrated a population relative risk, standardized for age and sex, of 10 for relatives of patients with Crohn's disease as well as for relatives of patients with ulcerative colitis [39]. An increased relative risk of the other of the two diagnoses in first-degree relatives of patients with Crohn's disease and patients with ulcerative colitis was also observed, λ_R (lambda) 4.4 and 1.8, respectively. Similar data have been generated by case–control studies. Peeters et al. reported a prevalence of IBD of 14.5 % in first-degree relatives of patients affected by Crohn's disease, corresponding to a relative risk of 13 [48]. The observed prevalence rates were 3.3 % in siblings, 1.6 % in parents, and 2.0 % in offspring. Similar data have been reported from Oxford, with a prevalence of IBD among first-degree relatives of 2.6 %, corresponding to a population relative risk of developing IBD of 15, based on an estimated prevalence of IBD of 170/100,000 in the background population [59]. On the whole, the estimated relative risk to a sibling λ_S (lambda) of a patient with Crohn's disease is 13–36 and for ulcerative colitis 7–17 [25]. This compares with a λ_S (lambda) of 9 for schizophrenia, 15 for type 1 diabetes, and 500 for cystic fibrosis [25].

However, from a clinical perspective the relative risk of developing IBD is of less interest. It is the absolute risk of IBD in first-degree relatives and especially in the patient's offspring that is requested. There are limited studies addressing this clinically relevant question, and quoted absolute risks differ between the studies. The studies give a lifetime risk of developing IBD for first-degree relatives of a Crohn's disease patient of 4.8–5.2 % for non-Jews and 7.8 % for Jews [28, 47, 48]. The corresponding figures for first-degree relatives of a patient with ulcerative colitis are 1.6 % for non-Jews and 5.2 % for Jews [28]. Similarly, the age-corrected risk for offspring of a Crohn's disease patient developing IBD is 0–10.4 % in non-Jews and 7.4–15.8 % in Jews. The equivalent figures for offspring of a patient with ulcerative colitis are 11 % and 2.9–7.4 %, respectively [28, 47, 48].

The risk of IBD in an offspring increases dramatically if both parents suffer from IBD. Case series have estimated the risk of IBD in the offspring to be 33–52 %, depending on follow-up [60, 61].

Phenotypic Similarities Within Families with IBD

IBD has traditionally been categorized as Crohn's disease, ulcerative colitis, or colitis unclassified (IBD-U), based on clinical, endoscopic, radiological, and histological criteria [62, 63]. However, there are great heterogeneities within the three different diagnoses, suggesting the existence of subgroups based on clinical characteristics like age at diagnosis, location or extent of inflammation, disease behavior, disease activity, and response to therapy. Several attempts have been made to define phenotypic subgroups based on these different clinical characteristics, especially within Crohn's disease, but more recently also within ulcerative colitis.

The initial attempts to classify Crohn's disease were based on anatomic distribution of the inflammation in the gastrointestinal tracts [64]. Subsequent studies showed that the location of inflammation in Crohn's disease influenced the response to medical and surgical therapy [64–68]. Greenstein et al. reported that disease behavior was another distinct subgroup, independent of anatomic distribution [69]. Recurrent postoperative disease was shown to follow the same behavior as the primary disease, and the disease behavior also influenced the risk of postoperative recurrence [69, 70]. Perforating disease with fistulas and/or abscesses was more aggressive and separated from the more benign non-perforating disease. Based on these observational data, a formal classification of subgroups in Crohn's disease was proposed in 1992 [71]. The objectives were to standardize the description of study populations in clinical trials and to support the association of different possible etiological factors with specific clinical phenotypes. The proposal was refined in 1998, when age at diagnosis was incorporated and the Vienna classification was presented [72]. The classification was developed further in 2005 when the Montreal classification, also including ulcerative colitis, was introduced [73].

Epidemiological studies from the mid-1990s of familial IBD, that is, of families with multiple affected individuals, suggest that there could be a genetic basis for these different subgroups. In general, a high degree of clinical similarities of IBD has been observed within multiplex families, and data are especially striking for Crohn's disease. Bayless et al. observed concordance rates of 86 % for location of inflammation and 82 % for disease behavior in a group of 133 patients with Crohn's disease within 60 families [44]. In a family study from Oxford, 76 % of siblings were concordant for disease extent and 84 % for extraintestinal manifestations [74]. The phenotypes were less similar in parent–child pairs than within sibling pairs. In a French series of 176 patients with Crohn's disease within 72 families, a concordance rate of 56 % was observed for disease location [75]. The concordance rates increased with the number of affected relatives within the family, and in families with more than two affected individuals, 83 % were concordant for disease location. Similarly, concordance rates of 58 % for location, 44 % for disease behavior, and 42 % for number of bowel resections were observed in a Belgian study [48]. Using the κ (kappa) statistics, they showed a strong agreement for age at diagnosis and initial disease location within generations. On the other side, Lee and Lennard-Jones could not observe any concordance in extent, type of disease, or extraintestinal manifestations in a large study of 67 families with three or more affected first-degree relatives [76]. However, sporadic cases were not included in the study, and the results might have been biased by the use of historical controls.

Less is known about concordance within families with multiple affected members with ulcerative colitis. Satsangi et al. observed agreement in disease extent in 53 % of 27 parent–child pairs and in 69 % in 35 sibling pairs [74]. Similar findings have been reported from an Italian study encompassing 64 families, where concordance rates of 33 % for disease extent, 47 % for need for corticosteroids, and 34 % for relapse were observed [77]. These figures are lower than the observed concordance rates within families with Crohn's disease in the two studies, which supports a smaller contribution of genetics to disease susceptibility in ulcerative colitis than in Crohn's disease. The influence of genetics in disease characteristics of Crohn's disease is further supported by a study of married couples with IBD. In the 17 couples where both partners suffered from Crohn's disease, no significant similarities were identified for disease location or behavior [61]. However, it cannot be ruled out that the observed concordance in disease characteristics within families with IBD is an effect of shared environment within the families, rather than of genetic predisposition.

Familial and Sporadic IBD

Based on the findings in family studies, it has even been proposed that familial IBD could be a different entity than sporadic disease, but the evidence for phenotypic differences between these two groups is sparse, and the literature rather contradictory. An earlier age at onset for familial cases of IBD than for patients without any family history of IBD is probably the most consistent observation. In a US study of 552 consecutive patients with Crohn's disease, 17 % had positive family history of IBD [49]. This subgroup of patients was younger at diagnosis, 21.8 vs. 26.7 years in the group of sporadic Crohn's disease. Similar findings have been reported in a French study, where the median age at diagnosis of Crohn's disease was 22 years in the familial form and 26.5 years in the sporadic form [75]. The even younger age at diagnosis in multiplex families with more than two affected individuals in the French cohort adds further support to this observation. Similarly, Carbonnel et al. observed younger age in familial than in sporadic Crohn's disease, although the difference did not reach statistical significance [50]. On the other side, Lee and Lennard-Jones reported younger age at diagnosis in familial Crohn's disease only in the second generation in their large study of families with three or more first-degree relatives [76]. However, due to absence of controls, that is, sporadic cases, the findings are difficult to interpret.

Beyond younger age at diagnosis, predominance of female cases has also been reported in familial Crohn's disease. A predominance of women is normally observed in Crohn's disease, but the observed female-to-male ratio of 1.23:1 to 1.68:1 outnumbers the ratio in sporadic cases as well as the expected ratio [45, 75, 76]. In a study where both a parent and a child had IBD, transmission of the disease was more common from mother to child than among father to child, 93 vs. 42 families, respectively [78]. Similarly, Zelinkova et al. reported 55 transmissions from mother to child and 32 transmissions from father to child in their cohort of 608 cases with familial IBD [79]. The female imprinting was specifically related to Crohn's disease, and also a higher female-to-female transmission compared with female-to-male transmission was observed. The authors proposed that a female sex-specific epigenetic inheritance pattern for Crohn's disease is a major contributing factor in the family-specific risk in Crohn's disease.

Possible differences in disease location, behavior, extraintestinal manifestations, and disease severity between familial and sporadic Crohn's disease have been explored in numerous studies. However, the definitions used differ between the studies, and interpretation of the data is difficult, since the majority of the studies are based on univariate analyzes, and associations between different clinical characteristics exist. Higher frequencies of ileal and ileocolonic disease, and less frequent occurrence of pure colonic disease, have been reported from both Belgian and French cohorts with familial Crohn's disease [75, 80]. Similarly, a positive family history of IBD was less frequently reported in a series of Italian patients with colonic Crohn's disease than in those with ileal or ileocolonic disease [81]. However, these findings might be secondary to the observed earlier age at diagnosis in the familial form, since consistent associations between age at diagnosis and disease location have been reported. Historically, a younger age at diagnosis was associated with ileal and ileocolonic disease. In contrast, pure colonic disease was more common in patients with late onset. However, the recently reported increasing incidence of colonic Crohn's disease in the pediatric population challenges this dogma [15]. So far, there is no widely accepted index of severity, although the LéMann score has recently been proposed [82]. Instead, pseudomarkers of disease severity, such as requirements for medical or surgical therapy, have been used to evaluate the possible influence of a family history on severity of Crohn's disease. Polito et al. observed an association between family history for Crohn's disease and risk of surgery for abscess or perforation [49]. There was no difference in disease behavior in patients with family history of IBD and patients without family history of IBD. In contrast, Carbonnel et al. did not observe any difference in incidence or extent of Crohn's disease-related resections between familial and sporadic Crohn's disease [50]. In addition, based on log rank test, the estimated time to prescription of immunosuppressive drugs and first intestinal resection was similar in familial and sporadic cases.

There are fewer data on possible phenotypic differences between familial and sporadic ulcerative colitis. However, younger age at diagnosis in familial cases has also been reported in ulcerative colitis. In a study primarily analyzing possible differences between Jews and non-Jews, the mean age at diagnosis of ulcerative colitis was 23.3 years in patients with positive family history of IBD and 28.6 years in patients with the sporadic form [28]. Similarly, a preponderance of women has also been reported in ulcerative colitis, with a female-to-male ratio of 1.4:1 in familial cases [76]. In contrast, among all cases, ulcerative colitis seems to be slightly more common among males, if any gender-related difference exists [53]. However, comparisons with the total ulcerative colitis population might be misleading since environmental factors, such as smoking, influence the risk of developing the disease.

In conclusion, the observed possible differences between familial and sporadic IBD are difficult to evaluate [83]. In general, the studies are limited by their retrospective design. Family history as such might introduce differential bias, including recruitment bias and reporting bias, especially recall bias. The observed difference in age at diagnosis between familial and sporadic cases is the most significant finding, and other possible phenotypic differences could be secondary phenomena. However, with a few exceptions, the family studies have not standardized their findings according to age [28, 47, 48], a systematic weakness that might jeopardize the overall results.

Genetic Anticipation

Based on the observed earlier age at onset within families with IBD than in cases of sporadic disease, it has been suggested that this might be due to genetic anticipation, a pattern of inheritance in which a hereditary illness strikes earlier, and often more severely, in succeeding generations. Genetic anticipation is molecularly explained by amplification of trinucleotide repeats. The phenomenon has been observed in diseases like Fragile X [84], myotonic dystrophy [85], and Huntington's disease [86], where the length and instability of the repeats are associated with an age of onset and the severity of the disease. Based on the consistent finding of an earlier age at onset in children than in parents in families with Crohn's disease, it has been suggested since the mid-1990s that this observation might be due to genetic anticipation [87–90]. However, this theory has been heavily questioned because of possible biases. First, the results have not been adjusted for age and thereby observation period; that is, healthy individuals within the younger generation might develop IBD later in life. Second, since the studies were retrospective, parents were interviewed at an older age, with the risk of report bias and especially recall bias. Third, the median age at diagnosis of IBD in the affected parents was higher than among the sporadic cases with Crohn's disease, pointing towards selection bias and lower inclusion of parents with early onset of disease. The importance of these plausible biases is highlighted by the reanalysis of the initial study. By adjusting the analysis to include patients who were at least 40 years of age at the time of study entry and had the diagnosis before the age of 40 years, the intergeneration difference in age of diagnosis was reduced from >10 years to around 1 year [87, 91].

Twin Studies

Twin studies provide a powerful tool to disentangle the relative contribution of genetics and environmental factors to the etiology of complex diseases like IBD. Monozygotic twins are genetically identical and share the intrauterine environment and, to a high extent, but necessarily not all, external factors during childhood. Dizygotic twins share environment to the same extent as monozygotic pairs, but on average, only half of the genes. Thus, the influence of genetics in disease pathogenesis would be reflected by a higher rate of concordance, both twins being affected, within monozygotic twin pairs than within dizygotic twin pairs. In contrast, a disease caused purely by exposure to environmental factors would predict equivalent rates of concordance in monozygotic and dizygotic twin pairs.

In 1988 Tysk et al. published the first unbiased study showing a higher concordance rate in monozygotic twin pairs than in dizygotic twin pairs with Crohn's disease, reflecting the influence of genetics in the disease pathogenesis [92]. Since then, data on twins with IBD have been reported from the United Kingdom,

Denmark, and Germany [93–97]. The design of the studies differs between countries, with the Scandinavian studies being based on the national twin registry in each country [92, 93]. In contrast, the British study was set up by an appeal for twins with IBD in a newsletter distributed to members of the National Association for Crohn's and Colitis, an IBD patient support group in the United Kingdom with approximately 20,000 members [94, 95, 97]. In addition, gastroenterologists within the London IBD Forum were asked to refer twins with IBD attending their outpatient clinics. Similarly, the German cohort was established by several calls for twins with IBD, using advertisements as well as the nationwide newsletter of the German Crohn's and Colitis Association [96]. The concordance rate in twin pairs can be calculated in different ways. The pair concordance rate simply reflects the proportion of concordant pairs. However, the pair concordance varies with the thoroughness of ascertainment, which makes comparisons between different studies difficult. Thus, to be able to compare reported concordance rates in different twin studies, the proband concordance should be used. Tysk et al. observed proband concordance rates of 58 % and 4 % in monozygotic and dizygotic twins with Crohn's disease, respectively, reflecting the pronounced genetic predisposition [92]. The corresponding figures for twins with ulcerative colitis were 6 % and 0 %, respectively. Orholm et al. later confirmed these findings in the Danish cohort, where proband concordance rates of 58 % and 0 % were observed in monozygotic and dizygotic twins with Crohn's disease, respectively [93]. Similarly, the corresponding rates in twins with ulcerative colitis were 18 % and 4 %, respectively. Follow-ups of the two Scandinavian cohorts, extending the observation period in previous healthy twin siblings, had only marginal effects on the concordance rates, since just a few additional twins had been diagnosed during the extended observation (Table 2.1) [98, 99]. In general, these additionally diagnosed twins were symptomatic, but without sufficient evidence for a definite diagnosis at the time of the initial studies [98, 99]. The population-based data from the Scandinavian twin registries are supported by the German and British twin studies (Table 2.2) [94–97], although the British study does not include any information on proband concordance rates. Recently, the extraordinarily high concordances in monozygotic twins have been questioned. The very short observation periods between the year of birth in national twin cohorts and the year of study might have biased the inclusion towards twins with early onset disease and possibly with a more aggressive disease course, disease phenotypes that could be associated with a more pronounced genetic predisposition [100]. In contrast to some of the previous family studies in IBD, none of the original twin studies have standardized their findings according to age. Similarly, the follow-ups of the Scandinavian cohorts have studied twin pairs included in the original publications only, not extending their analyzes to potential new twin pairs within the total background twin population in each country. Appropriate concordance rates can be achieved either by standardizing previous results by age or by extending the observation periods. By rerunning the Swedish hospital discharge register with the Swedish twin registry and restricting the analyzes to twins born during the original studied period, that is, 1886–1958, the proband concordance rates in monozygotic and dizygotic twins with Crohn's disease were corrected to 38 % and 2 %,

Table 2.2 Concordance rates in different twin cohorts

	Crohn's disease				Ulcerative colitis			
	Monozygotic twins		Dizygotic twins		Monozygotic twins		Dizygotic twins	
Cohorts	Pair concordance (%)	Proband concordance (%)	Pair concordance (%)	Proband concordance (%)	Pair concordance (%)	Proband concordance (%)	Pair concordance (%)	Proband concordance (%)
Swedish, 1988 [92] ($n=80$)	44.4	58.3	3.8	3.8	6.3	6.6	0	0
Swedish, follow-up, 2000 [98] ($n=80$)	50.0	62.5	3.8	3.8	18.8	18.8	0	0
Swedish, 2011 [100] ($n=229$)	27.3[a]	38.5[a]	2.0[a]	2.0[a]	14.6[a]	14.6[a]	6.1[a]	8.0[a]
Danish, 2000 [93] ($n=103$)	50.0	58.3	0	0	14.3	18.2	4.5	4.5
Danish follow-up, 2005 [99] ($n=103$)	55.6	63.6	3.6	3.6	14.3	18.2	4.5	4.5
British, 1996 [95] ($n=150$)	20.0	Not published	6.5	Not published	15.8	Not published	2.9	Not published
British, 1998 [94] ($n=249$)	33.3	Not published	9.9	Not published	13.4	Not published	4.4	Not published
British, 2012 [97] ($n=250$)	25.6	Not published	7.3	Not published	19.1	Not published	1.5	Not published
German, 2008 y ($n=189$)	35.5	52.4	3.5	6.7	16.2	27.9	1.6	3.1

n number of twin pairs
[a]Analyzes restricted to twin pairs born in 1886–1958 ($n=179$)

respectively [100]. The corresponding figures for twins with ulcerative colitis were 15 % and 8 %, respectively.

Although the observed concordance slightly varies between the four countries, probably due to the differences in study design, the overall picture is similar (Table 2.2). The concordance is higher in monozygotic than in dizygotic pairs, and the difference is more pronounced in Crohn's disease than in ulcerative colitis [101]. In addition, monozygotic twin pairs with one twin suffering from Crohn's disease and one from ulcerative colitis are very uncommon, and only a few pairs have been reported in the literature [100, 102].

The heritability, providing an estimate of the relative contribution of genetics to disease etiology, can be calculated from the difference in concordance rates between monozygotic and dizygotic twin pairs. Utmost high heritability for Crohn's disease has been reported in the twin studies. However, the methods for calculating the heritability vary with the studies, and the confidence intervals when reported are wide. Thus, reported heritability for Crohn's disease and ulcerative colitis, respectively, needs to be interpreted with caution and to be considered as rough estimates. Undoubtedly, the heritability of Crohn's disease is high and within the same range as other diseases with a pronounced genetic contribution, like type 1 diabetes. Yet, in spite of the fact that monozygotic twins share each other's genome, fewer than 50 % of the pairs with Crohn's disease are concordant, reflecting the influence of environment. A higher relative risk for concordant disease has also been observed in dizygotic twins than in ordinary siblings [103]. This would point towards the importance of shared internal intrauterine factors and/or shared external childhood environment, although the conclusion should be treated with some caution, since the results are based on small numbers and a less robust method.

Concordance in clinical characteristics has also been observed within twin pairs where both twins are affected by IBD. This was first reported in the follow-up of the original Swedish twin study [98]. In spite of the limited number of concordant monozygotic twins with CD, a remarkable phenotypic similarity was observed statistically within twin pairs, considering age at diagnosis, location of the disease, and progress in extent of inflammation. In six out of nine pairs, the diagnosis was made 2 years or less apart in each twin, but in three pairs the time interval was considerably longer. The onset of symptoms showed similar time aspects. The location of disease at diagnosis according to the Vienna classification was identical in seven of nine pairs, but differed slightly in 2. During the observation period, the extent of the inflammation in each twin could progress or remain within the same part of the gastrointestinal tract. In eight pairs, the twins showed the same pattern in this respect. The probability for the observed overall concordance in the nine monozygotic twin pairs was 6.5×10^{-9}, which strongly speaks against an association by chance. Although the similarity in disease behavior was of borderline significance only, concordance in disease behavior has been confirmed more recently in the combined Swedish–Danish twin cohort [104]. The observation of concordance in disease behavior also includes the entity perianal disease. In the Scandinavian study, 15 of 17 monozygotic twin pairs with CD were concordant for presence or absence of perianal disease [104].

Beyond the similarity in clinical characteristics at diagnosis of Crohn's disease, phenotypic concordance has also been reported longitudinally during the disease course. In total, 11 of 16 monozygotic pairs concordant for Crohn's disease had identical disease location, and 13 pairs had identical behavior of disease 10 years after diagnosis in the population-based combined Scandinavian cohort. Only one of the 16 monozygotic twin pairs concordant for CD and observed for ≥10 years had been living together during the course of the disease. These findings point towards a pronounced genetic impact on clinical characteristics. However, beyond *NOD2/CARD15*, studies based on molecular genetics, including genome-wide association studies, have not shown any firm genotype–phenotype associations [55]. This absence of genotype–phenotype associations could suggest that the phenotypic similarity within monozygotic twin pairs concordant for Crohn's disease is due to shared internal intrauterine factors and/or shared external childhood environment rather than to genetic predisposition. Actually, the latter might be supported by serologic data, where it has been suggested that commensal bacteria trigger and perpetuate a more complicated disease behavior in individuals predisposed to CD [105]. Also still to be explored is the possible influence of disease tolerance, that is, differences in susceptibility to tissue damage, in contrast to disease resistance [106]. Comparison of the degree of phenotypic similarity within monozygotic and dizygotic twin pairs where both twins are affected by Crohn's disease could disentangle the possible influence of genetics vs. the environment in disease characteristics. Unfortunately, concordant dizygotic twin pairs with Crohn's disease are rare, and so far, no comparisons have been performed.

In contrast to the high degree of similarity within pairs with Crohn's disease, phenotypic concordance has been observed for age at diagnosis and symptomatic onset only in monozygotic twins where both twins are affected by ulcerative colitis [98, 99, 104]. In the combined Scandinavian twin cohort, only four of nine monozygotic twin pairs had identical extent of inflammation at diagnosis, and only three were concordant for extent 10 years after diagnosis. Seven of nine pairs (78 %) were concordant for "colectomy or not," as they had not undergone surgery.

Associated Syndromes and Diseases with Well-Recognized Genetic Susceptibility

The observed associations of IBD with genetically determined syndromes, including Turner syndrome [107], Hermansky–Pudlak syndrome [108], glycogen storage disease Ib [109], cystic fibrosis [110], and pachydermoperiostosis [111], also provide epidemiological evidence for a role of genetics in IBD. An increased prevalence of IBD has also been observed in other inflammatory disorders with strong evidence of genetic susceptibility, like ankylosing spondylitis [112], psoriasis [113], multiple sclerosis [114], and celiac disease [115]. In more recent years, molecular studies have confirmed that a number of immune-mediated and inflammatory diseases have susceptibility genes in common [50, 55].

References

1. Economou M, Pappas G (2008) New global map of Crohn's disease: genetic, environmental, and socioeconomic correlations. Inflamm Bowel Dis 14(5):709–720
2. Molodecky NA, Soon IS, Rabi DM, Ghali WA, Ferris M, Chernoff G et al (2012) Increasing incidence and prevalence of the inflammatory bowel diseases with time, based on systematic review. Gastroenterology 142(1):46–54
3. Binder V (1998) Genetic epidemiology in inflammatory bowel disease. Dig Dis 16(6): 351–355
4. Tysk C, Järnerot G (1992) Ulcerative proctocolitis in Orebro, Sweden. A retrospective epidemiologic study, 1963–1987. Scand J Gastroenterol 27(11):945–950
5. Langholz E, Munkholm P, Nielsen OH, Kreiner S, Binder V (1991) Incidence and prevalence of ulcerative colitis in Copenhagen county from 1962 to 1987. Scand J Gastroenterol 26(12):1247–1256
6. Lindberg E, Järnerot G (1991) The incidence of Crohn's disease is not decreasing in Sweden. Scand J Gastroenterol 26(5):495–500
7. Srivastava ED, Mayberry JF, Morris TJ, Smith PM, Williams GT, Roberts GM et al (1992) Incidence of ulcerative colitis in Cardiff over 20 years: 1968–87. Gut 33(2):256–258
8. Thomas GA, Millar-Jones D, Rhodes J, Roberts GM, Williams GT, Mayberry JF (1995) Incidence of Crohn's disease in Cardiff over 60 years: 1986–1990 an update. Eur J Gastroenterol Hepatol 7(5):401–405
9. Loftus CG, Loftus EV Jr, Harmsen WS, Zinsmeister AR, Tremaine WJ, Melton LJ III et al (2007) Update on the incidence and prevalence of Crohn's disease and ulcerative colitis in Olmsted County, Minnesota, 1940–2000. Inflamm Bowel Dis 13(3):254–261
10. Kyle J (1992) Crohn's disease in the northeastern and northern Isles of Scotland: an epidemiological review. Gastroenterology 103(2):392–399
11. Lapidus A (2006) Crohn's disease in Stockholm County during 1990–2001: an epidemiological update. World J Gastroenterol 12(1):75–81
12. Gunesh S, Thomas GA, Williams GT, Roberts A, Hawthorne AB (2008) The incidence of Crohn's disease in Cardiff over the last 75 years: an update for 1996–2005. Aliment Pharmacol Ther 27(3):211–219
13. Chouraki V, Savoye G, Dauchet L, Vernier-Massouille G, Dupas JL, Merle V et al (2011) The changing pattern of Crohn's disease incidence in northern France: a continuing increase in the 10- to 19-year-old age bracket (1988–2007). Aliment Pharmacol Ther 33(10):1133–1142
14. Molinie F, Gower-Rousseau C, Yzet T, Merle V, Grandbastien B, Marti R et al (2004) Opposite evolution in incidence of Crohn's disease and ulcerative colitis in Northern France (1988–1999). Gut 53(6):843–848
15. Hildebrand H, Finkel Y, Grahnquist L, Lindholm J, Ekbom A, Askling J (2003) Changing pattern of paediatric inflammatory bowel disease in northern Stockholm 1990–2001. Gut 52(10):1432–1434
16. Sood A, Midha V, Sood N, Bhatia AS, Avasthi G (2003) Incidence and prevalence of ulcerative colitis in Punjab, North India. Gut 52(11):1587–1590
17. Wright JP, Froggatt J, O'Keefe EA, Ackerman S, Watermeyer S, Louw J et al (1986) The epidemiology of inflammatory bowel disease in Cape Town 1980–1984. S Afr Med J 70(1):10–15
18. Yang SK, Hong WS, Min YI, Kim HY, Yoo JY, Rhee PL et al (2000) Incidence and prevalence of ulcerative colitis in the Songpa-Kangdong District, Seoul, Korea, 1986–1997. J Gastroenterol Hepatol 15(9):1037–1042
19. Yao T, Matsui T, Hiwatashi N (2000) Crohn's disease in Japan: diagnostic criteria and epidemiology. Dis Colon Rectum 43(10 suppl):S85–S93
20. Yang SK, Yun S, Kim JH, Park JY, Kim HY, Kim YH et al (2008) Epidemiology of inflammatory bowel disease in the Songpa-Kangdong district, Seoul, Korea, 1986–2005: a KASID study. Inflamm Bowel Dis 14(4):542–549

21. Calkins BM, Mendeloff AI (1986) Epidemiology of inflammatory bowel disease. Epidemiol Rev 8:60–91
22. Kurata JH, Kantor-Fish S, Frankl H, Godby P, Vadheim CM (1992) Crohn's disease among ethnic groups in a large health maintenance organization. Gastroenterology 102(6): 1940–1948
23. Ogunbi SO, Ransom JA, Sullivan K, Schoen BT, Gold BD (1998) Inflammatory bowel disease in African-American children living in Georgia. J Pediatr 133(1):103–107
24. Montgomery SM, Morris DL, Pounder RE, Wakefield AJ (1999) Asian ethnic origin and the risk of inflammatory bowel disease. Eur J Gastroenterol Hepatol 11(5):543–546
25. Ahmad T, Satsangi J, McGovern D, Bunce M, Jewell DP (2001) Review article: the genetics of inflammatory bowel disease. Aliment Pharmacol Ther 15(6):731–748
26. Sandler E (1994) Epidemiology of inflammatory bowel disease. In: Targan SR, Shanahan F (eds) Inflammatory bowel disease: from bench to bedside. Williams & Wilkins, Baltimore, pp 5–30
27. Roth MP, Petersen GM, McElree C, Feldman E, Rotter JI (1989) Geographic origins of Jewish patients with inflammatory bowel disease. Gastroenterology 97(4):900–904
28. Yang H, McElree C, Roth MP, Shanahan F, Targan SR, Rotter JI (1993) Familial empirical risks for inflammatory bowel disease: differences between Jews and non-Jews. Gut 34(4): 517–524
29. Niv Y, Abuksis G, Fraser GM (1999) Epidemiology of Crohn's disease in Israel: a survey of Israeli kibbutz settlements. Am J Gastroenterol 94(10):2961–2965
30. Odes HS, Fraser D, Hollander L (1989) Epidemiological data of Crohn's disease in Israel: etiological implications. Public Health Rev 17(4):321–335
31. Allchin WH (1909) Ulcerative colitis. Proc Roy Soc Med 2:59–75
32. Crohn BB (1934) The broadening conception of regional ileitis. Am J Dig Dis 1:97–99
33. Spriggs EI (1934) Chronic ulceration of the colon. Q J Med 27:549–578
34. Moltke O (1936) Familial occurrences of non-specific suppurative coloproctitis. Acta Med Scand 78(suppl 77):426–432
35. Jackman RF, Bargen JA (1942) Familial occurrence of chronic ulcerative colitis. Am J Dig Dis 9:147–149
36. Morris PJ (1965) Familial ulcerative colitis. Gut 36:176–178
37. Sherlock P, Bell BM, Steinberg H, Almy TP (1963) Familial occurrence of regional enteritis and ulcerative colitis. Gastroenterology 45:413–420
38. Kirsner JB, Spencer JA (1963) Family occurrences of ulcerative colitis, regional enteritis, and ileocolitis. Ann Intern Med 59:133–144
39. Orholm M, Munkholm P, Langholz E, Nielsen OH, Sorensen IA, Binder V (1991) Familial occurrence of inflammatory bowel disease. N Engl J Med 324(2):84–88
40. Binder V, Weeke E, Olsen JH, Anthonisen P, Riis P (1966) A genetic study of ulcerative colitis. Scand J Gastroenterol 1(1):49–56
41. Farmer RG, Michener WM, Mortimer EA (1980) Studies of family history among patients with inflammatory bowel disease. Clin Gastroenterol 9(2):271–277
42. Mayberry JF, Rhodes J, Newcombe RG (1980) Familial prevalence of inflammatory bowel disease in relatives of patients with Crohn's disease. Br Med J 280(6207):84
43. Monsen U, Brostrom O, Nordenvall B, Sörstad J, Hellers G (1987) Prevalence of inflammatory bowel disease among relatives of patients with ulcerative colitis. Scand J Gastroenterol 22(2):214–218
44. Bayless TM, Tokayer AZ, Polito JM II, Quaskey SA, Mellits ED, Harris ML (1996) Crohn's disease: concordance for site and clinical type in affected family members–potential hereditary influences. Gastroenterology 111(3):573–579
45. Monsen U, Bernell O, Johansson C, Hellers G (1991) Prevalence of inflammatory bowel disease among relatives of patients with Crohn's disease. Scand J Gastroenterol 26(3): 302–306
46. Weterman IT, Pena AS (1984) Familial incidence of Crohn's disease in The Netherlands and a review of the literature. Gastroenterology 86(3):449–452

47. Roth MP, Petersen GM, McElree C, Vadheim CM, Panish JF, Rotter JI (1989) Familial empiric risk estimates of inflammatory bowel disease in Ashkenazi Jews. Gastroenterology 96(4):1016–1020
48. Peeters M, Nevens H, Baert F, Hiele M, de Meyer AM, Vlietinck R et al (1996) Familial aggregation in Crohn's disease: increased age-adjusted risk and concordance in clinical characteristics. Gastroenterology 111(3):597–603
49. Polito JM II, Childs B, Mellits ED, Tokayer AZ, Harris ML, Bayless TM (1996) Crohn's disease: influence of age at diagnosis on site and clinical type of disease. Gastroenterology 111(3):580–586
50. Carbonnel F, Macaigne G, Beaugerie L, Gendre JP, Cosnes J (1999) Crohn's disease severity in familial and sporadic cases. Gut 44(1):91–95
51. Probert CS, Jayanthi V, Hughes AO, Thompson JR, Wicks AC, Mayberry JF (1993) Prevalence and family risk of ulcerative colitis and Crohn's disease: an epidemiological study among Europeans and south Asians in Leicestershire. Gut 34(11):1547–1551
52. Russel MG, Pastoor CJ, Janssen KM, van Deursen CT, Muris JW, van Wijlick EH et al (1997) Familial aggregation of inflammatory bowel disease: a population-based study in South Limburg, The Netherlands. The South Limburg IBD Study Group. Scand J Gastroenterol Suppl 223:88–91
53. Loftus EV Jr (2004) Clinical epidemiology of inflammatory bowel disease: Incidence, prevalence, and environmental influences. Gastroenterology 126(6):1504–1517
54. Bridger S, Lee JC, Bjarnason I, Jones JE, Macpherson AJ (2002) In siblings with similar genetic susceptibility for inflammatory bowel disease, smokers tend to develop Crohn's disease and non-smokers develop ulcerative colitis. Gut 51(1):21–25
55. Lees CW, Barrett JC, Parkes M, Satsangi J (2011) New IBD genetics: common pathways with other diseases. Gut 60(12):1739–1753
56. Anderson CA, Boucher G, Lees CW, Franke A, D'Amato M, Taylor KD et al (2011) Meta-analysis identifies 29 additional ulcerative colitis risk loci, increasing the number of confirmed associations to 47. Nat Genet 43(3):246–252
57. Franke A, McGovern DP, Barrett JC, Wang K, Radford-Smith GL, Ahmad T et al (2010) Genome-wide meta-analysis increases to 71 the number of confirmed Crohn's disease susceptibility loci. Nat Genet 42(12):1118–1125
58. Nimmo ER, Prendergast JG, Aldhous MC, Kennedy NA, Henderson P, Drummond HE et al (2012) Genome-wide methylation profiling in Crohn's disease identifies altered epigenetic regulation of key host defense mechanisms including the Th17 pathway. Inflamm Bowel Dis 18(5):889–899
59. Satsangi J, Rosenberg WMC, Jewell DP (1994) The prevalence of inflammatory bowel disease in relatives of patients with Crohn's disease. Eur J Gastroenterol Hepatol 6:413–416
60. Bennett RA, Rubin PH, Present DH (1991) Frequency of inflammatory bowel disease in offspring of couples both presenting with inflammatory bowel disease. Gastroenterology 100(6):1638–1643
61. Laharie D, Debeugny S, Peeters M, Van Gossum A, Gower-Rousseau C, Belaiche J et al (2001) Inflammatory bowel disease in spouses and their offspring. Gastroenterology 120(4): 816–819
62. Lennard-Jones JE (1989) Classification of inflammatory bowel disease. Scand J Gastroenterol Suppl 170:2–6; discussion 16–19
63. Stange EF, Travis SP, Vermeire S, Beglinger C, Kupcinkas L, Geboes K et al (2006) European evidence based consensus on the diagnosis and management of Crohn's disease: definitions and diagnosis. Gut 55(suppl 1):i1–i15
64. Farmer RG, Hawk WA, Turnbull RB Jr (1975) Clinical patterns in Crohn's disease: a statistical study of 615 cases. Gastroenterology 68(4 pt 1):627–635
65. Farmer RG, Whelan G, Fazio VW (1985) Long-term follow-up of patients with Crohn's disease. Relationship between the clinical pattern and prognosis. Gastroenterology 88(6): 1818–1825

66. Franchimont DP, Louis E, Croes F, Belaiche J (1998) Clinical pattern of corticosteroid dependent Crohn's disease. Eur J Gastroenterol Hepatol 10(10):821–825
67. Griffiths AM, Wesson DE, Shandling B, Corey M, Sherman PM (1991) Factors influencing postoperative recurrence of Crohn's disease in childhood. Gut 32(5):491–495
68. Whelan G, Farmer RG, Fazio VW, Goormastic M (1985) Recurrence after surgery in Crohn's disease. Relationship to location of disease (clinical pattern) and surgical indication. Gastroenterology 88(6):1826–1833
69. Greenstein AJ, Lachman P, Sachar DB, Springhorn J, Heimann T, Janowitz HD et al (1988) Perforating and non-perforating indications for repeated operations in Crohn's disease: evidence for two clinical forms. Gut 29(5):588–592
70. Aeberhard P, Berchtold W, Riedtmann HJ, Stadelmann G (1996) Surgical recurrence of perforating and nonperforating Crohn's disease. A study of 101 surgically treated Patients. Dis Colon Rectum 39(1):80–87
71. Sachar DB, Andrews HA, Farmer RG, Pallone F, Pena AS, Prantera C et al (1992) Proposed classification of patient subgroups in Crohn's disease Gastroenterology International. Gut 5:141–154
72. Gasche C, Scholmerich J, Brynskov J, D'Haens G, Hanauer SB, Irvine EJ et al (2000) A simple classification of Crohn's disease: report of the Working Party for the World Congresses of Gastroenterology, Vienna 1998. Inflamm Bowel Dis 6(1):8–15
73. Silverberg MS, Satsangi J, Ahmad T, Arnott ID, Bernstein CN, Brant SR et al (2005) Toward an integrated clinical, molecular and serological classification of inflammatory bowel disease: Report of a Working Party of the 2005 Montreal World Congress of Gastroenterology. Can J Gastroenterol 19(suppl A):5–36
74. Satsangi J, Grootscholten C, Holt H, Jewell DP (1996) Clinical patterns of familial inflammatory bowel disease. Gut 38(5):738–741
75. Colombel JF, Grandbastien B, Gower-Rousseau C, Plegat S, Evrard JP, Dupas JL et al (1996) Clinical characteristics of Crohn's disease in 72 families. Gastroenterology 111(3):604–607
76. Lee JC, Lennard-Jones JE (1996) Inflammatory bowel disease in 67 families each with three or more affected first-degree relatives. Gastroenterology 111(3):587–596
77. Annese V, Andreoli A, Astegiano M, Campieri M, Caprilli R, Cucchiara S et al (2001) Clinical features in familial cases of Crohn's disease and ulcerative colitis in Italy: a GISC study. Italian study group for the disease of colon and rectum. Am J Gastroenterol 96(10):2939–2945
78. Akolkar PN, Gulwani-Akolkar B, Heresbach D, Lin XY, Fisher S, Katz S et al (1997) Differences in risk of Crohn's disease in offspring of mothers and fathers with inflammatory bowel disease. Am J Gastroenterol 92(12):2241–2244
79. Zelinkova Z, Stokkers PC, van der Linde K, Kuipers EJ, Peppelenbosch MP, van der Woude CP (2012) Maternal imprinting and female predominance in familial Crohn's disease. J Crohns Colitis 6(7):771–776
80. Franchimont D, Belaiche J, Louis E, Simon S, GrandBastien B, Gower-Rousseau C et al (1997) Familial Crohn's disease: a study of 18 families. Acta Gastroenterol Belg 60(2): 134–137
81. Cottone M, Brignola C, Rosselli M, Oliva L, Belloli C, Cipolla C et al (1997) Relationship between site of disease and familial occurrence in Crohn's disease. Dig Dis Sci 42(1): 129–132
82. Pariente B, Cosnes J, Danese S, Sandborn WJ, Lewin M, Fletcher JG et al (2011) Development of the Crohn's disease digestive damage score, the Lemann score. Inflamm Bowel Dis 17(6):1415–1422
83. Peeters M, Cortot A, Vermeire S, Colombel JF (2000) Familial and sporadic inflammatory bowel disease: different entities? Inflamm Bowel Dis 6(4):314–320
84. Fu YH, Kuhl DP, Pizzuti A, Pieretti M, Sutcliffe JS, Richards S et al (1991) Variation of the CGG repeat at the fragile X site results in genetic instability: resolution of the Sherman paradox. Cell 67(6):1047–1058

85. Harley HG, Brook JD, Rundle SA, Crow S, Reardon W, Buckler AJ et al (1992) Expansion of an unstable DNA region and phenotypic variation in myotonic dystrophy. Nature 355(6360):545–546
86. Anon (1993) A novel gene containing a trinucleotide repeat that is expanded and unstable on Huntington's disease chromosomes. The Huntington's Disease Collaborative Research Group. Cell 72(6):971–983
87. Polito JM II, Rees RC, Childs B, Mendeloff AI, Harris ML, Bayless TM (1996) Preliminary evidence for genetic anticipation in Crohn's disease. Lancet 347(9004):798–800
88. Grandbastien B, Peeters M, Franchimont D, Gower-Rousseau C, Speckel D, Rutgeerts P et al (1998) Anticipation in familial Crohn's disease. Gut 42(2):170–174
89. Bayless TM, Picco MF, LaBuda MC (1998) Genetic anticipation in Crohn's disease. Am J Gastroenterol 93(12):2322–2325
90. Bengtson MB, Solberg C, Aamodt G, Jahnsen J, Moum B, Sauar J et al (2009) Clustering in time of familial IBD separates ulcerative colitis from Crohn's disease. Inflamm Bowel Dis 15(12):1867–1874
91. Picco MF, Goodman S, Reed J, Bayless TM (2001) Methodologic pitfalls in the determination of genetic anticipation: the case of Crohn disease. Ann Intern Med 134(12):1124–1129
92. Tysk C, Lindberg E, Järnerot G, Floderus-Myrhed B (1988) Ulcerative colitis and Crohn's disease in an unselected population of monozygotic and dizygotic twins. A study of heritability and the influence of smoking. Gut 29(7):990–996
93. Orholm M, Binder V, Sorensen TI, Rasmussen LP, Kyvik KO (2000) Concordance of inflammatory bowel disease among Danish twins. Results of a nationwide study. Scand J Gastroenterol 35(10):1075–1081
94. Subhani J, Montgomery SM, Pounder RE, Wakefield AJ (1998) Concordance rates of twins and siblings in inflammatory bowel disease (IBD). Gut 42(suppl I):A40–A48
95. Thompson NP, Driscoll R, Pounder RE, Wakefield AJ (1996) Genetics versus environment in inflammatory bowel disease: results of a British twin study. BMJ 312(7023):95–96
96. Spehlmann ME, Begun AZ, Burghardt J, Lepage P, Raedler A, Schreiber S (2008) Epidemiology of inflammatory bowel disease in a German twin cohort: results of a nationwide study. Inflamm Bowel Dis 14(7):968–976
97. Ng SC, Woodrow S, Patel N, Subhani J, Harbord M (2012) Role of genetic and environmental factors in British twins with inflammatory bowel disease. Inflamm Bowel Dis 18(4): 725–736
98. Halfvarson J, Bodin L, Tysk C, Lindberg E, Järnerot G (2003) Inflammatory bowel disease in a Swedish twin cohort: a long-term follow-up of concordance and clinical characteristics. Gastroenterology 124(7):1767–1773
99. Jess T, Riis L, Jespersgaard C, Hougs L, Andersen PS, Orholm MK et al (2005) Disease concordance, zygosity, and NOD2/CARD15 status: follow-up of a population-based cohort of Danish twins with inflammatory bowel disease. Am J Gastroenterol 100(11):2486–2492
100. Halfvarson J (2011) Genetics in twins with Crohn's disease: less pronounced than previously believed? Inflamm Bowel Dis 17(1):6–12
101. Brant SR (2011) Update on the heritability of inflammatory bowel disease: the importance of twin studies. Inflamm Bowel Dis 17(1):1–5
102. Breslin NP, Todd A, Kilgallen C, O'Morain C (1997) Monozygotic twins with Crohn's disease and ulcerative colitis: a unique case report. Gut 41(4):557–560
103. Bengtson MB, Aamodt G, Vatn MH, Harris JR (2010) Concordance for IBD among twins compared to ordinary siblings–a Norwegian population-based study. J Crohns Colitis 4(3): 312–318
104. Halfvarson J, Jess T, Bodin L, Järnerot G, Munkholm P, Binder V et al (2007) Longitudinal concordance for clinical characteristics in a Swedish-Danish twin population with inflammatory bowel disease. Inflamm Bowel Dis 13(12):1536–1544
105. Dubinsky MC, Lin YC, Dutridge D, Picornell Y, Landers CJ, Farrior S et al (2006) Serum immune responses predict rapid disease progression among children with Crohn's disease: immune responses predict disease progression. Am J Gastroenterol 101(2):360–367

106. Medzhitov R, Schneider DS, Soares MP (2012) Disease tolerance as a defense strategy. Science 335(6071):936–941
107. Weinrieb IJ, Fineman RM, Spiro HM (1976) Turner syndrome and inflammatory bowel disease. N Engl J Med 294(22):1221–1222
108. Schinella RA, Greco MA, Cobert BL, Denmark LW, Cox RP (1980) Hermansky-Pudlak syndrome with granulomatous colitis. Ann Intern Med 92(1):20–23
109. Couper R, Kapelushnik J, Griffiths AM (1991) Neutrophil dysfunction in glycogen storage disease Ib: association with Crohn's-like colitis. Gastroenterology 100(2):549–554
110. Lloyd-Still JD (1994) Crohn's disease and cystic fibrosis. Dig Dis Sci 39(4):880–885
111. Compton RF, Sandborn WJ, Yang H, Lindor NM, Tremaine WJ, Davis MD et al (1997) A new syndrome of Crohn's disease and pachydermoperiostosis in a family. Gastroenterology 112(1):241–249
112. Jayson MI, Bouchier IA (1968) Ulcerative colitis in patients with ankylosing spondylitis. Proc R Soc Med 61(4):340–341
113. Yates VM, Watkinson G, Kelman A (1982) Further evidence for an association between psoriasis, Crohn's disease and ulcerative colitis. Br J Dermatol 106(3):323–330
114. Minuk GY, Lewkonia RM (1986) Possible familial association of multiple sclerosis and inflammatory bowel disease. N Engl J Med 314(9):586
115. Cottone M, Cappello M, Puleo A, Cipolla C, Filippazzo MG (1989) Familial association of Crohn's and coeliac diseases. Lancet 2(8658):338
116. Halme L, Turunen U, Helio T, Paavola P, Walle T, Miettinen A et al (2002) Familial and sporadic inflammatory bowel disease: comparison of clinical features and serological markers in a genetically homogeneous population. Scand J Gastroenterol 37(6):692–698
117. Freeman HJ (2002) Familial Crohn's disease in single or multiple first-degree relatives. J Clin Gastroenterol 35(1):9–13

Chapter 3
Insights from Recent Advances in Animal Models of Inflammatory Bowel Disease

Emiko Mizoguchi, Atsushi Mizoguchi, and Atul K. Bhan

Abstract Experimental animal models of intestinal inflammation that mimic inflammatory bowel disease (Nat Genet 2010, 42(4), 332–337) have become increasingly common over the last 2 decades. These include experimentally induced and genetically altered models as well as models of spontaneous intestinal inflammation. None of these models exactly replicate all features of Crohn's disease and ulcerative colitis, the two major subtypes of human IBD, but they have furthered our understanding and define the pathogenesis underlining intestinal inflammation. Of particular note is the recognition that enteric microorganisms play a central role in the maintenance of normal mucosal immune homeostasis and the development of pathologic innate and adaptive immune responses that lead to mucosal inflammation and disease. In this review, we summarize the data gathered from commonly utilized animal models of IBD. In addition, we review experimental models developed to define the role of recently identified human IBD susceptibility genes. These genes may regulate host responsiveness against microbial and environmental factors and control immune effector molecules during inflammation. These models not only promise to increase our understanding of the pathogenesis of IBD but also provide rationale for new therapeutic strategies.

E. Mizoguchi, M.D., Ph.D. (✉) • A. Mizoguchi, M.D., Ph.D.
Gastroinestinal Unit, Massachusetts General Hospital, GRJ 825D, Boston, MA 02114, USA
e-mail: emizoguchi@partners.org; amizoguchi@partners.org

A.K. Bhan, M.B.B.S., M.D.
Department of Pathology, Massachusetts General Hospital, 55 Fruit St., Warren 501, Boston, MA 02114, USA
e-mail: abhan@partners.org

M. D'Amato and J.D. Rioux (eds.), *Molecular Genetics of Inflammatory Bowel Disease*, DOI 10.1007/978-1-4614-8256-7_3,

Abbreviations

Ab	Antibody
APC	Antigen-presenting cell
ATG	Autophagy-related protein
BCR	B cell receptor
BMDC	Bone marrow-derived dendritic cells
BrdU	5-Bromo-2′-deoxyuridine
CD	Crohn's disease
cdcs1	Cytokine deficiency-induced colitis susceptibility-1
CECs	Colonic epithelial cells
DC	Dendritic cells
Defcr	Defensin-related cryptdin
DKO	Double knockout
DTH	Delayed-type hypersensitivity
ER	Endoplasmic reticulum
FM	Follicular mature
GF	Germfree
GWAS	Genome-wide association studies
HM	Hypomorphic
IBD	Inflammatory bowel disease
IEL	Intraepithelial lymphocytes
IFN	Interferon
Igμ	Immunoglobulin mu
IL or Il	Interleukin
KO	Knockout
LPS	Lipopolysaccharide
LRR	Leucine-rich repeat
LTi	Lymphoid tissue inducer
MAPK	Mitogen-activated protein kinase
MDP	Muramyl dipeptide
MHC	Major histocompatibility complex
MLN	Mesenteric lymph nodes
MyD88	Myeloid differentiation primary response gene 88
MZ	Marginal zone
NADPH	Nicotinamide adenine dinucleotide phosphate
NoD2	Nucleotide-binding oligomerization domain-containing protein 2
PD	Parkinson's disease
PKCθ	Protein kinase C theta
Rag, V(D) J	Recombination activation gene
SNP	Single-nucleotide polymorphism
SPF	Specific pathogen-free
STAT	Signal transducer and activator of transcription
TCRα	T cell receptor alpha-chain
TH	T helper
TLR	Toll-like receptor

TNBS	2,4,6-Trinitrobenzene sulfonic acid
TNF	Tumor necrosis factor
TRIF	TIR-domain-containing adapter-inducing interferon-β
TRUC	T-bet and Rag-2 DKO
UC	Ulcerative colitis
XBP-1	X-box-binding protein-1
γc	Common gamma chain
γδ	Gamma delta

Introduction

Inflammatory bowel disease (IBD), including Crohn's disease (CD) and ulcerative colitis (UC), is a group of intestinal chronic inflammatory conditions that affect individuals throughout life [1, 2]. Although the mechanisms involved in the pathogenesis of IBD are complex, continuous microbial stimulation resulting in pathogenic immune responses in a genetically susceptible host likely plays a key role in the development of IBD [3, 4]. Recent human genome-wide association studies (GWAS) have identified 140 risk loci/genes in CD and 133 risk loci/genes in UC [3, 5–7]. Polymorphisms in genes, which regulate intestinal homeostasis and immunity, including *IL10R* and its signaling components, are strongly associated with the pathogenesis of human IBD [3, 8–10]. Therefore, analysis of animal models that manipulate these IBD susceptibility identified by GWAS is likely to enhance our understanding of their role in IBD pathogenesis.

More than 60 different animal models of IBD have been established to study intestinal inflammation [11]. These include genetically manipulated animal models [conventional/conditional knockout (KO), knockin, or transgenic mice] that spontaneously develop colitis and/or ileitis [11–13]. Other models use chemical compounds or adaptive cell transfer to induce intestinal inflammation to investigate different phases of intestinal inflammation [14–16]. These experimental models have provided a way to dissect mechanisms that may underlie the pathogenesis of human IBD. Here, we summarize findings in commonly utilized as well as IBD susceptibility gene-manipulated animal models (Table 3.1) of intestinal inflammation and assess their value in human IBD.

Spontaneous Colitis Models

C3H/HeJBir Mouse Strain

C3H/HeJBir (C3Bir) mice, a substrain of inbred C3H/HeJ mice generated at the Jackson Laboratory, spontaneously develop inflammation in the cecum and right colon [17]. C3H/HeJ mice have a missense mutation in the third exon of the Toll-like receptor-4 gene (*Tlr4*) and therefore express a defective response to the biologic

Table 3.1 Categories of murine models of colitis

Category	Animal models	Sections in this chapter
Spontaneous colitis models	C3H/HeJBir	"*C3H/HeJBir* Mouse Strain"
	SAMP/Yit	"SAMP/Yit Fc Mouse Strain"
Induced models of colitis	Acetic acid	
	DSS	"DSS-Induced Colitis"
	TNBS	"TNBS-Induced Colitis"
	Oxazolone	"Oxazolone-Induced Colitis"
	CD45RBhi cell transfer	"CD45RB^{hi} Cell Transferred Model"
	CD3ε transgenic model	
	CD8 transfer	
	Anti-CD40 mAb-induced	"Anti-CD40 mAb-Induced Colitis"
Bacterial infectious colitis	*Salmonella*	
	Citrobacter	
	AIEC	
	ECOVA	
Genetically manipulated colitis (conventional KO only)	*Il-2* KO	"*Il2* KO and *Il2R* KO Mice"
	Il-2R KO	"*Il2* KO and *Il2R* KO Mice"
	Il-10 KO	"*IL-10* KO Mice"
	Tcrα KO	"*Tcra* KO Mice"
	Gαi2 KO	"*Gαi2* KO Mice"
	Tbet x *Rag* DKO (TRUC)	"*T-bet/Rag-2* DKO (TRUC) Mice"
	Wasp KO	"*Wasp* KO Mice"
	Integrin αV KO	
	Jak3 KO	
	A20 KO	
	Tgfβ KO	
	Mdr1a KO	
	Tlr5 KO	
	Muc2 KO	
	Ship1 KO	
	Keratin 8 KO	
	Gpx1/2 KO	
IBD susceptibility gene related models	*Il-10* KO	"*IL-10* KO Mice"
	Il-10R KO	
	Nod2 KO	"Nod2 KO Mice"
	Atg5 KO	"*Atg5* KO and *Atg16l1* Mutant Mice"
	Il23R KO	"IL23 receptor KO Mice"
	Stat3 KO	"Conditional *Stat3* KO Mice"
	Xbp1 KO	"*Xbp1* KO Mice"
	Lrrk2 KO	"*Lrrk2* KO Mice"
	P40phox KO	"*p40*^{phox} KO Mice"

Selected models of colitis (as noted above) have been described in the chapter

effects of lipopolysaccharide (LPS) [18, 19]. C3Bir mice also express this unresponsiveness [20]. The inflammation involving the right colon and cecum peaks at around 3–6 weeks of age with resolution by 12 weeks of age. However, a mild recurrence of colitis is sporadically observed after 1 year of age [17]. Using Western blot analysis, sera from colitic C3Bir mice show a unique reactivity to antigen extracts from cecal contents (bacteria), while sera from colitic C3H/HeJ mice do not show such reactivity [21].

Histologically, focal inflammation involving the cecum extends to the right colon with presence of mixed acute and chronic inflammation, colonic crypt hyperplasia, and submucosal scarring. In severe cases, crypt abscess formation is observed [22]. Enteric microflora plays a central role in the pathogenesis of intestinal inflammation in these mice [21]. Cell transfer studies with C3Bir $CD4^+$ T cells into combined immunodeficiency mice (*scid*, which have absent or atypical T cells and B cells) suggest that only cecal bacterial antigen-reactivated C3Bir $CD4^+$ T cells are the predominant pathogenic population to induce colitis [23].

The Il-10-deficient C3Bir (*C3Bir.Il10* KO) mice developed much more severe colitis in the cecum and colon as early as 4 weeks of age, as compared to *B6.Il10* KO mice [24]. Severe inflammation with ulceration, epithelial hyperplasia, and exuberant exudates can be seen in the colons of *C3Bir.Il10* KO [25]. Genetic analysis revealed that *Cdcs1* (cytokine deficiency-induced colitis susceptibility-1) on chromosome 3 has a strong linkage to the control of multiple colitogenic subphenotypes in C3Bir mice. The C3Bir *Cdcs1* susceptibility allele confers a reduced responsiveness to Tlr2, Tlr5, Tlr9, and Nod2-mediated bacteria sensor signaling. Compensatory mechanisms of adaptive immune responses in turn induce the hyperactivation of nuclear factor-kappa B (NF-κB) signaling in macrophages [25].

SAMP/YitFc Mouse Strain

A colony of the senescence-accelerated mouse (SAM1–10) was developed at Kyoto University after an unintentional outcross between a non-AKR strain and AKR/J strain [26]. Subsequently, a SAMP/Yit substrain was generated from the SAMP1 line that exhibited severe intestinal inflammation (restricted to the ileum and cecum with a skip pattern) and skin lesions [27]. A distinct substrain (SAMP1/YitFc) displays unique features of intestinal inflammation including perianal disease with fistula formation in a subset of mice (approximately 5 %). These mice developed ileitis as early as 10 weeks of age with elevated interferon gamma (IFN-γ) and tumor necrosis factor alpha (TNF-α) levels. The incidence of skin lesions was inversely correlated with the occurrence of ileitis [28, 29]. By 30 weeks of age, both SAMP/Yit and SAMP1/YitFc mice develop transmural and discontinuous inflammation in the terminal ileum with 100 % penetrance [27, 28]. In SAMP1/YitFc mice, the earliest histological change entails a dramatic expansion of epithelial cells of the secretory lineage (e.g., Paneth, goblet, and intermediate cells) [30]. In addition, global expansion of the crypt epithelial stem cell population can be seen as disease severity progresses [30, 31]. In inflamed areas, SAMP mice show an increased number of apoptotic intestinal epithelial cells (IRCs), which are normalized after anti-TNF treatment [32]. These epithelial alterations are considered to be the primary defects that trigger ileitis in this model [33]. The mice also develop *Helicobacter*-negative, immune-mediated gastritis [34].

The spontaneous ileitis in SAMP1/YitFc mice resembles human Crohn's disease [35]. In a genome-wide scan performed in two cohorts of F2 mice (SAMP1Yit/Fc x C57Bl/6J), a single SAMP-derived susceptibility locus (*Ibdq1*) was identified on chromosome 9 (Chr9) [35]. Both IL-10 receptor alpha (*Il10ra*) and IL-18 (*Il18*) genes, which regulate inflammatory responses, are located on *Ibdq1*. Comparative sequencing detected two single-nucleotide insertions at positions 732 and 3098 of introns 1 and 3 of the *Il10ra* gene in SAMP/YitFc strain as compared to AKR/J or C57Bl6/J mice, but no polymorphisms were detected for *Il18* among the three mouse strains [35]. However, recent studies have identified association of polymorphisms in the promoters of *Il18* [36] and the *Il18* gene haplotype-2 [37], suggesting the involvement of an undetected alteration in the *Il18* locus of the SAMP1/YitFc mice [35].

Both Th1 and Th2 responses are exaggerated in SAMP/YitFc mice. In particular, Il13 and Il5 expression levels are significantly (>25-fold) increased in the inflamed mucosa between 9 and 16 weeks of age [38, 39]. The Th2 pathway appears to contribute to the amplification of the inflammatory process during the chronic stage in SAMP mice [38] as the administration of anti-IL-4 monoclonal antibody after 15 weeks of age (with established disease) significantly decreases ileitis. Administration of anti-IL-15 antibody also ameliorates SAMP ileitis [39]. SAMP B cells also contribute to ileitis by producing immunoglobulins that specifically recognize antigenic epitopes resulting in immune complex formation in the gut mucosa [40]. A variety of experimental treatments including anti-TNF [32, 41] have been shown effective in treating the ileitis in SAMP mice studies.

Commonly Utilized Induced Models of Colitis

These models include colitis induced by the administration of chemicals or by the transfer of cells or antibodies. Although these models do not represent the spontaneous nature of human IBD, the induced intestinal inflammation may reflect various pathophysiological aspects of human IBD (Table 3.1). Furthermore, these models are often used to reveal susceptibility to mucosal injury and inflammation in genetically altered mice that may not develop spontaneous intestinal inflammation. Selected induced models of colitis are summarized in this chapter.

DSS-Induced Colitis

The dextran sodium sulfate (DSS; $C_6H_7Na_3O_{14}S_3$)-provoked colitis is one of the most widely used chemically induced colitis models. Approximately 2–5 % DSS (molecular weight of approximately 40 kDa) is orally administered in the drinking water to induce acute or chronic colitis in mice by inducing direct hyperosmotic damage to epithelial cells [42]. One cycle of 3–5 % DSS administration for 5–7 days, followed by regular drinking water intake, results in extensive colonic injury with complete crypt depletion in the acute phase and subsequently regeneration of epithelial cells in the recovery phase followed by regeneration of epithelial cells in the recovery phase. The C57Bl/6 strain is relatively resistant to this colitis, while the C3H/HeJ strain is highly susceptible [43].

After 5–10 % DSS administration for 8–9 continuous days, Balb/c mice gradually develop weight loss and diarrhea with presence of occult blood. The colon shows multiple mucosal erosive lesions with inflammatory cell infiltration of mononuclear cells (lymphocytes, macrophages) and polymorphonuclear cells [44]. Occasional crypt abscesses and regenerative epithelium are also seen in the colonic mucosa. After termination of DSS treatment, the colonic mucosa recovers gradually from the acute colonic injury. Therefore, DSS-induced colitis has been used for studying epithelial wound healing and regulatory immune cells during the recovery phase [45]. Chronic colitis can be induced by treating mice with repeated (3–5) cycles of DSS; each cycle consists of DSS administration for 7 days followed by drinking water for the subsequent 10 consecutive days [44]. In chronic DSS colitis, lymphoid follicular formation accompanied by regenerative and dysplastic changes of the mucosal epithelium is frequently seen in the left side of the large intestine and the transverse colon [44]. Long-term DSS administration with repeated cycles produces colorectal dysplasia and/or adenocarcinoma, which is reported to be similar to human IBD-associated colon cancer [43].

Acute DSS colitis can be induced in the absence of acquired immunity as *scid* mice are also susceptible. In the inflammatory lesions of DSS-treated *scid* mice, abundant proinflammatory cytokines such as IL-1β, TNF-α, and IL-6 are detected. These cytokines are mainly produced by activated macrophages after exposure to luminal bacterial components or products. Since the colitis is ameliorated by treatment with selected antibiotics (metronidazole, or vancomycin-imipenem combination) [46], an antibiotic-sensitive component of luminal bacteria is involved in the pathogenesis of DSS-induced colitis. However, mice kept in a germfree (GF) facility or treated with wide-spectrum antibiotics (mixture of vancomycin, neomycin, metronidazole, and ampicillin) develop more severe and even lethal DSS-induced colitis accompanied by massive intestinal bleeding [47, 48]. This finding is supported by the observation that *MyD88* KO and *Tlr4* KO mice developed lethal colitis after DSS administration [48]. Therefore, signaling through TLRs, in particular TLR2, TLR4, and TLR9, is important in maintaining epithelial homeostasis and protection from direct epithelial injury induced by DSS. This is accomplished by directly inducing the expression of several factors, including heat shock proteins, TNF, IL-6, keratinocyte-derived chemokine-1 (KS), and/or type I IFN, which are involved in tissue repair and cytoprotection [48, 49]. The concentration of DSS, treatment periods, genetic background of mice, and microbial environment of facilities can influence the results of the experiments.

TNBS-Induced Colitis

TNBS (2,4,6-trinitrobenzene sulfonic acid; $C_6H_3N_3O_9S$) is a nitroaryl oxidizing acid. Simultaneous rectal administration of TNBS (between 0.5 and 6 mg per mouse) and ethanol (between 35 and 50 %) induces colonic inflammation [50]. A high dose of TNBS causes an acute necrotizing enterocolitis in mice [51]. C3H/HeJ, SJL/J, and Balb/c mice are highly susceptible to TNBS as compared to C57Bl/6 and DBA/2 mice [52, 53].

Acute colitis induced by rectally administered TNBS/ethanol leads to massive transmural inflammation associated with loose stools, rectal prolapse, and weight loss. After 2–3 days of TNBS administration, discrete foci of acute necrosis and mucosal inflammatory cell infiltration with focal basal cryptitis are observed. The acute inflammation is followed by mononuclear cell infiltration, which lasts for a variable amount of time [22]. The inflammation, which peaks a few days after TNBS administration, is mainly mediated by Th1-type immune responses with increased production of Th1 cytokines (e.g., IFN-γ, IL-2) and IL-12p70 (IL-12p40/p35) in the affected colon. However, the colitis is mediated by Th2 cytokine (e.g., IL-4) in Balb/c strain [54]. Generally, mice are sacrificed in the first or third week after TNBS administration depending on the stage of inflammation to be evaluated [55]. Chronic TNBS-associated colitis, induced by several weekly intrarectal administrations of TNBS in ethanol, is associated with IL23 (IL-12p40/p19) and IL-17 production by lamina propria mononuclear cells. Pathogenic IL23, but not IL-12p70, is associated with this type of colitis [56]. Zhang et al. demonstrated the importance of IL-17 during the development of chronic inflammation in TNBS-induced colitis, by utilizing IL-17 receptor A (*Il17ra*) KO mice [57]. Furthermore, administration of IL-17RA IgG1 fusion protein ameliorated colitis. IL-13 (a Th2-type cytokine), which triggers TGF-β-dependent tissue fibrosis, is also involved in this model [58].

By sensitizing mice with TNBS 6–7 days before the rectal administration of TNBS, the role of delayed hypersensitivity (DTH) reaction on colitis development can be studied [51]. The reaction is mediated by "hapten-modified self-antigen" followed by a local immune response through the activation of T cells and macrophages [59].

The TNBS-induced colitis model has been widely utilized for drug screening and examination of the effect of IBD susceptibility genetic variants in animal models [22, 56]. The relevance of the TNBS colitis model to CD is reflected in the experiments showing that *Nod2* (a key CD susceptibility gene) protects mice from TNBS colitis [60]. CD-associated frameshift-mutated *Nod2* in mice leads to increased sensitivity to colitis, compared to control mice carrying wild-type *Nod2* [61].

Oxazolone-Induced Colitis

Oxazolone (4-ethoxymethylene-2-phenyloxazol), a haptenating agent, has been used to elicit intestinal inflammation in mice. Intrarectal administration of 6 mg of oxazolone in 150 μL of 50 % ethanol in SJL mice leads to rapid onset of colitis marked by diarrhea, weight loss, and development of hemorrhagic colitis in the distal colon with patchy ulceration, submucosal edema, and presence of mucosal inflammatory infiltrate (lymphocytes and neutrophils) [62]. The inflammatory process is resolved by 10–12 days. Anti-IL-4, but not anti-IL-12, antibody leads to amelioration of colitis. In C57Bl/10 mice, pre-sensitization with 3 % oxazolone by skin painting 5 days before rectal challenge with 1 % oxazolone leads to marked edema of the colonic wall with mucosal ulceration and dense infiltration of superficial mucosa by neutrophils [63]. IL-13 but not IL-4 increases in the inflamed tissues, and mice treated with IL-13Rα2-Fc are protected from colitis. NK T cells are essential for the induction of colitis. Interestingly, lamina propria T cells from UC patients may produce more IL-13 than cells from healthy controls with a clear expansion of IL-13-producing CD161+ nonclassical NK T cells [64].

Studies in *Ifng* KO and *Il4* KO mice indicate that both Th1 and Th2 cytokines play a crucial pathogenic role in oxazolone colitis in C57BL/6 and BALB/c mice [65]. Recent studies indicate that IL-6 is also essential for the induction of oxazolone colitis; IL-6 production is regulated by the nuclear factor of activated T cells [66]. Oxazolone colitis can be suppressed by TNF-α by promoting local glucocorticoid synthesis [67].

CD45RB^{hi} Cell Transferred Model

CD4+ T cell population can be subdivided by CD45 antigen expression in the rat, mouse, and human [68]; naïve CD4^{+} T cells are included within the CD45RB^{high} fraction, whereas antigen-primed CD4^{+} memory T cells are in the CD45RB^{low} fraction. These two populations have direct lineage relationship [69]. Adoptive transfer of CD4^{+} CD45RB^{high} T cells (from the spleen of WT mice) into *scid* mice results in colitis [70, 71]. Severe colitis is found approximately 6–12 weeks after reconstitution of mice with $1–5 \times 10^5$ CD4^{+} CD45RB^{high} T cells. The characteristic features of colitis include transmural mononuclear cell infiltration, crypt hyperplasia, and goblet cell depletion. The induction of colitis is strain independent and can also be seen in nonobese diabetic mice [72], *Rag1* KO mice, *Cd3* KO mice, or athymic nude rats [73].

This model allows examination of the earliest immunological events associated with the induction/perpetuation of colitis [74]. Reconstitution of immunodeficient mice with CD4^{+} CD45RB^{high} T cells shows polarized Th1 cells with increased production of IFN-γ and TNF-α in the inflamed colon. The colitis is inhibited by treatment of CD45RB^{high} cells with anti-IFN-γ or anti-TNF-α monoclonal antibodies [71, 75]. In addition to Th1 cells, Th17 cells may also have a colitogenic effect in this model [76, 77]. However, other studies have shown that Th17 cells may not play any role in IBD [78–80]. In mice with colitis, CD4^{+} T cells isolated from the spleen, mesenteric lymph nodes (MLN), and colonic lamina propria produce both Th1 and Th17 types of cytokines [74, 81]. IL-6 trans-signaling is also required for the development of colitis [82] as IL-6 inhibits the generation of inducible regulatory T cells from naïve T cells [83]. IL-4 plays a role in the pathogenesis of CD45RB^{hi} colitis by blocking the generation of TGF-β-induced Foxp3^{+} Treg cells and by inducing a population of IL-9^{+} IL-10^{+} Foxp3(−) Th9 cells, which have no regulatory function despite producing abundant IL-10 [84].

This model has been very useful in the identification of regulatory T cell subsets involved in the suppression of colitis development. *Scid* mice that received CD45RB^{high} and CD45RB^{low} fractions together do not show severe colitis because of the presence of regulatory T cells within the CD45RB^{low} fraction [70, 71]. The regulatory effect of CD45RB^{low} cells is inhibited by anti-TGF-β, suggesting the critical role of TGF-β-producing CD4^{+} T (Th3) cells in the regulation of this form of colitis [85]. IL-10 drives the generation of a CD4^{+} T regulatory cell 1 (Tr1), which produce high levels of IL-10, low levels of IL-2, and no IL-4. Tr1 cells actively suppress pathogenic immune responses in this model in an antigen-dependent manner [86]. Furthermore, IL-10 produced by CD11b^{+} myeloid cells is a critical regulator since Treg cells could not maintain Foxp3 expression and regulatory activity in IL-10-deficient *Rag1* KO mice [87]. However, this observation is challenged by a recent finding that Foxp3 expression can be preserved in IL-10R-deficient Treg cells [88].

Anti-CD40 mAb-Induced Colitis

CD40 is a type I membrane glycoprotein and is expressed in many cell types including B cells, dendritic cells, monocytes/macrophages, fibroblasts, and activated endothelial cells [89–92]. The ligation of CD40 and CD154 (CD40L) activates the pathogenic signals in intestinal inflammation including TNBS-induced colitis [93] and CD45RB^{high} T cell transferred into *scid* mice [94, 95]; inhibition of CD40-CD40L interactions prevents or ameliorates the colonic inflammation. By generating a positive feedback loop, CD40^{+} APCs and CD40L^{+} T cells are further activated after CD40-CD40L ligation [92, 96].

A new colitis model was developed by injecting agonistic anti-CD40 monoclonal antibody FGK45 into *Rag1* KO or *scid* mice leading to body weight loss, diarrhea, and anal inflammation within the first 4 days after antibody treatment [97]. On day 7, the mice showed splenomegaly, hepatopathy, and lymphadenopathy (MLN) with colonic wall thickening. The colitis is characterized by epithelial hyperplasia with goblet cell depletion, epithelial cell damage, and inflammatory cell infiltration in the lamina propria. The colitis starts resolving by day 10 with complete resolution by day 21. Therefore, activation of CD40-CD40L pathway in the innate immune cells is sufficient to induce colitis in the absence of T and B cells and is mediated by TNF-α, IL-12p40, and IL23 (p19). Interestingly, IL23p19 controls CD40-induced colitis, and IL-12p35 is involved in the systemic manifestation of this colitis, including wasting disease and serum proinflammatory cytokine production. IL-12 (p40/p35) and IL23p19 initiate the release of proinflammatory cytokines by dendritic cells [TNF-α, IFN-γ, and IL-6] and by NK cells [IFN-γ] [97]. The major source of

IL23 is CD11c$^+$ DCs after anti-CD40 or bacterial stimulations [98–100]. Human and murine dendritic cells and macrophages not only produce IL23 but also express IL23R, which support the activation of autocrine loop within the innate immune response [97, 101, 102]. Thus, IL23 is a critical cytokine which initiates a primary response in the mucosal inflammatory cytokine cascade in this model.

Anti-CD40 monoclonal antibody-mediated colitis cannot be induced in mice lacking the orphan nuclear receptor RORγt. The RORγt-deficient mice do not have LTi (lymphoid tissue inducer) cells, lymph nodes, or intestinal lymphoid clusters [103, 104]. A subsequent study revealed that RORγt-deficient NKR (natural killer cell receptor)-expressing LTi cells, which produce IFN-γ in IL23-dependent manner and upregulate perforin and granzyme B expression, are required for colitis development in anti-CD40 monoclonal antibody-treated *Rag1* KO mice [105]. In contrast, RORγt$^+$ NKR-LTi cells produce IL-22, which may be involved in repairing tissue damage. The expression of RORγt in innate lymphoid cells determines their distinct function.

All in all, anti-CD40 monoclonal antibody-mediated colitis is a useful model to analyze the role of innate immune system in intestinal inflammation.

Genetically Manipulated Colitis Models

Il2 KO and Il2R KO Mice

IL-2 is a key cytokine in the regulation of immune and inflammatory responses and is necessary for the development of Treg population [106, 107]. IL-2 specifically binds with an IL-2 receptor complex consisting of IL-2Rα (CD25), IL-1Rβ (CD122), and common gamma chain (γc) (CD132) and subsequently activates important signaling (e.g., PI3K/Akt, Ras/MAPK, JAK/STAT) cascades for maintaining the immune system [108]. *Il2* KO mice [109] develop chronic colitis within 10–25 weeks of age when raised under SPF conditions [110]. Approximately 50 % of *Il2* KO mice die before 9 weeks of age with severe splenomegaly, lymphadenopathy, and anemia preceding colitis, and the remaining 50 % of mice developing colitis with 100 % penetrance [110]. Clinical signs of severe colitis in *Il2* KO mice include enlarged colon, rectal prolapse, splenomegaly, and lymphadenopathy. The chronic inflammation in the colon is characterized by crypt hyperplasia and transmural mononuclear cellular infiltration of lymphocytes and plasma cells [110]. T cells, but not B cells or autoantibodies, are necessary for the development of this disease [111]. *Il2* KO mice raised under GF conditions continue to develop splenomegaly, lymphadenopathy, and mild and focal colitis, manifested by mild infiltration with mononuclear cells and the loss of goblet cells, despite absence of clinical signs of disease (e.g., diarrhea, rectal prolapse, weight loss) or mortality up to 46 weeks of age [112]. When mono-colonized with *E. coli* mpk, *Il2* KO mice raised in GF conditions develop chronic colitis when mono-colonized with

E. coli mpk [113]. In contrast, mice colonized with *Bacteroides vulgatus* mpk remain healthy by promoting differentiation of semi-mature dendritic cells in the colonic lamina propria [114].

IL-2 specifically binds to the high-affinity IL-2 receptor (IL-2R), composed of three component chains, IL-2Rα, IL-2Rβ, and IL-2Rγ [115–117]. Both IL-2Rβ and IL-2Rγ chains form a low-affinity IL-2 receptor and induce signal transduction upon ligation with IL-2. In contrast, IL-2Rα cannot function independently but forms high-affinity IL-2R associated with the other two chains [118]. Both *Il2ra* KO [119] and *Il2rg* KO [120] mice develop colitis spontaneously. Colitis, characterized by colonic shortening, mucosal hypertrophy, diarrhea, rectal bleeding, and rectal prolapse, in *Il2rg* KO mice manifest within 4 months of age. Histologically, colonic epithelial hyperplasia, loss of goblet cells, crypt distortion, and macrophage/lymphocyte infiltration in colonic lamina propria are observed [120]. *Il2ra* KO mice develop massive enlargement of MLN and the spleen around 4–6 weeks of age due to the expansion of $CD4^+/CD8^+$ T cells and IgM^+B220^+ B cells. Approximately 25 % of *Il2ra* KO mice die due to a severe anemia at approximately 8–20 weeks of age. The majority of the surviving mice begin to develop severe colitis at 12–16 weeks of age. The colitis is characterized by a marked thickening of the colonic mucosa and epithelial destruction with mucosal ulceration, infiltration by lymphocytes/neutrophils, and presence of crypt abscesses [119].

IL-10 *KO Mice*

IL-10 is an important anti-inflammatory, immunoregulatory cytokine that is mainly produced by monocytes/macrophages, T cells, B cells, thymocytes, and keratinocytes upon activation [121, 122]. IL-10 inhibits antigen-specific T cell responses by downregulating the MHC (major histocompatibility complex) class II molecule on the surface of monocytes in an autoregulatory manner [123, 124]. *Il10* KO mice in a conventional facility demonstrate growth retardation, anemia, and chronic enterocolitis with normal development of B and T cells [125]. After 3 months of age, underweight *Il10* KO mice develop chronic enterocolitis involving the entire intestinal tract with marked regenerative crypt hyperplasia in the colon [125]. The colitis in *Il10* KO mice of C57Bl/6 background is much milder than the colitis seen in mice of C3H or Balb/c background. As the colitis can be accelerated after administration of nonsteroidal anti-inflammatory drugs (NSAID) such as piroxicam [126], NSAID treatment is useful in inducing rapid and reproducible colitis in *Il10* KO mice.

Under SPF conditions, the inflammation in *Il10* KO mice is in the proximal part of the colon but not in the small intestine. Development of colitis in *Il10* KO mice has been shown to be dependent on *Helicobacter* species [127]. T cells, in particular $TCR\alpha\beta^+CD4^+CD8\alpha^-$ and $CD4^+$ $CD8\alpha^+$ T cells, but not B cells, mediate chronic colitis in *Il10 KO* mice [128].

The chronic colitis in *Il10* KO mice is initiated by the IFN-γ-producing Th1 cells driven by IL-12 produced by antigen-presenting cells (APCs). Early treatment with anti-IFN-γ mAb or anti-IL-12p40 mAb (which neutralizes both IL-12 and IL23) prevents colitis [129, 130]. In contrast, anti-IL-12p40, but not anti-IFN-γ mAb, reverses ongoing colitis in *Il10* KO mice as well as *Rag*-KO mice reconstituted with IL-10 KO CD4+ T cells [130]. IL23 (IL-12p19/p40), not IL-12p70 (IL-12p35/p40), is essential for the development of chronic colitis in *Il10* KO mice, and the colitis is initiated by a unique subset of tissue-homing memory T cells, specifically activated by the proinflammatory mediators IL-17 and IL-6 [131]. IFN-γ has an anti-inflammatory effect in the initiation phase of this colitis as IFN-γ-deficient *Il10* KO mice demonstrate significantly increased colonic inflammation as compared to *Il10* KO mice [132]. IFN-γ exerts its regulatory function by targeting the colonic $CD11b^+$ cells, which are a primary source of IL23 during the development of colitis in *Il10* KO mice. Although earlier studies suggested that the colitis in *Il10* KO is "Th1 mediated," more recent studies indicate that the colitis is Th17 mediated [56]. Interestingly, Th1 (IFN-γ) may play a protective role during the development of chronic colitis in *Il10* KO mice [56].

TLR signaling in effecters $CD4^+$ T cells plays distinct roles depending on the nature of TLR ligand/TLR interaction in *Il10* KO mice. IL-10-deficient *Tlr9* KO mice do not develop colitis by 8 months of age. In contrast, IL-10-deficient *Tlr4* KO mice develop accelerated colitis as early as 8 weeks of age, suggesting that TLR4-mediated signaling has a protective role in the colitis of *Il10* KO mice [127, 133]. LPS stimulation of TLR4-expressing $CD4^+$ T cells inhibits MAPK (mitogen-activated protein kinase) p42/p44 activation upon subsequent TCR stimulation by interacting with MAP kinase phosphate 3, suggesting that TLR4 signaling plays an inhibitory role in TCR-dependent colitogenic $CD4^+$ T cell responses independent of TLR4 expression on innate immune cells [133]. However, chronic colitis in *Il10* KO mice seems to be partially mediated by MyD88 (myeloid differentiation primary response gene 88)-dependent TLR signaling since the presence of colitogenic $CD11c^+$ MHC class II high cells in the MLN is completely abolished upon MyD88 deficiency in Il10 × *MyD88* double knockout (DKO) mice [134, 135].

Recent GWAS have identified *Il10* as a key susceptibility gene for both UC and CD [3, 5, 6, 136]. Of note, the *Il10* gene is associated with other immune-mediated disorders (e.g., sarcoidosis, Behçet's disease, type 1 diabetes mellitus, systemic lupus erythematosus) and is part of the "shared loci," which are enriched for many genes, including *IL23*, *IL21*, *IL7R*, and *IFNG*, involved in Th1 and Th17 cell differentiation [5]. Furthermore, recent reports show that loss of IL-10 signaling with IL-10 and/or IL-10R deficiency is associated with early onset of IBD and allogeneic hematopoietic stem cell transplantation can induce remission in patients with IL-10R deficiency [137, 138]. In summary, IL-10 is a critical cytokine capable of suppressing inflammatory immune responses.

Tcra *KO Mice*

T cell receptor (TCR) is a specific receptor expressed on the surface of T cells and is responsible for recognizing antigens presented by APC via MHC restriction. The majority of T cells express α and β chains, while a small number of T cells express γ and δ chains. *Tcra* KO mice spontaneously developed Th2-mediated colitis after 4–5 months of age, and about 60 % of the mice developed chronic colitis at 6 months of age under SPF conditions [139, 140]. The colitis is strain dependent; C57Bl/6 strain mice are more susceptible as compared to C3H/HeJ or Balb/c strains [140]. Like UC, the inflammation is restricted to the colonic mucosa with elongation of crypts, goblet cell depletion, and a mixed cellular infiltration into the lamina propria [140–142]; occasional crypt abscesses are present in severe colitis. Unlike UC, mucosal ulceration or erosion is not generally observed in *Tcra* KO mice [141]. However, ulceration becomes detectable in CD1d-deficient *Tcra* KO mice [143]. Interestingly, the lack of both IL-4 and B cells in *Tcra* KO mice resulted in the development of granulomatous inflammation in the mucosa and submucosa of the colon and in the ileocecal junction at approximately 24 weeks of age [144].

Several factors are involved in the pathogenesis of colitis in *Tcra* KO mice [56]. *Tcra* KO mice develop unique $CD4^+$ TCR $\alpha^-\beta^+$ T cells, which express TCRβ chains without TCRα chain and primarily recognize superantigens. These unique T cells are immunologically functional and actively produce IL-4, which results in spontaneous colitis development [139, 145, 146]. The colitis is associated with the presence of restricted diversity of $V\beta 8.2^+$ T cell subsets, which are characterized by a specific TCR motif [147]. These T cells can survive under Th2 conditions with colonic epithelial cells (CECs), suggesting the requirement of self-antigen(s) for the survival of pathogenic Th2 cells in *Tcra* KO mice. In addition, TCRVα7.2 chain transgenic *Tcra* KO mice developed rapid onset of colitis as compared to *Tcra* KO mice. The TCRVα7.2 chain may have a chaperone function that extends the half-life of the newly synthesized TCRβ chain of pathogenic T cells, which clonally expand in *Tcra* KO mice [148]. Protein kinase C theta (PKCθ) is an important component in the intracellular signaling cascade and plays a fundamental role in *Tcra* KO mice; *Tcra* x *Prkcq* DKO mice develop a milder form of colitis, as compared to *Tcra* KO mice [149] with reduced proliferation and production of IL-17 as well as Th2 cytokines (e.g., IL-4, IL-13) but unaltered apoptosis. Gamma delta (γδ) T cells may play a pathogenic role in colitis development in *Tcra* KO mice [150]. TNF-α/TNFR2 [151], IL-6 [151], IL-7 [152, 153], IL-1 [142], lipoteichoic acid [154], galectin-4 [155], and chitinase 3-like 1 [156] are also involved in the pathogenesis of chronic colitis in *Tcra* KO mice.

Tcra KO mice fail to develop colitis under GF conditions [157]. Interestingly, the development of colitis is suppressed when *Tcra* KO mice are maintained for several generations in a conventional facility, as compared to a SPF facility [158, 159]. In contrast, B cell-deficient *Tcra* KO mice continue to develop severe colitis in conventional facilities, suggesting that B cells, in particular B-1 B cells, contribute to the suppression of colitis [159]. There is no evidence for primary epithelial barrier disruption, as determined by mannitol transmural flux, in *Tcra* KO mice between 6 and 25 weeks of age [160]. Similar to the *Il10* KO mouse model, treatment with piroxicam, which directly induces epithelial cell apoptosis and weakens the mucosal barrier function [161], accelerates the development of colitis in *Tcra* KO mice presumably by facilitating the invasion of luminal bacteria into the colonic mucosa

[162]. Of note, *Helicobacter hepaticus* infection is not necessary for intestinal disease in *Tcra* KO mice although this infection is sufficient to cause chronic proliferative colitis in *Tcrb* KO mice [163]. Interestingly, *Helicobacter* species infection shifts the cytokine profile of *Tcra* KO mice from Th2- to Th1-dominant responses [164]. In addition, *Tcra* KO mice maintained in SPF conditions develop spontaneous left-side colon, while *Helicobacter*-infected *Tcra* KO mice develop typhlitis (inflammation in cecum).

The colitis in *Tcra* KO mice can be regulated by several factors. In *Tcra* KO mice, resection of the cecal patch (equivalent to human appendectomy) before 3 weeks of age results in decreased number of MLN cells and significantly lower incidence of colitis (<3.3 %) at 6–7 months as compared to the sham-operated mice (>80 %) [165]. This suggests that the appendix lymphoid follicle (cecal patch) may be the priming site for pathogenic cells leading to the development of colitis in *Tcra* KO mice. Recent studies show that cigarette smoke (carbon monoxide) and heme oxygenase (HO-1) induction ameliorates active colitis in *Tcra* KO mice by suppressing the colonic production of IL-1β, TNF, and IL-4 [166]. Of interest, elemental diet-fed *Tcra* KO mice showed no pathogenic features of colitis, with reduced production of Th2-type cytokine, low incidence of *Bacteroides vulgatus* infection, and diversification of Vβ usage of $TCR\alpha^{-}\beta^{+}$ T cell population, as compared to regular diet-fed *Tcra* KO mice [167]. Oral administration of small-size (less than 10 μm in diameter) chitin, a polymer of *N*-acetylglucosamine, suppresses the development of *Tcra* KO mice [168]. Furthermore, regulatory B cells (Breg), which produce large amounts of IL-10 or IL-12p70, contribute to the suppression of colitis in these mice [143, 169]. Local delivery of Il-22 (an IL-10 cytokine member) gene in the colon enhanced STAT3 activation specifically within CECs and ameliorated mucosal inflammation by enhancing the restitution of mucus-producing goblet cells [170]. This result strongly suggests that the local *Il22* gene-delivery system could be a useful therapeutic strategy for treating UC.

Gαi2 *KO Mice*

G proteins are important signal transducing factors, which couple a large family of receptors to effectors, including phospholipase C, adenyl cyclase, and ion channels. They are composed of αβγ (alpha, beta, gamma) heterotrimers that are referred to by their α-subunits [171]. The α-subunit of G_{i2} (so-called Gαi2) has been listed as a potential IBD susceptibility gene based on linkage studies in human IBD [172, 173]. Rudolph et al. generated *Gαi2* KO mice by homologous recombination in embryonic stem cells, in which *Gαi2* was disrupted at the *Nco*I site in exon 3 [173]. After 16–20 weeks of age, every *Gαi2* KO mouse on 129/Sv background develops chronic active inflammation of the colon with mixed inflammatory cellular infiltration in the lamina propria [173]. Of note, nonpolypoid adenocarcinoma resembling neoplastic change seen with UC develops in approximately 30–40 % of *Gαi2* KO mice [173]. Studies so far suggest that dysfunction of Th1-polarized T cells potentially contributes to the development of chronic colitis as well as colonic adenocarcinoma in *Gαi2* KO mice [173]. Furthermore, *Gαi2* KO mice on a 129SvEv [125] background developed earlier and more severe colitis as compared to those on a C57Bl/6 background, accompanied by greater levels of IFN-γ, IL-6, IL-12p40, IL-17, and TNF-α in the colon [174]. The difference can be explained by the distinct signaling pathways in bone marrow-derived dendritic cells (BMDCs) between these two strains. BMDCs in the 129Sv *Gαi2* KO mice displayed increased MAPKp38 signal activation followed by TLR9 ligand (CpG) stimulation and less anti-inflammatory IL-10 production than the C57Bl/6 *Gαi2* KO mice. This result also supports previous observations that colitis in *Gαi2* KO mice is mediated mainly by Th1- but not Th2-type cytokines [175, 176]. Under both conventional and SPF conditions, *Gαi2* KO mice developed severe colitis and colonic adenocarcinoma, suggesting that the disease was mediated by normal flora. In addition, *Gαi2* KO mice have impaired TGF-β responses in peripheral T cells via decreased phosphorylation of Smad2 and Smad3 [177].

Dysregulated B cell subpopulations, including regulatory B cells, may be associated with the development of chronic colitis in these mice [178]. In particular, IL-10 producing CD1d ^{hi}, CD23^{hi}, and CD21^{intermediate} B cells, which are also observed in other inflammatory disorders [179], may be altered or diminished in *Gαi2* KO mice, suggesting an important role of Gαi2 protein in the development of immunoregulatory B cell populations.

T-bet/Rag-2 *DKO (TRUC) Mice*

T-bet protein, encoded by the *TBX21* gene in humans, is a member of the T-box transcription factor family, which orchestrates adaptive and innate immune systems by initiating proinflammatory Th1 lineage development from naïve T helper precursor cells [180, 181]. However, the precise role(s) of T-bet protein in autoimmunity or neoplastic diseases is not well understood [182, 183]. Garrett et al. examined T-bet deficiency in the innate immune compartment by generating *T-bet* and *Rag-2* DKO (named TRUC) mice [184]. Although *T-bet* KO mice do not develop spontaneous colitis, TRUC mice spontaneously developed colitis that resembles human UC after 4 weeks of age, characterized by rectal prolapse and continuous inflammation of the rectum/left-side colon with mucosal inflammatory infiltrate and ulceration [184]. Interestingly, the colitis in TRUC mice is transmittable to T-bet-sufficient wild-type mice both vertically (by cross fostering) and horizontally (by cohousing). In culture of colon explants from 4-week-old TRUC mice, TNF-α production was significantly higher as compared to those of *Rag-2* KO control mice, while there were no apparent difference in the levels of other inflammatory cytokines, including IFN-γ, IL-1α, IL-1β, IL-6, IL-10, IL-12, IL-13, or IL23 [184]. The major source of TNF-α in the colon was from the CD11c^{+} dendritic cells. T-bet may control the mucosal immune system by downregulating TNF-α production negatively in colonic dendritic cells at the initial stage of the colitis development. Since intestinal dendritic cells are constantly interacting with abundant microbes in the colonic lumen and maintaining intestinal homeostasis by controlling the clearance of entero-invasive pathogens [185, 186], perhaps T-bet deficiency in dendritic cells allows for the growth of potentially pathogenic organisms. Therefore, different antibiotics regimens were examined for potential therapeutic benefit. It was found that a combination of vancomycin, metronidazole, neomycin, and ampicillin, as well as selective treatment with metronidazole alone, significantly ameliorated colitis. This suggests that colitis in TRUC mice is dependent on the presence of certain microbes, in particular anaerobic pathogenic bacteria. A later study by the same group, utilizing 16S rRNA-based analysis in fecal samples, showed that *Klebsiella pneumoniae* and *Proteus mirabilis*, both gram-negative facultative organisms, correlate with the development of colitis in TRUC mice [187] and can induce disease given to wild-type recipients.

Interestingly, over 96 % of TRUC mice spontaneously develop colonic high-grade dysplasia and rectal adenocarcinoma by 6 months of age as a consequence of MyD88-independent intestinal inflammation [188]. Restoration of T-bet expression in colonic dendritic cells in TRUC mice reduces colonic inflammation and prevents colonic neoplastic development [188].

Wasp *KO Mice*

A model of IBD had been developed by deletion of gene that encodes for the Wiskott–Aldrich syndrome protein (WASP). Although this model is not widely utilized, we have included it as one of the Th2 type of colitis models in this chapter given its unique characteristic of being one of a few models with a human correlate. *Wasp* encodes a cytoplasmic protein involved in regulating actin cytoskeleton [189], which is defective or absent in patients with Wiskott–Aldrich syndrome, a subset of whom suffer from colitis. *Wasp* KO mice also develop T cell-mediated colitis by 6 months of age with markedly thickened colons seen grossly and crypt hyperplasia and a mixed lymphocytic and neutrophilic infiltrate seen histologically [190]. Lamina propria lymphocytes from *Wasp* KO mice secrete exaggerated levels of IFN-γ, IL-4, and IL-13 without much difference in IL-6 as compared to controls; there were no difference in IL-17 levels [190]. The colitis in *Wasp* KO mice is ameliorated by treatment with antibody against IL-4, but not to IFN-γ. Interestingly, *Wasp* KO mice have a decreased number of naturally occurring $CD4^+CD25^+Foxp3^+$ cells [88]. These Treg cells were found to be markedly defective in their ability to ameliorate colitis using $CD45RB^{hi}$ transfer model [191]; the initiating pathogenic deficiency appears to lie in the innate immune cell population. *Wasp* KO mice represent a model of colitis with a human correlate, potentially mediated by Th2 cytokines and associated with altered innate immune cell and regulatory T cell defects.

Genetically Manipulated Models Associated with the IBD Susceptibility Genes

Nod2 KO Mice

NOD2 (nucleotide-binding oligomerization domain 2) is a member of the NOD-leucine-rich repeat (LRR) protein family. *Nod2* gene codes an intracellular receptor of muramyl dipeptide (MDP), which is a moiety of bacterial cell wall peptidoglycan. In 2001, it was reported that *Nod2* variants confer an increased risk for development of CD [192, 193] and Blau syndrome, which is an autosomal dominant syndrome characterized by familial granulomatous arthritis, uveitis (iritis), and skin granulomas [194]. A mouse model, which carries a *Nod2*-2939insC (*Nod2*^{2839ic}) frameshift mutation similar to human CD-associated *Nod2*-3020insC (*Nod2*^{3020ic}) frameshift mutation, shows significantly more severe DSS-induced colitis with ulceration and increased infiltration of F4/80-positive macrophages, as compared to wild-type controls [195]. The DSS-induced acute injury in Nod2-deficient mice is exacerbated under GF conditions as compared to SPF conditions, suggesting that the *Nod2* gene plays a pivotal role in commensal flora-mediated immunoregulatory function during recovery from acute injury [48]. To assess the immunobiological function of Nod2, other groups examined *Nod2* KO mice containing deletion of exon 1 [196] or exon 3 [197]. In humans, these *Nod2* mutations show defective NF-κB activation after cell stimulation with bacterial products, including LPS and MDP [193, 198, 199]. In APCs of *Nod2* KO mice, IL-12p70 production was significantly increased in response to TLR2 ligation, suggesting that Nod2 signaling inhibits a potentially inflammatory Th1 response mediated by TLR2 signal [196]. The *Nod2* KO mice, which were generated by the deletion of exon 3, are healthy and fertile and demonstrate a normal lymphoid architecture and development in the thymus and spleen [197]. Since *Nod2* plays a pivotal role in the innate immune responses against host/microbial interactions, handling of *Listeria monocytogenes*, a gram-positive intracellular bacteria, was examined in wild-type and *Nod2* KO mice [197]. There was no significant difference survival between the wild-type and *Nod2* KO mice injected intravenously or intraperitoneally with *L. monocytogenes*. In contrast, *Nod2* KO mice challenged with *L. monocytogenes* via intragastric route showed a significantly increased bacterial burden in the spleen and liver, but not in Peyer's patches, as compared to wild-type mice. Furthermore, expression of anti-bacterial peptides [defensin, defensin-related cryptdin 4 (Defcr4), and Defcr-related sequence 10 (Defcr-rs10)], preferentially produced in intestinal Paneth cells, was significantly reduced in bacterial infected *Nod2* KO as compared to wild-type mice. This suggests that Nod2 plays a protective role specifically in Paneth cell, but not Peyer's patch-dependent route of bacterial infection in intestine. The expression of *Nod2* is dependent on the presence of commensal bacteria, since wild-type mice raised in GF conditions expressed significantly less Nod2 expression, with restoration of expression after infection with commensal bacteria [200]. Interestingly, this

Nod2 KO mouse strain is susceptible to granulomatous inflammation restricted to the ileocecal region in the context of *Helicobacter* infection [201]. In summary, the studies in *Nod2* KO mice support that mutations in the *Nod2* gene are important genetic risk factors in a subset of patients with CD.

Atg5 *KO* and Atg16l1 *Mutant Mice*

An autophagy gene, *ATG16L1*, has been identified as a susceptibility allele for CD by GWAS [202–204]. There was a significant increased risk for CD risk between markers rs22141880 (a SNP coding for T300A) in the *ATg16L1* gene and the established *NOD2* susceptibility variants. ATG16L1 protein is an essential component of autophagy, which is the major intracellular degradation system of a cell's own components (autophagy) [205, 206]. Autophagy is involved in the clearance of intracellular components such as apoptotic bodies and organelles as well as microbes (xenophagy), which results in protection against infectious intracellular pathogens [207]. In addition to ATG16L1, ATG5 is another essential autophagy protein, which is important for the biological functions of antibacterial peptide-containing Paneth cells.

Cadwell et al. generated two mouse lines with hypomorphic (HM) for the expression of the ATG16L1 protein (ATG16L1^{HM1} and ATG16L1^{HM2}) and a third-line knocking out *Atg5* expression in IECs specifically (*Atg*$^{flox/flox}$ *villin-Cre* mice) [8]. Abnormalities in both *Atg16l1^{HM}* and *Atg5* conditional KO mice were confirmed to only Paneth cells with a lack of lysosome staining in the mucus and disorganized/decreased numbers of granules [8]. Even with *L. monocytogenes* infection by oral gavage, the lack of any changes in the spleen, liver, and MLN indicate that Atg16l1 and Nod2 have a distinct function in maintaining the integrity of Paneth cells. Electron microscopic analysis also revealed that Paneth cells in *Atg16l1^{HM}* mice showed significantly increased numbers in cytoplasmic vesicles; a similar abnormality is present in CD patients [8]. Dendritic cells obtained from CD patients with *NOD2* or *ATG16L1* mutation showed functional defects in autophagy, bacterial processing/handling, and antigen presentation [208]. However, recent observations strongly suggest that intracellular sensors of NOD are critical for the autophagic responses [209–211].

Akira's laboratory generated *Atg16l1* mutant mice [212], which express deleted forms of the ATG16L1 protein, lacking the entire coiled-coil domain, that is essentially required for processing autophagy [206]. ATG16L1-deficient macrophages showed increased production of IL-1β at the posttranscriptional level, as compared to wild-type control, while both message and protein levels of TNF-α, IL-6, and IFN-β in response to LPS stimulation show no obvious difference between the two groups. The increased IL-1β production was due to the TRIF (Toll/IL-1 receptor domain-containing adaptor inducing IFN-β)-dependent activation of caspase 1. Interestingly, chimeric mice with ATG16L1-deficient hematopoietic cells are highly susceptible to DSS-induced colitis with severe ulceration, increased inflammatory cell infiltration, and increased serum levels of IL-1β/IL-18 and 100 % mortality, which can be ameliorated by neutralizing these cytokines [212]. Cadwell et al. elegantly demonstrated that murine norovirus-infected *Atg16l1* KO mice have aberrant DSS-induced changes with increased productions of TNF-α and IFN-γ as compared to wild-type mice [208]. Therefore, ATG16L1 is an essential factor in controlling endotoxin- as well as virus-induced inflammatory immune responses.

In summary, ATG16L1 and ATG5 play a central role in the secretion of granules in Paneth cells that may alter/exclude intestinal microorganisms efficiently. However, how *Atg16l1* polymorphisms affect the biological function of differentiated Paneth cells is still unclear [8].

IL23 Receptor KO Mice

IL23 is a heterodimeric cytokine comprising of p40 and p19 subunits [101]. The IL23 receptor is also composed of two subunits, IL-12Rβ1 and IL23R, and is mainly expressed on T cells and innate immune cells [102]. IL23 is produced by dendritic cells and macrophages in response to pathogenic bacteria such as *Mycobacterium tuberculosis* [213], *Streptococcus pyogenes* [214], and *Klebsiella pneumoniae* [215]. IL23 stimulates macrophages to produce TNF-α and serves as a maintenance factor of Th17 T cells producing IL-17A and IL-17F.

The ligation of IL23 receptor is involved in the pathogenesis of many inflammatory disorders, including those involving the joints [216], brain [217], and intestine [131, 218, 219]. Therefore, IL23 is a key cytokine in several autoimmune diseases. Recent GWAS have demonstrated that polymorphisms of the *IL23* receptor are negatively associated with the development of both UC and CD [220]. By analyzing 14,500 non-synonymous SNPs from 1,000 cases of autoimmune disorders and breast cancer, Burton et al. identified that the *IL23R* locus has an initial association with ankylosing spondylitis. This disease shows occasional strong association with CD development [221]. Following high-throughput re-sequencing of DNA pools, the protective effects of low-frequency coding variants (p.Arg381Gln, p.Gly149Arg and p.Val362Ile) against IBD were confirmed [10]. This piece of data added to the already known fact that *IL12B* (IL-12p40) has been identified as an IBD-associated gene [222].

Recently, Powrie et al. showed that IL23R signaling in intestinal T cells suppresses the IL-10 production by T cells as well as generation of FoxP3+ cells induced by adaptive transferring the $CD4^{+}CD45RB^{hi}$ (naïve) T cells (isolated from wild-type or *Il23R* KO mice) into *Rag1* KO mice [223]. The majority of mice reconstituted with *Il23R* KO $CD4^{+}$ T cells, developed milder colitis than those transferred with wild-type $CD4^{+}$ T cells. Interestingly, IL23R-deficient T cells do not accumulate in the colon after the adaptive transfer into *Rag1* KO mice. In contrast, these cells accumulate in the spleen and liver, suggesting that IL23 signaling in T cells is specifically necessary for effector T cell accumulation in the colon rather than systemic lymphoid organs [223]. Interestingly, *IL23R* and *NOd2* genes can encode truncated variants that inhibit their signaling pathways [224, 225]. Th17 cells isolated from subjects with an *IL23R R381Q* gene variant show reduced production of IL-17A in response to IL23, suggesting the importance of IL23-related pathways in both CD and UC [226].

Conditional Stat3 *KO Mice*

STAT3 (signal transducer and activator of transcription 3) is a member of the STAT protein family, which acts as transcription activators after being phosphorylated by receptor-associated kinase [227]. STAT3 was the first molecule identified as a member of this family, which is efficiently activated by IL-6 family cytokines [227, 228] and is known to be involved in cell survival and proliferation [229]. STAT3 proteins form either a homodimer or heterodimer when combined with STAT1 and rapidly translocate into the cell nucleus after activation [230, 231]. STAT3 is involved in a broad spectrum of innate and adaptive immune functions, including epithelial regeneration and Th17 differentiation. The importance of STAT3 in IBD is recognized after recent human GWAS as one of the genes associated with increased susceptibility to both CD and UC [5, 6, 220, 232].

Since conventional *Stat3* KO mice are embryonically lethal [233], the function of *Stat3* gene has to be analyzed in a cell or tissue-specific gene KO systems. Takeda et al. generated mice, in which STAT3 is deficient specifically in macrophages and neutrophils (*LysMcre/Stat* $^{flox/-}$) where cells have one floxed *Stat3* allele and one disrupted *Stat3* allele [234]. These deficient mice exhibited mortality to endotoxin shock with increased production of proinflammatory cytokines including TNF-α, IL-6, IL-1, and IFN-γ within 24 h after the injection of a small amount of LPS (20 μg), while the wild-type mice survived over 4 days after the injection. Furthermore, *LysMcre/Stat* $^{flox/-}$ mice spontaneously developed leukocytosis, anemia, and colitis at the age of 20 weeks. The markedly thickened colonic wall showed reduced crypts, goblet cell depletion, regenerative epithelium, mixed cellular infiltration in the lamina propria, frequent crypt abscesses, and occasional mucosal ulcers. The development of colitis is strongly associated with a decreased production of IL-10 by macrophages and enhanced production of IFN-γ by Th1 cells. Interestingly, the conditional *Stat3* KO mice do not develop colitis when crossed with *Rag-1* KO mice [235], suggesting the requirement of intact STAT3 signaling in the adaptive immune compartment.

To analyze the effect of STAT3 activation in multiple cell types by triggering type I IFN, mice carrying a *STAT3floxed* allele were crossed with *MX-Cre* transgenic mice expressing Cre recombinase under the control of IFN-responsive *Mx1* promoter (*MX+*; *STAT3*$^{fl/fl}$) [236]. In *MX+; STAT3*$^{fl/fl}$ mice, most of the enteric epithelial STAT3 signals disappeared. These mice developed a severe wasting syndrome with an aggressive form of colitis within 2–3 weeks after the injection of synthetic double-stranded RNA (pIpC), which almost completely deleted STAT3 expressions in liver, bone marrow, and adipose tissues. These mice produced high amounts of IL-6, IL-12p40, IFN-γ, and IL-10 as compared to *MX-; STAT3$^{fl/fl}$* mice. Treatment with the anti-IL-12p40 antibody, but not neutralization of CD4^{+} or NK cells, or treatment with oral antibiotics (2.5 mg/mL streptomycin/bacitracin in drinking water 7 days prior to pIpC injection) prevented the development of colitis, suggesting a crucial role of STAT3 in the maintenance of intestinal homeostasis. In addition, Pickert et al. showed that specific deletion of *STAT3* in epithelial cells increases susceptibility of mice to DSS-induced colitis [45], suggesting that STAT3 activation in innate versus adaptive immune responses plays distinctly different roles in the pathogenesis of colitis.

To further dissect the functional role of STAT3 on T cells during the development of colitis, Durant et al. utilized the CD45RB^{high} CD4^{+} T cell transfer into *Rag-2* KO colitis mouse model. The recipient *Rag-2* KO mice were reconstituted with naïve CD4^{+} T cells from control (*Stat3$^{flox/flox}$*) or *Stat3* KO (*cd4 Cre; Stat3$^{flox/flox}$*) mice. *Rag-2* KO recipient mice that received *Stat3* KO T cells had no colonic inflammation, while the mice that received control T cells developed marked colitis at 9 weeks of age [237]. Interestingly, chromatin immunoprecipitation and massive parallel sequencing analysis revealed that STAT3 directly binds to multiple survival gene promoters, including *Bcl2*, *ler3*, *Fos*, *Jun*, and *Fos12*, suggesting that STAT3 directly regulates genes that are involved in the survival as well as proliferation of CD4^{+} T cells.

Xbp1 *KO Mice*

The transcription factor XBP-1 (X-box-binding protein-1) is a key component for the stress response in the endoplasmic reticulum (ER) and is required for ER expansion, the development of highly secretory cells, and adaptation of tumor cells to stressful (e.g., low glucose, hypoxia) conditions [238–240]. ER stress is known to be increased in the IECs isolated from patients with IBD, and inflammation-induced ER stress is efficiently inhibited by IL-10 in vivo [241]. XBP-1 is expressed in the IECs and *Xbp1* deletion results in increased ER stress as well as the exacerbation of DSS-induced colitis [242]. To further analyze the role of XBP-1 on colonic epithelial homeostasis, intestinal epithelial-specific *Xbp1* KO (*Xbp$^{flox/flox}$ villin-Cre*) mice were generated [243]. About 60 % of *Xbp1* KO mice and 30 % of *Xbp1+/−* mice spontaneously developed small intestinal inflammation in association with ER stress, suggesting a pivotal role for mono-allelic expression of *Xbp1* in inducing the organ-specific inflammation [244]. The small intestinal inflammation was characterized by polymorphonuclear infiltration in the lamina propria with occasional crypt abscesses and mucosal ulceration. Paneth cells were completely absent and goblet cells were reduced in the small intestine. In contrast, the morphology and function of absorptive epithelium and enteroendocrine cells were intact [244] and the colon did not demonstrate any abnormalities. Examination of a German patient cohort (1103 controls, 550 CD, and 539 UC patients) found that the *Xbp1* variant rs35873774 had the strongest association with both CD and UC among the 20 candidate SNPs studied [244]. In summary, XBP-1 is one of the susceptibility factors for IBD, and its abnormality in IECs is associated with spontaneous ileitis development.

Lrrk2 *KO Mice*

LRRK (leucine-rich repeat kinase 2) is a large (285 kDa) protein, which contains a Ras of complex GTPase domain, a C-terminal of Ras of complex domain and an MAPK kinase domain. Point mutations in *Lrrk2* are the most common genetic cause of both familial and apparently sporadic forms of Parkinson's disease (PD) [245, 246]. A PD-related G2019S substitution in the kinase domain of LRRK2 enhances the phosphorylation of putative protein kinases ezrin, radixin, and the moesin (ERM) family proteins, which links the actin cytoskeleton with membrane proteins [247, 248]. The G2019S substitution likely increases kinase activity in LRRK2 [249, 250]. In addition, *LRRK2* PD-associated mutations induce an alteration of cell death and autophagy function [251, 252]. Of note, LRRK2 forms a protein complex with heat shock protein 90 (HSP90). Since inhibition of the Hsp90 chaperone function dramatically decreases the stability of LRRK2, Hsp90 may suppress the accumulation of mutant *LRRK2*, which is strongly related to pathogenic activities in neurons [253]. Tong et al. generated *Lrrk2* KO mice to analyze its physiological role in vivo. *Lrrk2* KO mice appear normal, but gross morphological abnormalities in the kidney (altered size, weight, color, and texture) become evident at 3–4 months of age [254]. Increased accumulation of autofluorescent granules in proximal renal tubules becomes obvious in *Lrrk2* KO mice with increasing age, although kidney filtration function is intact [255]. The impaired autophagy function in *Lrrk2* KO kidneys is observed by accumulation of lipofuscin granules, altered levels of LC3-1/II, a reliable marker for autophagy, and increased number of apoptotic cells [254, 255].

A recent GWAS identified a single-nucleotide polymorphism (SNP) rs11175593 as a risk factor for CD. The *LRRK2* gene is located downstream from this SNP [232]. *LRRK2* is mainly expressed in immune cells including B cells, monocytes, and dendritic cells, based on the microarray data using *LRRK2* probes from the Genomics Institute of the Novartis Research Foundation and RIKEN data sets [256]. The expression of *LRRK2* is significantly upregulated in the lamina propria cells of inflamed intestinal tissues obtained from CD patients, as compared to the cells from noninflamed areas from the same CD patients or from inflamed tissues from UC patients [256]. The same group also demonstrated that *LRRK2* is an IFN-γ target gene upregulated during bacterial infection, and it is one of the activators of the NF-κB pathway: LRRK2-induced NF-κB activation is IKK dependent but is independent of LRRK2 kinase activity. These results suggest that LRRK2 might be involved in the regulation of mucosal immune responses by activating the NF-κB signaling pathway, which is relevant to CD pathogenesis. Utilization of *Lrrk2* KO mice in chemically induced or bacterial infectious colitis models could provide further knowledge regarding the role of LRRK2 on regulation of the immune response to pathogen or epithelial injury.

p40^{phox} *KO Mice*

In phagocytes, reactive oxygen species (ROS), generated by NADPH oxidase, play a pivotal role in regulating proinflammatory signaling and in killing pathogens by phagocytosis. The NADPH oxidase complex is composed of five subunits including gp91^{phox}, p47^{phox}, p22^{phox}, p67^{phox}, and p40^{phox} [257]. Recent genetic studies have revealed an association between increased susceptibility for CD and *NCF4* (which encodes p40^{phox}) and *NCF2* (which encodes p67^{phox}) polymorphism [203, 258]. Functional studies have confirmed that impaired p40^{phox} promotes intestinal inflammation with impaired ROS production [259]. In addition, neutrophils isolated from *P40*^{phox} KO mice have severe defects in NADPH oxidase regulation as well as oxidant-dependent in killing of *Staphylococcus aureus,* both in vitro and in vivo [260]. The *p40*^{phox} KO mice showed enhanced intestinal inflammation during the acute and recovery phases of DSS colitis through upregulation of the chemokine receptor 1 and downregulation of enzymes for glycan modifications: these results were obtained by using an integrative bioinformatics approach [261].

Conclusions/Future Prospective

Although the intestinal inflammation seen in animal models does not exactly replicate human IBD, studies utilizing these models have provided important insights into the pathogenesis of IBD. One principle concept is the central role of the intestinal microbiota in the development of the innate and adaptive immune responses that can result in intestinal inflammation. Migratory patterns of immune cells and the effects of mucosal immune dysregulation and alteration of mucosal epithelial barrier during initiation and maintenance of intestinal inflammation can also be learned from these models. Furthermore, the development of animal models that incorporate variants of human IBD susceptibility genes has provided a means to directly examine the role of these genes in intestinal inflammation. Application of new technology will refine and extend our current knowledge thus far obtained from these models.

Acknowledgement The authors are grateful to Terry Danford Lott, Cindy W. Lau, and Doris A. Regal for their excellent secretarial assistance in preparing this manuscript and Drs. Daren Low and Deanna Nguyen for critically reading this manuscript.

Grant Supports: This work has been supported by the National Institute of Health and grants from the Eli and Edythe L. Broad Medical Foundation, the American Gastroenterological Association Foundation, and the Crohn's and Colitis Foundation of America.

Financial Disclosure: All authors have no disclosures in regard to any potential conflicts that may be relevant to this manuscript.

References

1. Baumgart DC, Sandborn WJ (2007) Inflammatory bowel disease: clinical aspects and established and evolving therapies. Lancet 369(9573):1641–1657, Epub 2007/05/15
2. Xavier RJ, Podolsky DK (2007) Unravelling the pathogenesis of inflammatory bowel disease. Nature 448(7152):427–434, Epub 2007/07/27
3. Khor B, Gardet A, Xavier RJ (2011) Genetics and pathogenesis of inflammatory bowel disease. Nature 474(7351):307–317, Epub 2011/06/17
4. Sartor RB (2008) Microbial influences in inflammatory bowel diseases. Gastroenterology 134(2):577–594, Epub 2008/02/05
5. Anderson CA, Boucher G, Lees CW, Franke A, D'Amato M, Taylor KD et al (2011) Meta-analysis identifies 29 additional ulcerative colitis risk loci, increasing the number of confirmed associations to 47. Nat Genet 43(3):246–252, Epub 2011/02/08
6. Franke A, McGovern DP, Barrett JC, Wang K, Radford-Smith GL, Ahmad T et al (2010) Genome-wide meta-analysis increases to 71 the number of confirmed Crohn's disease susceptibility loci. Nat Genet 42(12):1118–1125, Epub 2010/11/26
7. Jostins L, Ripke S, Weersma RK, Duerr RH, McGovern DP, Hui KY et al (2012) Host-microbe interactions have shaped the genetic architecture of inflammatory bowel disease. Nature 491(7422):119–124, Epub 2012/11/07
8. Cadwell K, Liu JY, Brown SL, Miyoshi H, Loh J, Lennerz JK et al (2008) A key role for autophagy and the autophagy gene Atg16l1 in mouse and human intestinal Paneth cells. Nature 456(7219):259–263, Epub 2008/10/14
9. Glocker EO, Kotlarz D, Boztug K, Gertz EM, Schaffer AA, Noyan F et al (2009) Inflammatory bowel disease and mutations affecting the interleukin-10 receptor. N Engl J Med 361(21): 2033–2045, Epub 2009/11/06
10. Momozawa Y, Mni M, Nakamura K, Coppieters W, Almer S, Amininejad L et al (2011) Resequencing of positional candidates identifies low frequency IL-23R coding variants protecting against inflammatory bowel disease. Nat Genet 43(1):43–47, Epub 2010/12/15
11. Mizoguchi A, Mizoguchi E (2010) Animal models of IBD: linkage to human disease. Curr Opin Pharmacol 10(5):578–587, Epub 2010/09/24
12. Saleh M, Elson CO (2011) Experimental inflammatory bowel disease: insights into the host-microbiota dialog. Immunity 34(3):293–302, Epub 2011/03/26
13. Strober W, Fuss I, Mannon P (2007) The fundamental basis of inflammatory bowel disease. J Clin Invest 117(3):514–521
14. Kawada M, Arihiro A, Mizoguchi E (2007) Insights from advances in research of chemically induced experimental models of human inflammatory bowel disease. World J Gastroenterol 13(42):5581–5593, Epub 2007/10/24
15. Mizoguchi A, Mizoguchi E, Bhan AK (2003) Immune networks in animal models of inflammatory bowel disease. Inflamm Bowel Dis 9(4):246–259, Epub 2003/08/07
16. Wirtz S, Neufert C, Weigmann B, Neurath MF (2007) Chemically induced mouse models of intestinal inflammation. Nat Protoc 2(3):541–546, Epub 2007/04/05
17. Sundberg JP, Elson CO, Bedigian H, Birkenmeier EH (1994) Spontaneous, heritable colitis in a new substrain of C3H/HeJ mice. Gastroenterology 107(6):1726–1735, Epub 1994/12/01

18. Mergenhagen SE, Pluznik DH (1984) Defective responses to lipid A in C3H/HeJ mice: approaches to an understanding of lipid A-cell interaction. Rev Infect Dis 6(4):519–523, Epub 1984/07/01
19. Poltorak A, He X, Smirnova I, Liu MY, Van Huffel C, Du X et al (1998) Defective LPS signaling in C3H/HeJ and C57BL/10ScCr mice: mutations in Tlr4 gene. Science 282(5396): 2085–2088, Epub 1998/12/16
20. Elson CO, Cong Y, Brandwein S, Weaver CT, McCabe RP, Mahler M et al (1998) Experimental models to study molecular mechanisms underlying intestinal inflammation. Ann N Y Acad Sci 859:85–95, Epub 1999/02/03
21. Brandwein SL, McCabe RP, Cong Y, Waites KB, Ridwan BU, Dean PA et al (1997) Spontaneously colitic C3H/HeJBir mice demonstrate selective antibody reactivity to antigens of the enteric bacterial flora. J Immunol 159(1):44–52, Epub 1997/07/01
22. Elson CO, Sartor RB, Tennyson GS, Riddell RH (1995) Experimental models of inflammatory bowel disease. Gastroenterology 109(4):1344–1367, Epub 1995/10/01
23. Cong Y, Brandwein SL, McCabe RP, Lazenby A, Birkenmeier EH, Sundberg JP et al (1998) CD4+ T cells reactive to enteric bacterial antigens in spontaneously colitic C3H/HeJBir mice: increased T helper cell type 1 response and ability to transfer disease. J Exp Med 187(6):855–864, Epub 1998/04/04
24. Bristol IJ, Farmer MA, Cong Y, Zheng XX, Strom TB, Elson CO et al (2000) Heritable susceptibility for colitis in mice induced by IL-10 deficiency. Inflamm Bowel Dis 6(4):290–302, Epub 2001/01/10
25. Beckwith J, Cong Y, Sundberg JP, Elson CO, Leiter EH (2005) Cdcs1, a major colitogenic locus in mice, regulates innate and adaptive immune response to enteric bacterial antigens. Gastroenterology 129(5):1473–1484, Epub 2005/11/16
26. Higuchi K (1997) Genetic characterization of senescence-accelerated mouse (SAM). Exp Gerontol 32(1–2):129–138, Epub 1997/01/01
27. Matsumoto S, Okabe Y, Setoyama H, Takayama K, Ohtsuka J, Funahashi H et al (1998) Inflammatory bowel disease-like enteritis and caecitis in a senescence accelerated mouse P1/Yit strain. Gut 43(1):71–78, Epub 1998/10/15
28. Kosiewicz MM, Nast CC, Krishnan A, Rivera-Nieves J, Moskaluk CA, Matsumoto S et al (2001) Th1-type responses mediate spontaneous ileitis in a novel murine model of Crohn's disease. J Clin Invest 107(6):695–702, Epub 2001/03/20
29. Rivera-Nieves J, Bamias G, Vidrich A, Marini M, Pizarro TT, McDuffie MJ et al (2003) Emergence of perianal fistulizing disease in the SAMP1/YitFc mouse, a spontaneous model of chronic ileitis. Gastroenterology 124(4):972–982, Epub 2003/04/03
30. Vidrich A, Buzan JM, Barnes S, Reuter BK, Skaar K, Ilo C et al (2005) Altered epithelial cell lineage allocation and global expansion of the crypt epithelial stem cell population are associated with ileitis in SAMP1/YitFc mice. Am J Pathol 166(4):1055–1067, Epub 2005/03/29
31. Reuter BK, Pizarro TT (2009) Mechanisms of tight junction dysregulation in the SAMP1/YitFc model of Crohn's disease-like ileitis. Ann N Y Acad Sci 1165:301–307, Epub 2009/06/23
32. Marini M, Bamias G, Rivera-Nieves J, Moskaluk CA, Hoang SB, Ross WG et al (2003) TNF-alpha neutralization ameliorates the severity of murine Crohn's-like ileitis by abrogation of intestinal epithelial cell apoptosis. Proc Natl Acad Sci U S A 100(14):8366–8371, Epub 2003/07/02
33. Olson TS, Reuter BK, Scott KG, Morris MA, Wang XM, Hancock LN et al (2006) The primary defect in experimental ileitis originates from a nonhematopoietic source. J Exp Med 203(3):541–552, Epub 2006/03/01
34. Reuter BK, Pastorelli L, Brogi M, Garg RR, McBride JA, Rowlett RM et al (2011) Spontaneous, immune-mediated gastric inflammation in SAMP1/YitFc mice, a model of Crohn's-like gastritis. Gastroenterology 141(5):1709–1719, Epub 2011/06/28

35. Pizarro TT, Pastorelli L, Bamias G, Garg RR, Reuter BK, Mercado JR et al (2011) SAMP1/YitFc mouse strain: a spontaneous model of Crohn's disease-like ileitis. Inflamm Bowel Dis 17(12):2566–2584, Epub 2011/05/11
36. Haas SL, Andreas Koch W, Schreiber S, Reinhard I, Koyama N, Singer MV et al (2005) 137 (G/C) IL-18 promoter polymorphism in patients with inflammatory bowel disease. Scand J Gastroenterol 40(12):1438–1443, Epub 2005/12/01
37. Takagawa T, Tamura K, Takeda N, Tomita T, Ohda Y, Fukunaga K et al (2005) Association between IL-18 gene promoter polymorphisms and inflammatory bowel disease in a Japanese population. Inflamm Bowel Dis 11(12):1038–1043, Epub 2005/11/25
38. Bamias G, Martin C, Mishina M, Ross WG, Rivera-Nieves J, Marini M et al (2005) Proinflammatory effects of TH2 cytokines in a murine model of chronic small intestinal inflammation. Gastroenterology 128(3):654–666, Epub 2005/03/15
39. Takedatsu H, Mitsuyama K, Matsumoto S, Handa K, Suzuki A, Takedatsu H et al (2004) Interleukin-5 participates in the pathogenesis of ileitis in SAMP1/Yit mice. Eur J Immunol 34(6):1561–1569, Epub 2004/05/27
40. Olson TS, Bamias G, Naganuma M, Rivera-Nieves J, Burcin TL, Ross W et al (2004) Expanded B cell population blocks regulatory T cells and exacerbates ileitis in a murine model of Crohn disease. J Clin Invest 114(3):389–398, Epub 2004/08/03
41. Burns RC, Rivera-Nieves J, Moskaluk CA, Matsumoto S, Cominelli F, Ley K (2001) Antibody blockade of ICAM-1 and VCAM-1 ameliorates inflammation in the SAMP-1/Yit adoptive transfer model of Crohn's disease in mice. Gastroenterology 121(6):1428–1436, Epub 2001/12/01
42. Schwartz L, Abolhassani M, Pooya M, Steyaert JM, Wertz X, Israel M et al (2008) Hyperosmotic stress contributes to mouse colonic inflammation through the methylation of protein phosphatase 2A. Am J Physiol Gastrointest Liver Physiol 295(5):G934–G941, Epub 2008/08/30
43. Dieleman LA, Ridwan BU, Tennyson GS, Beagley KW, Bucy RP, Elson CO (1994) Dextran sulfate sodium-induced colitis occurs in severe combined immunodeficient mice. Gastroenterology 107(6):1643–1652, Epub 1994/12/01
44. Okayasu I, Hatakeyama S, Yamada M, Ohkusa T, Inagaki Y, Nakaya R (1990) A novel method in the induction of reliable experimental acute and chronic ulcerative colitis in mice. Gastroenterology 98(3):694–702, Epub 1990/03/01
45. Pickert G, Neufert C, Leppkes M, Zheng Y, Wittkopf N, Warntjen M et al (2009) STAT3 links IL-22 signaling in intestinal epithelial cells to mucosal wound healing. J Exp Med 206(7): 1465–1472
46. Rath HC, Schultz M, Freitag R, Dieleman LA, Li F, Linde HJ et al (2001) Different subsets of enteric bacteria induce and perpetuate experimental colitis in rats and mice. Infect Immun 69(4):2277–2285, Epub 2001/03/20
47. Kitajima S, Morimoto M, Sagara E, Shimizu C, Ikeda Y (2001) Dextran sodium sulfate-induced colitis in germ-free IQI/Jic mice. Exp Anim 50(5):387–395, Epub 2002/01/05
48. Rakoff-Nahoum S, Paglino J, Eslami-Varzaneh F, Edberg S, Medzhitov R (2004) Recognition of commensal microflora by toll-like receptors is required for intestinal homeostasis. Cell 118(2):229–241, Epub 2004/07/21
49. Katakura K, Lee J, Rachmilewitz D, Li G, Eckmann L, Raz E (2005) Toll-like receptor 9-induced type I IFN protects mice from experimental colitis. J Clin Invest 115(3):695–702, Epub 2005/03/15
50. Neurath MF, Fuss I, Kelsall BL, Stuber E, Strober W (1995) Antibodies to interleukin 12 abrogate established experimental colitis in mice. J Exp Med 182(5):1281–1290
51. te Velde AA, Verstege MI, Hommes DW (2006) Critical appraisal of the current practice in murine TNBS-induced colitis. Inflamm Bowel Dis 12(10):995–999, Epub 2006/10/03
52. Elson CO, Beagley KW, Sharmanov AT, Fujihashi K, Kiyono H, Tennyson GS et al (1996) Hapten-induced model of murine inflammatory bowel disease: mucosa immune responses and protection by tolerance. J Immunol 157(5):2174–2185, Epub 1996/09/01

53. Scheiffele F, Fuss IJ (2001) Induction of TNBS colitis in mice. Curr Protoc Immunol 49:15.9.1–15.9.4
54. Dohi T, Fujihashi K, Kiyono H, Elson CO, McGhee JR (2000) Mice deficient in Th1- and Th2-type cytokines develop distinct forms of hapten-induced colitis. Gastroenterology 119(3):724–733, Epub 2000/09/13
55. Hollenbach E, Vieth M, Roessner A, Neumann M, Malfertheiner P, Naumann M (2005) Inhibition of RICK/nuclear factor-kappaB and p38 signaling attenuates the inflammatory response in a murine model of Crohn disease. J Biol Chem 280(15):14981–14988, Epub 2005/02/05
56. Mizoguchi A (2012) Animal models of inflammatory bowel disease. Prog Mol Biol Transl Sci 105:263–320, Epub 2011/12/06
57. Zhang Z, Zheng M, Bindas J, Schwarzenberger P, Kolls JK (2006) Critical role of IL-17 receptor signaling in acute TNBS-induced colitis. Inflamm Bowel Dis 12(5):382–388
58. Fichtner-Feigl S, Fuss IJ, Young CA, Watanabe T, Geissler EK, Schlitt HJ et al (2007) Induction of IL-13 triggers TGF-beta1-dependent tissue fibrosis in chronic 2,4,6-trinitrobenzene sulfonic acid colitis. J Immunol 178(9):5859–5870, Epub 2007/04/20
59. Ishiguro K, Ando T, Maeda O, Watanabe O, Goto H (2010) Novel mouse model of colitis characterized by hapten-protein visualization. Biotechniques 49(3):641–648, Epub 2010/09/22
60. Yang Z, Fuss IJ, Watanabe T, Asano N, Davey MP, Rosenbaum JT et al (2007) *NOD2* transgenic mice exhibit enhanced MDP-mediated down-regulation of TLR2 responses and resistance to colitis induction. Gastroenterology 133(5):1510–1521, Epub 2007/10/05
61. Watanabe T, Asano N, Murray PJ, Ozato K, Tailor P, Fuss IJ et al (2008) Muramyl dipeptide activation of nucleotide-binding oligomerization domain 2 protects mice from experimental colitis. J Clin Invest 118(2):545–559, Epub 2008/01/12
62. Boirivant M, Fuss IJ, Chu A, Strober W (1998) Oxazolone colitis: a murine model of T helper cell type 2 colitis treatable with antibodies to interleukin 4. J Exp Med 188(10):1929–1939, Epub 1998/11/17
63. Heller F, Fuss IJ, Nieuwenhuis EE, Blumberg RS, Strober W (2002) Oxazolone colitis, a Th2 colitis model resembling ulcerative colitis, is mediated by IL-13-producing NK-T cells. Immunity 17(5):629–638, Epub 2002/11/16
64. Fuss IJ, Heller F, Boirivant M, Leon F, Yoshida M, Fichtner-Feigl S et al (2004) Nonclassical CD1d-restricted NK T cells that produce IL-13 characterize an atypical Th2 response in ulcerative colitis. J Clin Invest 113(10):1490–1497
65. Iijima H, Neurath MF, Nagaishi T, Glickman JN, Nieuwenhuis EE, Nakajima A et al (2004) Specific regulation of T helper cell 1-mediated murine colitis by CEACAM1. J Exp Med 199(4):471–482, Epub 2004/02/19
66. Weigmann B, Lehr HA, Yancopoulos G, Valenzuela D, Murphy A, Stevens S et al (2008) The transcription factor NFATc2 controls IL-6-dependent T cell activation in experimental colitis. J Exp Med 205(9):2099–2110, Epub 2008/08/20
67. Noti M, Corazza N, Mueller C, Berger B, Brunner T (2010) TNF suppresses acute intestinal inflammation by inducing local glucocorticoid synthesis. J Exp Med 207(5):1057–1066, Epub 2010/05/05
68. Thomas ML (1989) The leukocyte common antigen family. Annu Rev Immunol 7:339–369, Epub 1989/01/01
69. Lee WT, Yin XM, Vitetta ES (1990) Functional and ontogenetic analysis of murine CD45Rhi and CD45Rlo CD4+ T cells. J Immunol 144(9):3288–3295, Epub 1990/05/01
70. Morrissey PJ, Charrier K, Braddy S, Liggitt D, Watson JD (1993) CD4+ T cells that express high levels of CD45RB induce wasting disease when transferred into congenic severe combined immunodeficient mice. Disease development is prevented by cotransfer of purified CD4+ T cells. J Exp Med 178(1):237–244
71. Powrie F, Leach MW, Mauze S, Caddle LB, Coffman RL (1993) Phenotypically distinct subsets of CD4+ T cells induce or protect from chronic intestinal inflammation in C. B-17 scid mice. Int Immunol 5(11):1461–1471

72. Shimada A, Rohane P, Fathman CG, Charlton B (1996) Pathogenic and protective roles of CD45RB(low) CD4+ cells correlate with cytokine profiles in the spontaneously autoimmune diabetic mouse. Diabetes 45(1):71–78, Epub 1996/01/01
73. Powrie F, Mason D (1990) OX-22high CD4+ T cells induce wasting disease with multiple organ pathology: prevention by the OX-22low subset. J Exp Med 172(6):1701–1708, Epub 1990/12/01
74. Ostanin DV, Bao J, Koboziev I, Gray L, Robinson-Jackson SA, Kosloski-Davidson M et al (2009) T cell transfer model of chronic colitis: concepts, considerations, and tricks of the trade. Am J Physiol Gastrointest Liver Physiol 296(2):G135–G146, Epub 2008/11/27
75. Powrie F, Leach MW, Mauze S, Menon S, Caddle LB, Coffman RL (1994) Inhibition of Th1 responses prevents inflammatory bowel disease in scid mice reconstituted with CD45RBhi CD4+ T cells. Immunity 1(7):553–562, Epub 1994/10/01
76. Elson CO, Cong Y, Weaver CT, Schoeb TR, McClanahan TK, Fick RB et al (2007) Monoclonal anti-interleukin 23 reverses active colitis in a T cell-mediated model in mice. Gastroenterology 132(7):2359–2370, Epub 2007/06/16
77. Izcue A, Coombes JL, Powrie F (2006) Regulatory T cells suppress systemic and mucosal immune activation to control intestinal inflammation. Immunol Rev 212:256–271, Epub 2006/08/15
78. Esplugues E, Huber S, Gagliani N, Hauser AE, Town T, Wan YY et al (2011) Control of TH17 cells occurs in the small intestine. Nature 475(7357):514–518, Epub 2011/07/19
79. Hundorfean G, Neurath MF, Mudter J (2012) Functional relevance of T helper 17 (Th17) cells and the IL-17 cytokine family in inflammatory bowel disease. Inflamm Bowel Dis 18(1):180–186, Epub 2011/03/08
80. O'Connor W Jr, Kamanaka M, Booth CJ, Town T, Nakae S, Iwakura Y et al (2009) A protective function for interleukin 17A in T cell-mediated intestinal inflammation. Nat Immunol 10(6):603–609
81. Kawachi S, Morise Z, Jennings SR, Conner E, Cockrell A, Laroux FS et al (2000) Cytokine and adhesion molecule expression in SCID mice reconstituted with CD4+ T cells. Inflamm Bowel Dis 6(3):171–180, Epub 2000/08/29
82. Atreya R, Mudter J, Finotto S, Mullberg J, Jostock T, Wirtz S et al (2000) Blockade of interleukin 6 trans signaling suppresses T-cell resistance against apoptosis in chronic intestinal inflammation: evidence in Crohn disease and experimental colitis in vivo. Nat Med 6(5): 583–588
83. Kenny EE, Pe'er I, Karban A, Ozelius L, Mitchell AA, Ng SM et al (2012) A genome-wide scan of Ashkenazi Jewish Crohn's disease suggests novel susceptibility loci. PLoS Genet 8(3):e1002559, Epub 2012/03/14
84. Dardalhon V, Awasthi A, Kwon H, Galileos G, Gao W, Sobel RA et al (2008) IL-4 inhibits TGF-beta-induced Foxp3+ T cells and, together with TGF-beta, generates IL-9+ IL-10+ Foxp3(-) effector T cells. Nat Immunol 9(12):1347–1355, Epub 2008/11/11
85. Powrie F, Carlino J, Leach MW, Mauze S, Coffman RL (1996) A critical role for transforming growth factor-beta but not interleukin 4 in the suppression of T helper type 1-mediated colitis by CD45RB(low) CD4+ T cells. J Exp Med 183(6):2669–2674, Epub 1996/06/01
86. Groux H, O'Garra A, Bigler M, Rouleau M, Antonenko S, de Vries JE et al (1997) A CD4+ T-cell subset inhibits antigen-specific T-cell responses and prevents colitis. Nature 389(6652):737–742, Epub 1997/10/24
87. Murai M, Turovskaya O, Kim G, Madan R, Karp CL, Cheroutre H et al (2009) Interleukin 10 acts on regulatory T cells to maintain expression of the transcription factor Foxp3 and suppressive function in mice with colitis. Nat Immunol 10(11):1178–1184, Epub 2009/09/29
88. Chaudhry A, Samstein RM, Treuting P, Liang Y, Pils MC, Heinrich JM et al (2011) Interleukin-10 signaling in regulatory T cells is required for suppression of Th17 cell-mediated inflammation. Immunity 34(4):566–578, Epub 2011/04/23
89. Anderson DM, Maraskovsky E, Billingsley WL, Dougall WC, Tometsko ME, Roux ER et al (1997) A homologue of the TNF receptor and its ligand enhance T-cell growth and dendritic-cell function. Nature 390(6656):175–179, Epub 1997/11/21

90. Cella M, Scheidegger D, Palmer-Lehmann K, Lane P, Lanzavecchia A, Alber G (1996) Ligation of CD40 on dendritic cells triggers production of high levels of interleukin-12 and enhances T cell stimulatory capacity: T-T help via APC activation. J Exp Med 184(2): 747–752, Epub 1996/08/01
91. Ranheim EA, Kipps TJ (1993) Activated T cells induce expression of B7/BB1 on normal or leukemic B cells through a CD40-dependent signal. J Exp Med 177(4):925–935, Epub 1993/04/01
92. van Kooten C, Banchereau J (1997) Functions of CD40 on B cells, dendritic cells and other cells. Curr Opin Immunol 9(3):330–337, Epub 1997/06/01
93. Stuber E, Strober W, Neurath M (1996) Blocking the CD40L-CD40 interaction in vivo specifically prevents the priming of T helper 1 cells through the inhibition of interleukin 12 secretion. J Exp Med 183(2):693–698, Epub 1996/02/01
94. De Jong YP, Comiskey M, Kalled SL, Mizoguchi E, Flavell RA, Bhan AK et al (2000) Chronic murine colitis is dependent on the CD154/CD40 pathway and can be attenuated by anti-CD154 administration. Gastroenterology 119(3):715–723, Epub 2000/09/13
95. Liu Z, Geboes K, Colpaert S, Overbergh L, Mathieu C, Heremans H et al (2000) Prevention of experimental colitis in SCID mice reconstituted with CD45RBhigh CD4+ T cells by blocking the CD40-CD154 interactions. J Immunol 164(11):6005–6014, Epub 2000/05/23
96. Diehl L, den Boer AT, Schoenberger SP, van der Voort EI, Schumacher TN, Melief CJ et al (1999) CD40 activation in vivo overcomes peptide-induced peripheral cytotoxic T-lymphocyte tolerance and augments anti-tumor vaccine efficacy. Nat Med 5(7):774–779, Epub 1999/07/08
97. Uhlig HH, McKenzie BS, Hue S, Thompson C, Joyce-Shaikh B, Stepankova R et al (2006) Differential activity of IL-12 and in mucosal and systemic innate immune pathology. Immunity 25(2):309–318
98. Becker C, Wirtz S, Blessing M, Pirhonen J, Strand D, Bechthold O et al (2003) Constitutive p40 promoter activation and IL-23 production in the terminal ileum mediated by dendritic cells. J Clin Invest 112(5):693–706, Epub 2003/09/04
99. Krajina T, Leithauser F, Moller P, Trobonjaca Z, Reimann J (2003) Colonic lamina propria dendritic cells in mice with CD4+ T cell-induced colitis. Eur J Immunol 33(4):1073–1083, Epub 2003/04/03
100. Morelli AE, Zahorchak AF, Larregina AT, Colvin BL, Logar AJ, Takayama T et al (2001) Cytokine production by mouse myeloid dendritic cells in relation to differentiation and terminal maturation induced by lipopolysaccharide or CD40 ligation. Blood 98(5):1512–1523, Epub 2001/08/25
101. Oppmann B, Lesley R, Blom B, Timans JC, Xu Y, Hunte B et al (2000) Novel p19 protein engages IL-12p40 to form a cytokine, IL-23, with biological activities similar as well as distinct from IL-12. Immunity 13(5):715–725
102. Parham C, Chirica M, Timans J, Vaisberg E, Travis M, Cheung J et al (2002) A receptor for the heterodimeric cytokine IL-23 is composed of IL-12Rbeta1 and a novel cytokine receptor subunit, IL-23R. J Immunol 168(11):5699–5708, Epub 2002/05/23
103. Buonocore S, Ahern PP, Uhlig HH, Ivanov II, Littman DR, Maloy KJ et al (2010) Innate lymphoid cells drive interleukin-23-dependent innate intestinal pathology. Nature 464(7293): 1371–1375, Epub 2010/04/16
104. Sun Z, Unutmaz D, Zou YR, Sunshine MJ, Pierani A, Brenner-Morton S et al (2000) Requirement for RORgamma in thymocyte survival and lymphoid organ development. Science 288(5475):2369–2373, Epub 2000/07/06
105. Vonarbourg C, Mortha A, Bui VL, Hernandez PP, Kiss EA, Hoyler T et al (2010) Regulated expression of nuclear receptor RORgammat confers distinct functional fates to NK cell receptor-expressing RORgammat(+) innate lymphocytes. Immunity 33(5):736–751, Epub 2010/11/26
106. Sakaguchi S, Sakaguchi N, Asano M, Itoh M, Toda M (1995) Immunologic self-tolerance maintained by activated T cells expressing IL-2 receptor alpha-chains (CD25). Breakdown of a single mechanism of self-tolerance causes various autoimmune diseases. J Immunol 155(3):1151–1164, Epub 1995/08/01

107. Thornton AM, Shevach EM (1998) CD4+CD25+ immunoregulatory T cells suppress polyclonal T cell activation in vitro by inhibiting interleukin 2 production. J Exp Med 188(2):287–296, Epub 1998/07/22
108. Beadling C, Johnson KW, Smith KA (1993) Isolation of interleukin 2-induced immediate-early genes. Proc Natl Acad Sci U S A 90(7):2719–2723, Epub 1993/04/01
109. Schorle H, Holtschke T, Hunig T, Schimpl A, Horak I (1991) Development and function of T cells in mice rendered interleukin-2 deficient by gene targeting. Nature 352(6336):621–624, Epub 1991/08/15
110. Sadlack B, Merz H, Schorle H, Schimpl A, Feller AC, Horak I (1993) Ulcerative colitis-like disease in mice with a disrupted interleukin-2 gene. Cell 75(2):253–261, Epub 1993/10/22
111. Ma A, Datta M, Margosian E, Chen J, Horak I (1995) T cells, but not B cells, are required for bowel inflammation in interleukin 2-deficient mice. J Exp Med 182(5):1567–1572, Epub 1995/11/01
112. Schultz M, Tonkonogy SL, Sellon RK, Veltkamp C, Godfrey VL, Kwon J et al (1999) IL-2-deficient mice raised under germfree conditions develop delayed mild focal intestinal inflammation. Am J Physiol 276(6 Pt 1):G1461–G1472, Epub 1999/06/11
113. Waidmann M, Bechtold O, Frick JS, Lehr HA, Schubert S, Dobrindt U et al (2003) Bacteroides vulgatus protects against Escherichia coli-induced colitis in gnotobiotic interleukin-2-deficient mice. Gastroenterology 125(1):162–177, Epub 2003/07/10
114. Muller M, Fink K, Geisel J, Kahl F, Jilge B, Reimann J et al (2008) Intestinal colonization of IL-2 deficient mice with non-colitogenic B. vulgatus prevents DC maturation and T-cell polarization. PLoS One 3(6):e2376
115. Leonard WJ (1988) Immune regulation: recombinant lymphokines and their receptors. Science 240(4854):932, Epub 1988/05/13
116. Smith KA (1988) Interleukin-2: inception, impact, and implications. Science 240(4856): 1169–1176, Epub 1988/05/27
117. Waldmann TA, Tsudo M (1988) The role of the multichain IL-2 receptor complex in the control of normal and malignant T-cell proliferation. Prog Clin Biol Res 262:283–293, Epub 1988/01/01
118. Wang HM, Smith KA (1987) The interleukin 2 receptor. Functional consequences of its bimolecular structure. J Exp Med 166(4):1055–1069, Epub 1987/10/01
119. Willerford DM, Chen J, Ferry JA, Davidson L, Ma A, Alt FW (1995) Interleukin-2 receptor alpha chain regulates the size and content of the peripheral lymphoid compartment. Immunity 3(4):521–530, Epub 1995/10/01
120. Murata Y, Yamashita A, Saito T, Sugamura K, Hamuro J (2002) The conversion of redox status of peritoneal macrophages during pathological progression of spontaneous inflammatory bowel disease in Janus family tyrosine kinase 3(-/-) and IL-2 receptor gamma(-/-) mice. Int Immunol 14(6):627–636, Epub 2002/06/01
121. Fiorentino DF, Bond MW, Mosmann TR (1989) Two types of mouse T helper cell. IV. Th2 clones secrete a factor that inhibits cytokine production by Th1 clones. J Exp Med 170(6): 2081–2095
122. Moore KW, O'Garra A, de Waal MR, Vieira P, Mosmann TR (1993) Interleukin-10. Annu Rev Immunol 11:165–190, Epub 1993/01/01
123. Fiorentino DF, Zlotnik A, Mosmann TR, Howard M, O'Garra A (1991) IL-10 inhibits cytokine production by activated macrophages. J Immunol 147(11):3815–3822, Epub 1991/12/01
124. Hsieh CS, Heimberger AB, Gold JS, O'Garra A, Murphy KM (1992) Differential regulation of T helper phenotype development by interleukins 4 and 10 in an alpha beta T-cell-receptor transgenic system. Proc Natl Acad Sci U S A 89(13):6065–6069, Epub 1992/07/01
125. Kuhn R, Lohler J, Rennick D, Rajewsky K, Muller W (1993) Interleukin-10-deficient mice develop chronic enterocolitis. Cell 75(2):263–274, Epub 1993/10/22
126. Berg DJ, Zhang J, Weinstock JV, Ismail HF, Earle KA, Alila H et al (2002) Rapid development of colitis in NSAID-treated IL-10-deficient mice. Gastroenterology 123(5):1527–1542, Epub 2002/10/31

127. Matharu KS, Mizoguchi E, Cotoner CA, Nguyen DD, Mingle B, Iweala OI et al (2009) Toll-like receptor 4-mediated regulation of spontaneous Helicobacter-dependent colitis in IL-10-deficient mice. Gastroenterology 137(4):1380–1390.e1-3, Epub 2009/07/15
128. Davidson NJ, Leach MW, Fort MM, Thompson-Snipes L, Kuhn R, Muller W et al (1996) T helper cell 1-type CD4+ T cells, but not B cells, mediate colitis in interleukin 10-deficient mice. J Exp Med 184(1):241–251, Epub 1996/07/01
129. Berg DJ, Davidson N, Kuhn R, Muller W, Menon S, Holland G et al (1996) Enterocolitis and colon cancer in interleukin-10-deficient mice are associated with aberrant cytokine production and CD4(+) TH1-like responses. J Clin Invest 98(4):1010–1020, Epub 1996/08/15
130. Davidson NJ, Hudak SA, Lesley RE, Menon S, Leach MW, Rennick DM (1998) IL-12, but not IFN-gamma, plays a major role in sustaining the chronic phase of colitis in IL-10-deficient mice. J Immunol 161(6):3143–3149
131. Yen D, Cheung J, Scheerens H, Poulet F, McClanahan T, McKenzie B et al (2006) IL-23 is essential for T cell-mediated colitis and promotes inflammation via IL-17 and IL-6. J Clin Invest 116(5):1310–1316, Epub 2006/05/04
132. Sheikh SZ, Matsuoka K, Kobayashi T, Li F, Rubinas T, Plevy SE (2010) Cutting edge: IFN-gamma is a negative regulator of IL-23 in murine macrophages and experimental colitis. J Immunol 184(8):4069–4073, Epub 2010/03/17
133. Gonzalez-Navajas JM, Fine S, Law J, Datta SK, Nguyen KP, Yu M et al (2010) TLR4 signaling in effector CD4+ T cells regulates TCR activation and experimental colitis in mice. J Clin Invest 120(2):570–581, Epub 2010/01/07
134. Biswas A, Wilmanski J, Forsman H, Hrncir T, Hao L, Tlaskalova-Hogenova H et al (2011) Negative regulation of Toll-like receptor signaling plays an essential role in homeostasis of the intestine. Eur J Immunol 41(1):182–194, Epub 2010/12/25
135. Rakoff-Nahoum S, Hao L, Medzhitov R (2006) Role of toll-like receptors in spontaneous commensal-dependent colitis. Immunity 25(2):319–329, Epub 2006/08/02
136. Lees CW, Barrett JC, Parkes M, Satsangi J (2011) New IBD genetics: common pathways with other diseases. Gut 60(12):1739–1753, Epub 2011/02/09
137. Kotlarz D, Beier R, Murugan D, Diestelhorst J, Jensen O, Boztug K et al (2012) Loss of interleukin-10 signaling and infantile inflammatory bowel disease—implications for diagnosis and therapy. Gastroenterology 143:347–355, Epub 2012/05/03
138. Moran CJ, Walters TD, Guo CH, Kugathasan S, Klein C, Turner D et al (2013) IL-10R polymorphisms are associated with very-early-onset ulcerative colitis. Inflamm Bowel Dis 19:115–123
139. Mizoguchi A, Mizoguchi E, Bhan AK (1999) The critical role of interleukin 4 but not interferon gamma in the pathogenesis of colitis in T-cell receptor alpha mutant mice. Gastroenterology 116(2):320–326, Epub 1999/01/29
140. Mombaerts P, Mizoguchi E, Grusby MJ, Glimcher LH, Bhan AK, Tonegawa S (1993) Spontaneous development of inflammatory bowel disease in T cell receptor mutant mice. Cell 75(2):274–282, Epub 1993/10/22
141. Bhan AK, Mizoguchi E, Smith RN, Mizoguchi A (2000) Spontaneous chronic colitis in TCR alpha-mutant mice; an experimental model of human ulcerative colitis. Int Rev Immunol 19(1):123–138, Epub 2000/03/21
142. Mizoguchi E, Mizoguchi A, Bhan AK (1997) Role of cytokines in the early stages of chronic colitis in TCR alpha-mutant mice. Lab Invest 76(3):385–397, Epub 1997/03/01
143. Mizoguchi A, Mizoguchi E, Takedatsu H, Blumberg RS, Bhan AK (2002) Chronic intestinal inflammatory condition generates IL-10-producing regulatory B cell subset characterized by CD1d upregulation. Immunity 16(2):219–230, Epub 2002/03/01
144. Mizoguchi A, Ogawa A, Takedatsu H, Sugimoto K, Shimomura Y, Shirane K et al (2007) Dependence of intestinal granuloma formation on unique myeloid DC-like cells. J Clin Invest 117(3):605–615, Epub 2007/02/24
145. Iijima H, Takahashi I, Kishi D, Kim JK, Kawano S, Hori M et al (1999) Alteration of interleukin 4 production results in the inhibition of T helper type 2 cell-dominated inflammatory bowel disease in T cell receptor alpha chain-deficient mice. J Exp Med 190(5):607–615, Epub 1999/09/08

146. Mizoguchi A, Mizoguchi E, Chiba C, Spiekermann GM, Tonegawa S, Nagler-Anderson C et al (1996) Cytokine imbalance and autoantibody production in T cell receptor-alpha mutant mice with inflammatory bowel disease. J Exp Med 183(3):847–856, Epub 1996/03/01
147. Mizoguchi A, Mizoguchi E, Saubermann LJ, Higaki K, Blumberg RS, Bhan AK (2000) Limited CD4 T-cell diversity associated with colitis in T-cell receptor alpha mutant mice requires a T helper 2 environment. Gastroenterology 119(4):983–995, Epub 2000/10/21
148. Prinz I, Klemm U, Kaufmann SH, Steinhoff U (2004) Exacerbated colitis associated with elevated levels of activated CD4+ T cells in TCRalpha chain transgenic mice. Gastroenterology 126(1):170–181, Epub 2003/12/31
149. Nagahama K, Ogawa A, Shirane K, Shimomura Y, Sugimoto K, Mizoguchi A (2008) Protein kinase C theta plays a fundamental role in different types of chronic colitis. Gastroenterology 134(2):459–469, Epub 2007/12/25
150. Nanno M, Kanari Y, Naito T, Inoue N, Hisamatsu T, Chinen H et al (2008) Exacerbating role of gammadelta T cells in chronic colitis of T-cell receptor alpha mutant mice. Gastroenterology 134(2):481–490, Epub 2008/02/05
151. Mizoguchi E, Mizoguchi A, Takedatsu H, Cario E, de Jong YP, Ooi CJ et al (2002) Role of tumor necrosis factor receptor 2 (TNFR2) in colonic epithelial hyperplasia and chronic intestinal inflammation in mice. Gastroenterology 122(1):134–144, Epub 2002/01/10
152. Okada E, Yamazaki M, Tanabe M, Takeuchi T, Nanno M, Oshima S et al (2005) IL-7 exacerbates chronic colitis with expansion of memory IL-7Rhigh CD4+ mucosal T cells in mice. Am J Physiol Gastrointest Liver Physiol 288(4):G745–G754, Epub 2004/11/20
153. Yamazaki M, Yajima T, Tanabe M, Fukui K, Okada E, Okamoto R et al (2003) Mucosal T cells expressing high levels of IL-7 receptor are potential targets for treatment of chronic colitis. J Immunol 171(3):1556–1563, Epub 2003/07/23
154. Haruta I, Hashimoto E, Kobayashi M, Shiratori K (2007) Systemic multifocal epithelial inflammations associated with PBC-like bile duct damage in chronic colitis harboring TCR alpha -/- x AIM -/- mice: does lipoteichoic acid affect the pathogenesis of epithelial inflammation followed by fibrosis? Ann N Y Acad Sci 1107:300–307, Epub 2007/09/07
155. Hokama A, Mizoguchi E, Sugimoto K, Shimomura Y, Tanaka Y, Yoshida M et al (2004) Induced reactivity of intestinal CD4(+) T cells with an epithelial cell lectin, galectin-4, contributes to exacerbation of intestinal inflammation. Immunity 20(6):681–693, Epub 2004/06/11
156. Mizoguchi E (2006) Chitinase 3-like-1 exacerbates intestinal inflammation by enhancing bacterial adhesion and invasion in colonic epithelial cells. Gastroenterology 130(2):398–411, Epub 2006/02/14
157. Dianda L, Hanby AM, Wright NA, Sebesteny A, Hayday AC, Owen MJ (1997) T cell receptor-alpha beta-deficient mice fail to develop colitis in the absence of a microbial environment. Am J Pathol 150(1):91–97, Epub 1997/01/01
158. Gaskins HR, Vondrak-Juergens GL, McCracken BA, Woolsey JH (1997) Specific-pathogen-free conditions enhance inflammatory bowel disease in T-cell receptor knockout, but not C3H/HeJBir mice. Lab Anim Sci 47(6):650–655, Epub 1998/01/20
159. Shimomura Y, Mizoguchi E, Sugimoto K, Kibe R, Benno Y, Mizoguchi A et al (2008) Regulatory role of B-1 B cells in chronic colitis. Int Immunol 20(6):729–737, Epub 2008/04/01
160. Sydora BC, Tavernini MM, Doyle J, Fedorak RN (2006) A defect in epithelial barrier integrity is not required for a systemic response to bacterial antigens or intestinal injury in T cell receptor-alpha gene-deficient mice. Inflamm Bowel Dis 12(8):750–757, Epub 2006/08/19
161. Hale LP, Gottfried MR, Swidsinski A (2005) Piroxicam treatment of IL-10-deficient mice enhances colonic epithelial apoptosis and mucosal exposure to intestinal bacteria. Inflamm Bowel Dis 11(12):1060–1069, Epub 2005/11/25
162. Nishiyori A, Nagakura Y, Ichikawa K (2009) Piroxicam accelerates development of colitis in T-cell receptor alpha chain-deficient mice. Eur J Pharmacol 615(1–3):241–245, Epub 2009/05/19

163. Chin EY, Dangler CA, Fox JG, Schauer DB (2000) Helicobacter hepaticus infection triggers inflammatory bowel disease in T cell receptor alphabeta mutant mice. Comp Med 50(6):586–594, Epub 2001/02/24
164. Burich A, Hershberg R, Waggie K, Zeng W, Brabb T, Westrich G et al (2001) Helicobacter-induced inflammatory bowel disease in IL-10- and T cell-deficient mice. Am J Physiol Gastrointest Liver Physiol 281(3):G764–G778, Epub 2001/08/24
165. Mizoguchi A, Mizoguchi E, Chiba C, Bhan AK (1996) Role of appendix in the development of inflammatory bowel disease in TCR-alpha mutant mice. J Exp Med 184(2):707–715, Epub 1996/08/01
166. Sheikh SZ, Hegazi RA, Kobayashi T, Onyiah JC, Russo SM, Matsuoka K et al (2011) An anti-inflammatory role for carbon monoxide and heme oxygenase-1 in chronic Th2-mediated murine colitis. J Immunol 186(9):5506–5513, Epub 2011/03/30
167. Kishi D, Takahashi I, Kai Y, Tamagawa H, Iijima H, Obunai S et al (2000) Alteration of V beta usage and cytokine production of CD4+ TCR beta beta homodimer T cells by elimination of Bacteroides vulgatus prevents colitis in TCR alpha-chain-deficient mice. J Immunol 165(10):5891–5899, Epub 2000/11/09
168. Nagatani K, Wang S, Llado V, Lau CW, Li Z, Mizoguchi A et al (2012) Chitin microparticles for the control of intestinal inflammation. Inflamm Bowel Dis 18:1698–1710, Epub 2012/01/14
169. Mizoguchi A, Mizoguchi E, Smith RN, Preffer FI, Bhan AK (1997) Suppressive role of B cells in chronic colitis of T cell receptor alpha mutant mice. J Exp Med 186(10):1749–1756, Epub 1997/12/31
170. Sugimoto K, Ogawa A, Mizoguchi E, Shimomura Y, Andoh A, Bhan AK et al (2008) IL-22 ameliorates intestinal inflammation in a mouse model of ulcerative colitis. J Clin Invest 118(2):534–544
171. Birnbaumer L (1992) Receptor-to-effector signaling through G proteins: roles for beta gamma dimers as well as alpha subunits. Cell 71(7):1069–1072, Epub 1992/12/24
172. Hampe J, Lynch NJ, Daniels S, Bridger S, Macpherson AJ, Stokkers P et al (2001) Fine mapping of the chromosome 3p susceptibility locus in inflammatory bowel disease. Gut 48(2):191–197, Epub 2001/01/13
173. Rudolph U, Finegold MJ, Rich SS, Harriman GR, Srinivasan Y, Brabet P et al (1995) Ulcerative colitis and adenocarcinoma of the colon in G alpha i2-deficient mice. Nat Genet 10(2):143–150, Epub 1995/06/01
174. Pena JA, Thompson-Snipes L, Calkins PR, Tatevian N, Puppi M, Finegold MJ (2009) Alterations in myeloid dendritic cell innate immune responses in the Galphai2-deficient mouse model of colitis. Inflamm Bowel Dis 15(2):248–260, Epub 2008/11/28
175. Bjursten M, Hultgren HE (2005) Dietary antigen-specific T-cell responses: switch from an interleukin-10-dominated response in normal mice to a T-helper 1 cytokine profile in Galphai2-deficient mice prior to colitis. Scand J Immunol 61(1):29–35, Epub 2005/01/13
176. Hornquist CE, Lu X, Rogers-Fani PM, Rudolph U, Shappell S, Birnbaumer L et al (1997) G(alpha)i2-deficient mice with colitis exhibit a local increase in memory CD4+ T cells and proinflammatory Th1-type cytokines. J Immunol 158(3):1068–1077, Epub 1997/02/01
177. Wu JY, Jin Y, Edwards RA, Zhang Y, Finegold MJ, Wu MX (2005) Impaired TGF-beta responses in peripheral T cells of G alpha i2-/- mice. J Immunol 174(10):6122–6128, Epub 2005/05/10
178. Dalwadi H, Wei B, Schrage M, Spicher K, Su TT, Birnbaumer L et al (2003) B cell developmental requirement for the G alpha i2 gene. J Immunol 170(4):1707–1715, Epub 2003/02/08
179. Fillatreau S, Sweenie CH, McGeachy MJ, Gray D, Anderton SM (2002) B cells regulate autoimmunity by provision of IL-10. Nat Immunol 3(10):944–950, Epub 2002/09/24
180. Glimcher LH (2007) Trawling for treasure: tales of T-bet. Nat Immunol 8(5):448–450, Epub 2007/04/19
181. Maruyama K, Sugano S (1994) Oligo-capping: a simple method to replace the cap structure of eukaryotic mRNAs with oligoribonucleotides. Gene 138(1–2):171–174, Epub 1994/01/28

182. Neurath MF, Weigmann B, Finotto S, Glickman J, Nieuwenhuis E, Iijima H et al (2002) The transcription factor T-bet regulates mucosal T cell activation in experimental colitis and Crohn's disease. J Exp Med 195(9):1129–1143, Epub 2002/05/08
183. Weigmann B, Neurath MF (2002) T-bet and mucosal Th1 responses in the gastrointestinal tract. Gut 51(3):301–303, Epub 2002/08/13
184. Garrett WS, Lord GM, Punit S, Lugo-Villarino G, Mazmanian SK, Ito S et al (2007) Communicable ulcerative colitis induced by T-bet deficiency in the innate immune system. Cell 131(1):33–45, Epub 2007/10/10
185. Chieppa M, Rescigno M, Huang AY, Germain RN (2006) Dynamic imaging of dendritic cell extension into the small bowel lumen in response to epithelial cell TLR engagement. J Exp Med 203(13):2841–2852, Epub 2006/12/06
186. Niess JH, Brand S, Gu X, Landsman L, Jung S, McCormick BA et al (2005) CX3CR1-mediated dendritic cell access to the intestinal lumen and bacterial clearance. Science 307(5707):254–258, Epub 2005/01/18
187. Garrett WS, Gallini CA, Yatsunenko T, Michaud M, DuBois A, Delaney ML et al (2010) Enterobacteriaceae act in concert with the gut microbiota to induce spontaneous and maternally transmitted colitis. Cell Host Microbe 8(3):292–300, Epub 2010/09/14
188. Garrett WS, Punit S, Gallini CA, Michaud M, Zhang D, Sigrist KS et al (2009) Colitis-associated colorectal cancer driven by T-bet deficiency in dendritic cells. Cancer Cell 16(3):208–219, Epub 2009/09/08
189. Snapper SB, Rosen FS, Mizoguchi E, Cohen P, Khan W, Liu CH et al (1998) Wiskott-Aldrich syndrome protein-deficient mice reveal a role for WASP in T but not B cell activation. Immunity 9(1):81–91, Epub 1998/08/11
190. Nguyen DD, Maillard MH, Cotta-de-Almeida V, Mizoguchi E, Klein C, Fuss I et al (2007) Lymphocyte-dependent and Th2 cytokine-associated colitis in mice deficient in Wiskott-Aldrich syndrome protein. Gastroenterology 133(4):1188–1197, Epub 2007/09/04
191. Maillard MH, Cotta-de-Almeida V, Takeshima F, Nguyen DD, Michetti P, Nagler C et al (2007) The Wiskott-Aldrich syndrome protein is required for the function of CD4(+)CD25(+) Foxp3(+) regulatory T cells. J Exp Med 204(2):381–391, Epub 2007/02/14
192. Hugot JP, Chamaillard M, Zouali H, Lesage S, Cezard JP, Belaiche J et al (2001) Association of *NOD2* leucine-rich repeat variants with susceptibility to Crohn's disease. Nature 411(6837):599–603, Epub 2001/06/01
193. Ogura Y, Bonen DK, Inohara N, Nicolae DL, Chen FF, Ramos R et al (2001) A frameshift mutation in *NOD2* associated with susceptibility to Crohn's disease. Nature 411(6837): 603–606, Epub 2001/06/01
194. Miceli-Richard C, Lesage S, Rybojad M, Prieur AM, Manouvrier-Hanu S, Hafner R et al (2001) CARD15 mutations in Blau syndrome. Nat Genet 29(1):19–20, Epub 2001/08/31
195. Maeda S, Hsu LC, Liu H, Bankston LA, Iimura M, Kagnoff MF et al (2005) *Nod2* mutation in Crohn's disease potentiates NF-kappaB activity and IL-1beta processing. Science 307(5710):734–738, Epub 2005/02/05
196. Watanabe T, Kitani A, Murray PJ, Strober W (2004) *NOD2* is a negative regulator of Toll-like receptor 2-mediated T helper type 1 responses. Nat Immunol 5(8):800–808, Epub 2004/06/29
197. Kobayashi KS, Chamaillard M, Ogura Y, Henegariu O, Inohara N, Nunez G et al (2005) *Nod2*-dependent regulation of innate and adaptive immunity in the intestinal tract. Science 307(5710):731–734, Epub 2005/02/05
198. Bonen DK, Ogura Y, Nicolae DL, Inohara N, Saab L, Tanabe T et al (2003) Crohn's disease-associated *NOD2* variants share a signaling defect in response to lipopolysaccharide and peptidoglycan. Gastroenterology 124(1):140–146, Epub 2003/01/04
199. Gutierrez O, Pipaon C, Inohara N, Fontalba A, Ogura Y, Prosper F et al (2002) Induction of *Nod2* in myelomonocytic and intestinal epithelial cells via nuclear factor-kappa B activation. J Biol Chem 277(44):41701–41705
200. Petnicki-Ocwieja T, Hrncir T, Liu YJ, Biswas A, Hudcovic T, Tlaskalova-Hogenova H et al (2009) *Nod2* is required for the regulation of commensal microbiota in the intestine. Proc Natl Acad Sci U S A 106(37):15813–15818, Epub 2009/10/07

201. Biswas A, Liu YJ, Hao L, Mizoguchi A, Salzman NH, Bevins CL et al (2010) Induction and rescue of *Nod2*-dependent Th1-driven granulomatous inflammation of the ileum. Proc Natl Acad Sci U S A 107(33):14739–14744, Epub 2010/08/04
202. Hampe J, Franke A, Rosenstiel P, Till A, Teuber M, Huse K et al (2007) A genome-wide association scan of nonsynonymous SNPs identifies a susceptibility variant for Crohn disease in ATG16L1. Nat Genet 39(2):207–211, Epub 2007/01/04
203. Rioux JD, Xavier RJ, Taylor KD, Silverberg MS, Goyette P, Huett A et al (2007) Genome-wide association study identifies new susceptibility loci for Crohn disease and implicates autophagy in disease pathogenesis. Nat Genet 39(5):596–604, Epub 2007/04/17
204. Wellcome Trust Case Control C, Australo-Anglo-American Spondylitis C, Burton PR, Clayton DG, Cardon LR, Craddock N et al (2007) Association scan of 14,500 nonsynonymous SNPs in four diseases identifies autoimmunity variants. Nat Genet 39(11):1329–1337, Epub 2007/10/24.
205. Fujita N, Itoh T, Omori H, Fukuda M, Noda T, Yoshimori T (2008) The Atg16L complex specifies the site of LC3 lipidation for membrane biogenesis in autophagy. Mol Biol Cell 19(5):2092–2100, Epub 2008/03/07
206. Mizushima N, Kuma A, Kobayashi Y, Yamamoto A, Matsubae M, Takao T et al (2003) Mouse Apg16L, a novel WD-repeat protein, targets to the autophagic isolation membrane with the Apg12-Apg5 conjugate. J Cell Sci 116(Pt 9):1679–1688, Epub 2003/04/01
207. Kundu M, Thompson CB (2008) Autophagy: basic principles and relevance to disease. Annu Rev Pathol 3:427–455, Epub 2007/11/28
208. Marks DJ, Harbord MW, MacAllister R, Rahman FZ, Young J, Al-Lazikani B et al (2006) Defective acute inflammation in Crohn's disease: a clinical investigation. Lancet 367(9511):668–678, Epub 2006/03/01
209. Cooney R, Baker J, Brain O, Danis B, Pichulik T, Allan P et al (2010) *NOD2* stimulation induces autophagy in dendritic cells influencing bacterial handling and antigen presentation. Nat Med 16(1):90–97, Epub 2009/12/08
210. Juarez E, Carranza C, Hernandez-Sanchez F, Leon-Contreras JC, Hernandez-Pando R, Escobedo D et al (2012) *NOD2* enhances the innate response of alveolar macrophages to Mycobacterium tuberculosis in humans. Eur J Immunol 42(4):880–889, Epub 2012/04/26
211. Travassos LH, Carneiro LA, Ramjeet M, Hussey S, Kim YG, Magalhaes JG et al (2010) Nod1 and *Nod2* direct autophagy by recruiting ATG16L1 to the plasma membrane at the site of bacterial entry. Nat Immunol 11(1):55–62, Epub 2009/11/10
212. Saitoh T, Fujita N, Jang MH, Uematsu S, Yang BG, Satoh T et al (2008) Loss of the autophagy protein Atg16L1 enhances endotoxin-induced IL-1beta production. Nature 456(7219):264–268, Epub 2008/10/14
213. Verreck FA, de Boer T, Langenberg DM, Hoeve MA, Kramer M, Vaisberg E et al (2004) Human IL-23-producing type 1 macrophages promote but IL-10-producing type 2 macrophages subvert immunity to (myco)bacteria. Proc Natl Acad Sci U S A 101(13):4560–4565, Epub 2004/04/09
214. Veckman V, Miettinen M, Pirhonen J, Siren J, Matikainen S, Julkunen I (2004) Streptococcus pyogenes and Lactobacillus rhamnosus differentially induce maturation and production of Th1-type cytokines and chemokines in human monocyte-derived dendritic cells. J Leukoc Biol 75(5):764–771, Epub 2004/02/18
215. Happel KI, Zheng M, Young E, Quinton LJ, Lockhart E, Ramsay AJ et al (2003) Cutting edge: roles of Toll-like receptor 4 and IL-23 in IL-17 expression in response to Klebsiella pneumoniae infection. J Immunol 170(9):4432–4436, Epub 2003/04/23
216. Murphy CA, Langrish CL, Chen Y, Blumenschein W, McClanahan T, Kastelein RA et al (2003) Divergent pro- and antiinflammatory roles for IL-23 and IL-12 in joint autoimmune inflammation. J Exp Med 198(12):1951–1957, Epub 2003/12/10
217. Cua DJ, Sherlock J, Chen Y, Murphy CA, Joyce B, Seymour B et al (2003) Interleukin-23 rather than interleukin-12 is the critical cytokine for autoimmune inflammation of the brain. Nature 421(6924):744–748, Epub 2003/03/01

218. Hue S, Ahern P, Buonocore S, Kullberg MC, Cua DJ, McKenzie BS et al (2006) Interleukin-23 drives innate and T cell-mediated intestinal inflammation. J Exp Med 203(11):2473–2483
219. Kullberg MC, Jankovic D, Feng CG, Hue S, Gorelick PL, McKenzie BS et al (2006) IL-23 plays a key role in Helicobacter hepaticus-induced T cell-dependent colitis. J Exp Med 203(11):2485–2494, Epub 2006/10/13
220. Duerr RH, Taylor KD, Brant SR, Rioux JD, Silverberg MS, Daly MJ et al (2006) A genome-wide association study identifies IL-23R as an inflammatory bowel disease gene. Science 314(5804):1461–1463, Epub 2006/10/28
221. Burton PR, Clayton DG, Cardon LR, Craddock N, Deloukas P, Duncanson A et al (2007) Genome-wide association study of 14,000 cases of seven common disease and 3,00 shared controls. Nature 447:661–678
222. McGovern DP, Gardet A, Torkvist L, Goyette P, Essers J, Taylor KD et al (2010) Genomewide association identifies multiple ulcerative colitis susceptibility loci. Nat Genet 42(4):332– 337, Epub 2010/03/17
223. Ahern PP, Schiering C, Buonocore S, McGeachy MJ, Cua DJ, Maloy KJ et al (2010) Interleukin-23 drives intestinal inflammation through direct activity on T cells. Immunity 33(2):279–288, Epub 2010/08/25
224. Rosenstiel P, Huse K, Till A, Hampe J, Hellmig S, Sina C et al (2006) A short isoform of *NOD2*/CARD15, *NOD2*-S, is an endogenous inhibitor of *NOD2*/receptor-interacting protein kinase 2-induced signaling pathways. Proc Natl Acad Sci U S A 103(9):3280–3285, Epub 2006/02/24
225. Yu RY, Gallagher G (2010) A naturally occurring, soluble antagonist of human IL-23 inhibits the development and in vitro function of human Th17 cells. J Immunol 185(12):7302–7308, Epub 2010/11/16
226. Di Meglio P, Di Cesare A, Laggner U, Chu CC, Napolitano L, Villanova F et al (2011) The IL-23R R381Q gene variant protects against immune-mediated diseases by impairing IL-23-induced Th17 effector response in humans. PLoS One 6(2):e17160, Epub 2011/03/03
227. Akira S, Nishio Y, Inoue M, Wang XJ, Wei S, Matsusaka T et al (1994) Molecular cloning of APRF, a novel IFN-stimulated gene factor 3 p91-related transcription factor involved in the gp130-mediated signaling pathway. Cell 77(1):63–71, Epub 1994/04/08
228. Zhong Z, Wen Z, Darnell JE Jr (1994) Stat3: a STAT family member activated by tyrosine phosphorylation in response to epidermal growth factor and interleukin-6. Science 264(5155):95–98, Epub 1994/04/01
229. Bourillot PY, Aksoy I, Schreiber V, Wianny F, Schulz H, Hummel O et al (2009) Novel STAT3 target genes exert distinct roles in the inhibition of mesoderm and endoderm differentiation in cooperation with Nanog. Stem Cells 27(8):1760–1771, Epub 2009/06/23
230. Ihle JN (1996) STATs: signal transducers and activators of transcription. Cell 84(3):331–334, Epub 1996/02/09
231. Schindler C, Darnell JE Jr (1995) Transcriptional responses to polypeptide ligands: the JAK-STAT pathway. Annu Rev Biochem 64:621–651, Epub 1995/01/01
232. Barrett JC, Hansoul S, Nicolae DL, Cho JH, Duerr RH, Rioux JD et al (2008) Genome-wide association defines more than 30 distinct susceptibility loci for Crohn's disease. Nat Genet 40(8):955–962, Epub 2008/07/01
233. Takeda K, Noguchi K, Shi W, Tanaka T, Matsumoto M, Yoshida N et al (1997) Targeted disruption of the mouse Stat3 gene leads to early embryonic lethality. Proc Natl Acad Sci U S A 94(8):3801–3804, Epub 1997/04/15
234. Takeda K, Clausen BE, Kaisho T, Tsujimura T, Terada N, Forster I et al (1999) Enhanced Th1 activity and development of chronic enterocolitis in mice devoid of Stat3 in macrophages and neutrophils. Immunity 10(1):39–49, Epub 1999/02/19
235. Reindl W, Weiss S, Lehr HA, Forster I (2007) Essential crosstalk between myeloid and lymphoid cells for development of chronic colitis in myeloid-specific signal transducer and activator of transcription 3-deficient mice. Immunology 120(1):19–27, Epub 2007/01/20
236. Alonzi T, Newton IP, Bryce PJ, Di Carlo E, Lattanzio G, Tripodi M et al (2004) Induced somatic inactivation of STAT3 in mice triggers the development of a fulminant form of enterocolitis. Cytokine 26(2):45–56, Epub 2004/03/31

237. Durant L, Watford WT, Ramos HL, Laurence A, Vahedi G, Wei L et al (2010) Diverse targets of the transcription factor STAT3 contribute to T cell pathogenicity and homeostasis. Immunity 32(5):605–615, Epub 2010/05/25
238. Lee AH, Chu GC, Iwakoshi NN, Glimcher LH (2005) XBP-1 is required for biogenesis of cellular secretory machinery of exocrine glands. EMBO J 24(24):4368–4380, Epub 2005/12/20
239. Reimold AM, Iwakoshi NN, Manis J, Vallabhajosyula P, Szomolanyi-Tsuda E, Gravallese EM et al (2001) Plasma cell differentiation requires the transcription factor XBP-1. Nature 412(6844):300–307, Epub 2001/07/19
240. Shaffer AL, Shapiro-Shelef M, Iwakoshi NN, Lee AH, Qian SB, Zhao H et al (2004) XBP1, downstream of Blimp-1, expands the secretory apparatus and other organelles, and increases protein synthesis in plasma cell differentiation. Immunity 21(1):81–93, Epub 2004/09/04
241. Shkoda A, Ruiz PA, Daniel H, Kim SC, Rogler G, Sartor RB et al (2007) Interleukin-10 blocked endoplasmic reticulum stress in intestinal epithelial cells: impact on chronic inflammation. Gastroenterology 132(1):190–207, Epub 2007/01/24
242. Bertolotti A, Wang X, Novoa I, Jungreis R, Schlessinger K, Cho JH et al (2001) Increased sensitivity to dextran sodium sulfate colitis in IRE1beta-deficient mice. J Clin Invest 107(5):585–593, Epub 2001/03/10
243. Madison BB, Dunbar L, Qiao XT, Braunstein K, Braunstein E, Gumucio DL (2002) Cis elements of the villin gene control expression in restricted domains of the vertical (crypt) and horizontal (duodenum, cecum) axes of the intestine. J Biol Chem 277(36):33275–33283, Epub 2002/06/18
244. Kaser A, Lee AH, Franke A, Glickman JN, Zeissig S, Tilg H et al (2008) XBP1 links ER stress to intestinal inflammation and confers genetic risk for human inflammatory bowel disease. Cell 134(5):743–756, Epub 2008/09/09
245. Paisan-Ruiz C, Jain S, Evans EW, Gilks WP, Simon J, van der Brug M et al (2004) Cloning of the gene containing mutations that cause PARK8-linked Parkinson's disease. Neuron 44(4):595–600, Epub 2004/11/16
246. Zimprich A, Biskup S, Leitner P, Lichtner P, Farrer M, Lincoln S et al (2004) Mutations in LRRK2 cause autosomal-dominant parkinsonism with pleomorphic pathology. Neuron 44(4):601–607, Epub 2004/11/16
247. Bretscher A, Edwards K, Fehon RG (2002) ERM proteins and merlin: integrators at the cell cortex. Nat Rev Mol Cell Biol 3(8):586–599, Epub 2002/08/03
248. Mangeat P, Roy C, Martin M (1999) ERM proteins in cell adhesion and membrane dynamics: authors' correction. Trends Cell Biol 9(7):289, Epub 1999/06/17
249. Greggio E, Jain S, Kingsbury A, Bandopadhyay R, Lewis P, Kaganovich A et al (2006) Kinase activity is required for the toxic effects of mutant LRRK2/dardarin. Neurobiol Dis 23(2):329–341, Epub 2006/06/06
250. Smith WW, Pei Z, Jiang H, Dawson VL, Dawson TM, Ross CA (2006) Kinase activity of mutant LRRK2 mediates neuronal toxicity. Nat Neurosci 9(10):1231–1233, Epub 2006/09/19
251. Iaccarino C, Crosio C, Vitale C, Sanna G, Carri MT, Barone P (2008) Apoptotic mechanisms in mutant LRRK2-mediated cell death. Hum Mol Genet 16:1319–1326
252. Plowey ED, Cherra SJ 3rd, Liu YJ, Chu CT (2008) Role of autophagy in G2019S-LRRK2-associated neurite shortening in differentiated SH-SY5Y cells. J Neurochem 105(3):1048–1056, Epub 2008/01/10
253. Wang L, Xie C, Greggio E, Parisiadou L, Shim H, Sun L et al (2008) The chaperone activity of heat shock protein 90 is critical for maintaining the stability of leucine-rich repeat kinase 2. J Neurosci 28(13):3384–3391, Epub 2008/03/28
254. Tong Y, Yamaguchi H, Giaime E, Boyle S, Kopan R, Kelleher RJ 3rd et al (2010) Loss of leucine-rich repeat kinase 2 causes impairment of protein degradation pathways, accumulation of alpha-synuclein, and apoptotic cell death in aged mice. Proc Natl Acad Sci U S A 107(21):9879–9884, Epub 2010/05/12
255. Tong Y, Giaime E, Yamaguchi H, Ichimura T, Liu Y, Si H et al (2012) Loss of leucine-rich repeat kinase 2 causes age-dependent bi-phasic alterations of the autophagy pathway. Mol Neurodegener 7:2, Epub 2012/01/11

256. Gardet A, Benita Y, Li C, Sands BE, Ballester I, Stevens C et al (2010) LRRK2 is involved in the IFN-gamma response and host response to pathogens. J Immunol 185(9):5577–5585, Epub 2010/10/06
257. Lambeth JD (2004) NOX enzymes and the biology of reactive oxygen. Nat Rev Immunol 4(3):181–189, Epub 2004/03/25
258. Roberts RL, Hollis-Moffatt JE, Gearry RB, Kennedy MA, Barclay ML, Merriman TR (2008) Confirmation of association of IRGM and NCF4 with ileal Crohn's disease in a population-based cohort. Genes Immun 9(6):561–565, Epub 2008/06/27
259. Somasundaram R, Deuring JJ, van der Woude CJ, Peppelenbosch MP, Fuhler GM (2012) Linking risk conferring mutations in NCF4 to functional consequences in Crohn's disease. Gut 61:1097–1098, Epub 2011/10/24
260. Ellson CD, Davidson K, Ferguson GJ, O'Connor R, Stephens LR, Hawkins PT (2006) Neutrophils from p40phox-/- mice exhibit severe defects in NADPH oxidase regulation and oxidant-dependent bacterial killing. J Exp Med 203(8):1927–1937, Epub 2006/08/02
261. Conway KL, Goel G, Sokol H, Manocha M, Mizoguchi E, Terhorst C et al (2012) P40phox expression regulates meutrophil recruitment and function during the resolution phase of intestinal inflammation. J Immunol 189(17):3631–3640, Epub 2012/08/22

Part II
The Genetic and Molecular Makeup of Inflammatory Bowel Disease

Chapter 4
Complex Disease Genes and Their Discovery

Jeffrey C. Barrett and Mark J. Daly

Abstract The study of the genetic underpinning of heritable human diseases stretches back nearly a century. While thousands of mutations in single genes have been found that cause severe "Mendelian" disorders, attempts to find such single genes for complex diseases have been relatively unsuccessful. Instead it has become clear that complex diseases, like IBD, are affected by many (likely hundreds or even thousands) different genes as well as environmental factors. Here we describe the process by which that discovery was made, as well as the technological advances from small-scale candidate gene to genome-wide association studies. These approaches, especially when undertaken in large-scale collaborations, have unlocked thousands of complex disease genes, including 163 associated with IBD. Despite these exciting developments, the discovery of genes represents the first stage in translating that knowledge into biological understanding of disease and possible future treatments.

Background

It has long been appreciated that genetics plays an important role in susceptibility to a wide variety of complex human diseases. Indeed it has been almost 100 years since R. A. Fisher and others reconciled the discrete Mendelian inheritance of

J.C. Barrett, B.S., D.Phil. (✉)
Wellcome Trust Sanger Institute, Wellcome Trust Genome Campus, Hinxton, Cambridge, CB10 1HH, UK
e-mail: barrett@sanger.ac.uk

M.J. Daly, Ph.D.
Simches Research Center, Massachusetts General Hospital, 185 Cambridge Street, CPZN 6818, Boston, MA 02114, USA

Analytic and Translational Genetics Unit, Department of Medicine, Massachusetts General Hospital, 185 Cambridge Street, CPZN 6818, Boston, MA 02114, USA
e-mail: mjdaly@atgu.mgh.harvard.edu

M. D'Amato and J.D. Rioux (eds.), *Molecular Genetics of Inflammatory Bowel Disease*, DOI 10.1007/978-1-4614-8256-7_4,

individual genes with the continuous distribution of complex heritable traits, such as height [1]. The earliest geneticists at that time were realizing that while most traits were correlated among relatives and thereby appeared "heritable," only a few of them strictly followed Mendel's laws of inheritance. Instead, the majority of heritable traits and diseases quite evidently involved the action of many genes (as well as nongenetic or environmental factors). The suggestion that individual genetic variants might have relatively modest effects on these traits, and that if they were sufficiently numerous, would give rise to normal distributions in general populations is one which continues to reverberate through the most cutting-edge genetic studies of today.

Indeed, the history of complex disease genetics has been a story of reconciling the obvious family clustering of these diseases with an evolving understanding of the types of genetic variation that exist in human populations and how they affect disease risk. In this chapter we will describe how that process moved (over many decades) from relatively fruitless searches to find single genes explaining disease in individual families to international collaborations studying tens of thousands of patients with complex diseases at once. At each stage, a combination of dedicated clinical researchers, statistical analysts, and new technologies enabled new discoveries to be made. We will try to illustrate this process with examples from IBD and conclude by considering what biological lessons have been learned and what challenges remain.

Chasing a Successful Paradigm: Linkage Studies in Complex Disease

Even in the simplest, so-called "Mendelian" diseases, the earliest studies of inherited phenotypes often showed that two genetic traits cosegregated—that is, were correlated in their transmission from parents to children. This is of course the consequence of the genes being "linked" or closely located on the same chromosome, such that from generation to generation, the pair of genes is passed on intact from the same parental chromosome without intervening meiotic recombination. Sturtevant and Morgan 100 years ago conceived that this cosegregation of phenotypes could be used to create a linear map of the underlying order of the genes responsible for those phenotypes—thus creating the first linkage map [2] (in this instance in fruit flies). Mathematically, the principles of linkage analysis were worked out independent of even the definition of the structure of DNA [3], but it was not until the 1970s and 1980s that the molecular techniques to clone and sequence DNA permitted the experimental connection of genetic linkage maps of phenotypes to underlying DNA variation and thus to identify genes responsible for phenotype via linkage analysis, followed by detailed sequencing and functional studies to define the specific underlying causal gene and mutation—so-called positional cloning.

Use of this strategy to approach human disease was outlined at this time [4, 5] and first successfully demonstrated in the localization of the Huntington's disease

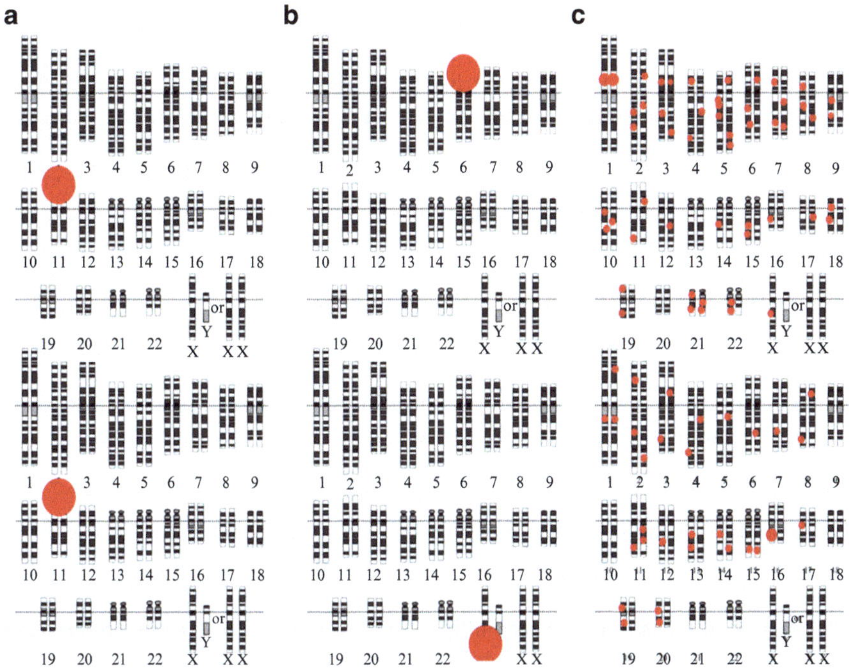

Fig. 4.1 Three illustrations of the spectrum of disease genes. Six images of imaginary "patient genomes" with *red circles* corresponding to risk alleles for a particular disease. Three columns (**a**–**c**) correspond to different disorders, with two patients of each for comparison. (**a**) Some single gene, or Mendelian, disorders, such as sickle cell disease, are caused by mutations in the same gene in all patients. (**b**) Other disorders, such as intellectual disability, are often caused by a single mutation in each patient, but can be in a variety of genes. (**c**) Complex diseases, like IBD, are instead affected by a large number of individually weak variants across the genome, none individually as strong as the Mendelian variants

gene in 1983 [6]—later coming to fruition in the identification of genes for cystic fibrosis and Duchenne muscular dystrophy as the 1980s drew to a close. Family-based linkage studies in humans are the most direct approach to analyzing the simple consequences of Mendelian inheritance from one generation to the next and the resulting sharing of relatively long segments of DNA *identical by descent*. Consider families with multiple individuals affected with a rare disease (see Fig. 4.1) such as Huntington's disease (which affects a handful of individuals per 100,000 populations [7]): it is very unlikely that such co-occurrence would happen by chance, so a genetic explanation is likely. If the family is large, with a sufficient number of affected individuals, a classic pattern of Mendelian inheritance might be clear, such as *recessive* (requiring two copies of a damaging mutation to be affected, one from each parent) or *dominant* (individuals with one copy of the mutation, from either parent, are affected—the pattern seen in Huntington's). Because the segments of shared chromosomes between nearby relatives are large (tens of megabases among first-degree relatives), it is possible to identify which parts of which chromosomes are shared between affected individuals with relatively few DNA markers. If one chromosomal segment is shared in a consistent way among all affected family members, but not those who are healthy, it likely carries a mutation that causes disease. The same region (containing the relevant gene) will be "linked" in this way in many different families, even if the individual mutations responsible vary across families.

For this reason, the earliest studies of complex disease genetics built upon the successes of family-based linkage studies in mapping Mendelian disease genes. The linkage approach was rapidly applied across a range of rare diseases suspected of Mendelian inheritance, leading to the identification of hundreds of disease genes throughout the 1990s. These successes led to the application of the same approach to more common, complex diseases, such as type 2 diabetes and inflammatory bowel disease. Much like Mendelian disease, these diseases were known to run in families, and studies of disease concordance in monozygous and dizygous twins rigorously established that genetics plays an important role in their etiology. With a few exceptions (see next chapters), however, linkage studies in these complex diseases did not lead to the discovery of major genetic risk factors.

The principal insight from this failure was the lesson learned from the earliest genetic studies in fruit flies, specifically that, in contrast to Mendelian disease, there was not a single gene (nor even a very small number) for most complex diseases. This was partially unsurprising, as multiply affected families with these diseases did not display the classic Mendelian patterns of inheritance, but instead appeared to be driven by combinations of many genetic factors each exerting a relatively weak effect—just as predicted by the biometrical models of Fisher. These realizations lead to the invention of more sophisticated statistical techniques for linkage analysis [8] aimed at discovering loci which only partially explain the disease state of family members.

A key observation was made [9] that the power of linkage studies falls rapidly with decreasing effect size of the associated genetic variant. If the genetic basis of complex disease was completely unlike Mendelian disease, and instead consisted of dozens or hundreds of small effects, then linkage would never, for practical purposes, be able to discover them. *Association studies*, where one simply compared the allele frequency of a particular variant between unrelated cases and controls, had the potential to discover these tiny effects. At the time of this publication, however, technology did not exist to make such studies possible, nor was there even a compelling estimate of how much genetic variation existed in the human population. By contrast to genome-wide linkage mapping, where the recombination map of humans had been described for a decade and could be conveniently assayed by fewer than 1,000 polymorphic markers, genome-wide association would need to wait.

Motivated by Biology: Candidate Gene Studies

The complex disease genetics community was thus faced with the twin realities that linkage mapping would be unlikely to discover risk loci for these diseases and that genome-wide application of the association study paradigm was still technologically impossible. One possible solution to this problem would be to prioritize genes for genetic study that seemed biologically plausible candidates for particular diseases. For instance, one of the few success stories in complex disease linkage mapping was the identification of a tandem repeat polymorphism in *INS* (the gene encoding the insulin protein) associated with type 1 diabetes [10]. This discovery fit neatly into the developing biological understanding of the disease and suggested that perhaps genetic discoveries could be made by first guessing the relevant candidates.

Unfortunately, three problems undermined this candidate gene approach. First, and perhaps most importantly, the ability of researchers to predict which genes would be associated with which diseases was poor: obvious connections like insulin and type 1 diabetes were not common. Second, the available patient collections typically numbered in dozens or low hundreds: too small to detect the very weak effects that would come to typify the contribution of common alleles to complex diseases. Finally even with the good fortune of picking the right gene, it was not possible to select SNPs that represented the diversity of variation within that gene in a systematic fashion. Just as it had been impossible to query the reference sequence of a gene before the draft human genome was finished, it was now impossible to look up how a particular gene commonly varied within a population of interest.

Maps of Common Human Genetic Variation

As noted above, a common feature of candidate gene studies was the selection of only a handful of SNPs in each gene being considered to test for association. This limitation was largely a result of a lack of comprehensive databases of genetic variation throughout the human genome. The first project aimed at producing such a database was the SNP Consortium [11], which undertook large-scale genome resequencing and identified over one million SNPs. This effort provided the substrate for a wide array of subsequent investigations into the number, distribution, and frequency of SNPs throughout the genome.

One research area transformed by this new abundance of variation data was the study of population genetics: the quantity and frequency of, and patterns of correlation among, genetic variation in different populations around the world provided empirical data with which to fit models of human demography and selection. Two forces increase variation in the genome: mutation, which introduces new variants, and recombination, which reshuffles the existing patterns. Random drift, evolutionary selection (either positive or negative), and human population history then shape this pool of variation into the patterns seen in modern humans. Previous population genetics work could be used to make very specific predictions about the extent of correlation between nearby SNPs (known as *linkage disequilibrium*, or LD) given certain assumptions about the history of humans and, crucially, that recombination occurred uniformly throughout the genome. Two simultaneous observations suggested, however, that recombination was instead clustered in punctate "hotspots"—the vast majority of historical human recombinations had happened in a relatively small fraction of the genome sequence. Molecular typing of multiple sperm from a single individual showed clustering of directly observable recombinations [12], and an analysis of the precise positions where LD decayed in a survey of general population variation suggested this process was consistently concentrated over many generations [13].

Prompted by these insights, the International HapMap [14] project was launched to create a genome variation reference for medical genetics in multiple human populations across the entire genome. The project was undertaken in two main phases which yielded a map of the frequencies and LD patterns of over 2.5 million SNPs in individuals of European, West African, and East Asian ancestry. The HapMap provided both a generic variation reference and revealed new specific insights into human population history, such as strong support for the out-of-Africa hypothesis of human migration and a realization that a huge fraction of variation is shared across the world. In addition, this large-scale collaboration contributed heavily to the development of high-throughput genotyping technologies. In the course of the project, it became possible to move from genotyping dozens of SNPs to thousands and then hundreds of thousands—technological advances which would prove to be just as transformative as the scientific discoveries of the project.

Nowhere were the implications of these data and technologies greater than for the study of the role of genetic variation in disease risk. It became clear that it was possible to select a small number of SNPs from a particular region of the genome that were highly correlated with all nearby SNPs. These "tag" SNPs could then be genotyped as an efficient means of capturing all the information contained in the full complement of SNPs in the region [15]. This tagging approach was quickly shown to be scalable genome wide, so that fewer than 500,000 carefully chosen SNPs could capture nearly all the common variation in populations of European descent [16]. The stage was set for a revolution in the discovery of genetic risk loci for common diseases.

Genome-Wide Association Studies

Several developments from the HapMap project presented new opportunities for disease gene mapping: an understanding of genome-wide LD patterns, algorithms and tools for selecting efficient tag SNP sets, and affordable technologies for genotyping hundreds of thousands of SNPs. Taken together, these offered the ability to genotype large groups of healthy individuals and cases of particular diseases in a way which captured nearly all the common variation in individuals of European ancestry on an affordable scale. It was also recognized that robust genome-wide statistics would likely only emerge from much larger sample collections than customarily used in candidate gene studies, and indeed even at this time both the lack of consistent marker maps and small samples with inconclusive statistical support were creating a cacophony of inconsistent candidate gene studies for many gene, disease pairings. These early *genome-wide association studie*s (GWAS) showed some early successes [17]; they also confirmed the increasing suspicion that individual common risk alleles generally exercised very weak effects on disease risk: few odds ratios were >1.2. Crohn's disease (as will be described in future chapters) benefited from a number of early GWAS discoveries, increasing the number of confirmed loci to a dozen [18–20].

It quickly became clear that data quality control of GWAS data was essential to producing interpretable and reproducible results [21]. While the genotyping platforms produced data that were extremely high quality on average, the sheer size of the datasets compared to earlier studies meant that even very low error rates could produce spurious associations. A suite of quality control metrics, including missing data rates, Hardy–Weinberg equilibrium, and overall heterozygosity quickly became standards in GWAS analysis, and geneticists became familiar with QQ-plots and other statistical tools as a rapid transition from genetic studies where each genotyping assay was manually inspected and scored to automated genome-wide typing technologies took place. It was also recognized that, even if genotyping data were perfect, false inference of association could arise if the ancestries of cases and controls were not well matched and the frequency differences characteristic of different populations were confounded with case–control status. Here a parallel set of methods emerged [22, 23] to measure and control for population structure in association studies that, like the QC standards, are still in wide use. Furthermore, the genetics community insisted on stringent statistical significance thresholds ($p<5\times10^{-8}$ being a common criterion for genome-wide significance [24]) and replication of any putative findings in independent samples [25] to generate ultimate assurance that novel genetic findings constituted truly durable insights into disease pathogenesis.

These rigorous guidelines for GWAS produced a substantial shift away from the contentious and generally irreproducible findings from linkage and candidate gene studies and rapidly produced a swath of bona fide associations to a wide variety of common diseases. Despite these early GWAS successes, however, it became apparent that associations from the first generation of studies explained only a very small fraction of the total genetic contribution to disease [26]. A variety of hypotheses were proposed to explain this so-called missing heritability, including a preponderance of rare variants, copy number variation, and complex interactions among risk loci [27]. None of these explanations lent themselves to straightforward post-GWAS experimentation nor had direct evidence that they explained the majority of what was not yet found. What was clear was that the confirmed findings from GWAS were both numerous and generally barely strong enough to have been detected, suggesting that many more results might lie just beneath the surface. Thus, the natural next step was for geneticists to set aside historical competitions in favor of combining GWAS datasets studying the same disease to investigate what additional associations might be discovered via collaboration.

Meta-analysis and the Importance of Sample Size

Early examples of GWAS *meta-analysis*, where individual scans were combined, often using summary association statistics from individual projects, began shortly after the initial GWAS publications [28, 29]. These studies, which typically consisted of a few thousand individuals, rapidly confirmed the suspicion that a large number of additional common alleles of small effect were waiting to be identified. Reassuringly, and in contrast to the experience of both complex disease linkage and candidate gene studies, GWAS meta-analyses also confirmed nearly all the previously published associations in the smaller scans.

The possibilities of this approach were most clearly realized by researchers studying quantitative traits, such as height or cholesterol levels, which had been measured in hundreds of thousands of individuals subjected to GWAS analysis. Unlike specific disease studies, which were limited by the incidence of diseases and the difficulty of recruiting large numbers of cases, these quantitative trait studies could draw samples collected for any number of different study designs, so long as the measurement of interest had been recorded in a consistent way. In the most recent meta-analysis of height GWAS [30], for instance, nearly 200 independent genomic loci showed significant association. A similar trend was observed across a wide variety of traits and diseases: as sample sizes increased, so did the number of associated loci. What often differed, however, was the number of samples required to make the earliest discoveries (i.e., find the biggest effects in that disease) and the rate at which loci subsequently accumulated. It is still unclear whether fundamental differences in genetic architecture, heterogeneity of diagnoses, or other factors might explain this locus discovery "coefficient." IBD once again reaped the benefits of these approaches via a series of successively larger meta-analyses culminating in the discovery of 163 independent loci [28, 31–33].

In addition to unleashing a torrent of individual associations across hundreds of diseases and traits, large meta-analysis sample sizes encouraged the application of newly developed statistical methods that analyzed the entire genome at once. Rather than focus on the most statistically significant associations, these methods [34] aimed to evaluate the total amount of phenotypic variance explained by common variation across the entire genome. These models suggested that a much larger fraction of total variance in many traits and risk of diseases could be explained by common variation than was explained by the genome-wide significant loci. It may be, therefore, that seeking to fully uncover the "missing heritability" is a fruitless effort since these hundreds or thousands of tiny effects will be impossible to pinpoint individually.

Biological Insights from Disease Gene Mapping

In parallel to the goal of trying to identify the specific regions of the genome associated with disease risk, or particular DNA variants which cause those associations or the overall contribution of common variation in general, disease gene mapping also provides an opportunity to better understand the biological mechanisms of health and disease. Indeed, given the difficulty in accurately predicting disease susceptibility from GWAS-type analyses [35], it is likely that new biological insights will be the most important long-term benefits of studying the genetics of complex diseases.

GWAS studies of fetal hemoglobin (HbF) levels in sickle cell disease (SCD) patients offer an informative example of this process. In some ways, SCD is the fundamental example of a Mendelian genetic disease, as the recessive mutation (in the hemoglobin beta gene) which causes the disease has been known for over 60 years [36]. It has also been long known that increased levels of HbF (encoded by a different gene and typically not expressed after birth) substantially reduce the severity of SCD. This observation led to a GWAS for HbF level [37], which identified a strong effect of variants near the *BCL11A* gene on persistence of HbF after birth. This discovery was followed by the remarkable discovery that reducing the activity of *BCL11A* could substantially alleviate SCD in mice [38]. A single new GWAS discovery opened a potential therapeutic avenue that had remained undetected despite decades of biologically motivated research into the relationship between HbF and SCD.

The discovery of *BCL11A* as a key regulator of HbF also serves as an illustration of the caution needed when predicting how quickly GWAS results will enable new diagnostics or treatments. They serve as a critical starting point, a biological truth that some functional unit in a particular part of the genome is related to disease risk. That piece of information alone offers little clue how to then move towards more complete knowledge of disease processes but certainly offers better prospects than aimlessly trying to make sense of those same processes in the context of the entirety of human biology.

Future Directions

The most substantial change to mapping complex disease genes at the present is the transition from GWAS-style data (where only a subset of common variation is studied) to complete genome sequencing, enabled by the plunging cost of sequencing compared to genotyping [39]. These technological shifts have the potential to open up the study of a wide spectrum of variation beyond the common alleles targeted by GWAS. It will be important, however, not to forget the lessons of that era principally that large sample sizes are critical to success. In addition to broadening the types of genetic variation which can be detected, sequencing-based studies (either directly or indirectly through imputation in projects like 1,000 Genomes) have the potential to more rapidly proceed from region of association to specific causal variants. While these discoveries only slightly affect the amount of variation explained in a disease or trait of interest, they have great potential to aid the biological inferences described above.

In a very real sense the progression of gene mapping described in this chapter is approaching its final stages: it will soon be possible to analyze the complete genome sequence of nearly all the patients seen with a particular disease. It is likely that even at that point, it will be impossible to perfectly predict an individual's risk of disease. Instead, we must ask whether we can use this limited information in a clinically useful way (in a similar sense to currently used risk measures, like cholesterol levels, which are strongly but imperfectly predictive of outcomes like heart disease) and simultaneously promote the utility of genetic association results in more fundamental biological studies of human health.

References

1. Fisher RA (1919) The correlation between relatives on the supposition of Mendelian inheritance. Trans R Soc Edinb 52:399–433
2. Sturtevant AH (1913) The linear arrangement of six sex-linked factors in Drosophila, as shown by their mode of association. J Exp Zool 14:43–59
3. Morton NE (1955) Sequential tests for the detection of linkage. Am J Hum Genet 7:277
4. Kan YW, Dozy AM (1978) Polymorphism of DNA sequence adjacent to human beta-globin structural gene: relationship to sickle mutation. Proc Natl Acad Sci 75:5631–5635
5. Botstein D, White RL, Skolnick M, Davis RW (1980) Construction of a genetic linkage map in man using restriction fragment length polymorphisms. Am J Hum Genet 32:314
6. Gusella JF et al (1983) A polymorphic DNA marker genetically linked to Huntington's disease. Nature 306:234–238
7. Wise J (2010) Prevalence of Huntington's disease is underestimated in UK. BMJ 340:c3516
8. Markianos K, Daly MJ, Kruglyak L (2001) Efficient multipoint linkage analysis through reduction of inheritance space. Am J Hum Genet 68(4):963–977
9. Risch N, Merikangas K (1996) The future of genetic studies of complex human diseases. Science 273:1516–1517
10. Undlien DE et al (1995) Insulin gene region-encoded susceptibility to IDDM maps upstream of the insulin gene. Diabetes 44:620–625
11. Sachidanandam R et al (2001) A map of human genome sequence variation containing 1.42 million single nucleotide polymorphisms. Nature 409:928–933
12. Jeffreys AJ, Kauppi L, Neumann R (2001) Intensely punctate meiotic recombination in the class II region of the major histocompatibility complex. Nat Genet 29:217–222
13. Daly MJ, Rioux JD, Schaffner SF, Hudson TJ, Lander ES (2001) High-resolution haplotype structure in the human genome. Nat Genet 29:229–232
14. The International HapMap Consortium (2005) A haplotype map of the human genome. Nature 437:1299–1320
15. Pe'er I et al (2006) Evaluating and improving power in whole-genome association studies using fixed marker sets. Nat Genet 38:663–667
16. Barrett JC, Cardon LR (2006) Evaluating coverage of genome-wide association studies. Nat Genet 38:659–662
17. Frayling TM et al (2007) A common variant in the FTO gene is associated with body mass index and predisposes to childhood and adult obesity. Science 316:889–894
18. Duerr RH et al (2006) A genome-wide association study identifies IL23R as an inflammatory bowel disease gene. Science 314:1461–1463
19. Rioux JD et al (2007) Genome-wide association study identifies new susceptibility loci for Crohn disease and implicates autophagy in disease pathogenesis. Nat Genet 39:596–604
20. Parkes M et al (2007) Sequence variants in the autophagy gene IRGM and multiple other replicating loci contribute to Crohn's disease susceptibility. Nat Genet 39:830–832
21. Anderson CA et al (2010) Data quality control in genetic case–control association studies. Nat Protoc 5:1564–1573

22. Devlin B, Roeder K (1999) Genomic control for association studies. Biometrics 55: 997–1004
23. Price AL et al (2006) Principal components analysis corrects for stratification in genome-wide association studies. Nat Genet 38:904–909
24. Pe'er I, Yelensky R, Altshuler D, Daly MJ (2008) Estimation of the multiple testing burden for genomewide association studies of nearly all common variants. Genet Epidemiol 32:381–385
25. WTCCC (2007) Genome-wide association study of 14,000 cases of seven common diseases and 3,000 shared controls. Nature 447:661–678
26. Lee SH, Wray NR, Goddard ME, Visscher PM (2011) Estimating missing heritability for disease from genome-wide association studies. Am J Hum Genet 88(3):294–305. doi:10.1016/j.ajhg.2011.02.002
27. Eichler EE et al (2010) Missing heritability and strategies for finding the underlying causes of complex disease. Nat Rev Genet 11:446–450
28. Barrett JC et al (2008) Genome-wide association defines more than 30 distinct susceptibility loci for Crohn's disease. Nat Genet 40:955–962
29. Voight BF et al (2010) Twelve type 2 diabetes susceptibility loci identified through large-scale association analysis. Nat Genet 42:579–589
30. Paper A. F. L. O. A. A. T. A. A. A. T. E. O. T (2010) Hundreds of variants clustered in genomic loci and biological pathways affect human height. Nature 467:832–838
31. Franke A et al (2010) Genome-wide meta-analysis increases to 71 the number of confirmed Crohn's disease susceptibility loci. Nat Genet 42(12):1118–1125. doi:10.1038/ng.717
32. Anderson CA et al (2011) Meta-analysis identifies 29 additional ulcerative colitis risk loci, increasing the number of confirmed associations to 47. Nat Genet 43:246–252
33. Jostins L, Ripke S, Barrett JC, Cho JH (2012) Host-microbe interactions have shaped the genetic architecture of inflammatory bowel disease. Nature 490:119–124
34. International Schizophrenia Consortium et al (2009) Common polygenic variation contributes to risk of schizophrenia and bipolar disorder. Nature 460:748–752
35. Jostins L, Barrett JC (2011) Genetic risk prediction in complex disease. Hum Mol Genet 20:R182–R188
36. Ingram VM (1957) Gene mutations in human haemoglobin: the chemical difference between normal and sickle cell haemoglobin. Nature 180:326–328
37. Menzel S et al (2007) A QTL influencing F cell production maps to a gene encoding a zinc-finger protein on chromosome 2p15. Nat Genet 39:1197–1199
38. Xu J et al (2011) Correction of sickle cell disease in adult mice by interference with fetal hemoglobin silencing. Science 334:993–996
39. Kilpinen H, Barrett JC (2013) How next-generation sequencing is transforming complex disease genetics. Trends Genet 29(1):23–30. doi:10.1016/j.tig.2012.10.001

Chapter 5
The Genetics of Crohn's Disease

Andre Franke and Miles Parkes

Abstract This chapter summarises progress in understanding the genetic basis of Crohn's disease (CD). It starts with a brief review of family studies for CD epidemiology and then summarises findings of the so-called "linkage era". Given the success of genome-wide association studies (GWAS) in terms of identifying CD susceptibility loci, the focus of this chapter is on the key GWAS studies and their main results. These have demonstrated association with multiple Th17 pathway components and strongly implicated defects in innate immunity, particularly in autophagy and the handling of intracellular bacteria, as playing key roles in CD pathogenesis. Besides GWAS for adult-onset CD, paediatric-onset GWAS are discussed. Although paediatric-onset CD presents with more extensive disease and rapid progression compared to adult-onset CD, genetic studies have shown marked molecular similarities between the two disease forms. Not only have single GWAS contributed to completing the molecular map of CD genetics, but also systematic cross-phenotype analyses and meta-analyses of several CD GWAS, both of which are discussed in the current chapter. Lastly, the first sequencing studies for CD as well as future challenges are described.

A. Franke, Ph.D. (✉)
Institute of Clinical Molecular Biology, Christian–Albrechts–University of Kiel, Schittenhelmstraße 12, 24105 Kiel, Schleswig-Holstein, Germany
e-mail: a.franke@mucosa.de

M. Parkes, M.A., M.B.B.S., F.R.C.P., D.M.
IBD Research Unit, Department of Medicine, University of Cambridge, Cambridge, UK
e-mail: miles.parkes@doctors.org.uk

M. D'Amato and J.D. Rioux (eds.), *Molecular Genetics of Inflammatory Bowel Disease*,
DOI 10.1007/978-1-4614-8256-7_5, © Springer Science+Business Media New York 2013

Introduction

This chapter summarises progress in understanding the genetic basis of Crohn's disease (CD). It starts with a brief review of family studies for CD epidemiology and then summarises findings of the so-called "linkage era". Given the success of genome-wide association studies (GWAS) in terms of identifying CD susceptibility loci, the focus of this chapter is on the key GWAS studies and their main results. These have demonstrated association with multiple Th17 pathway components and strongly implicated defects in innate immunity, particularly in autophagy and the handling of intracellular bacteria, as playing key roles in CD pathogenesis. Besides GWAS for adult-onset CD, paediatric-onset GWAS are discussed. Although paediatric-onset CD presents with more extensive disease and rapid progression compared to adult-onset CD, genetic studies have shown marked molecular similarities between the two disease forms. Not only have single GWAS contributed to completing the molecular map of CD genetics, but also systematic cross-phenotype analyses and meta-analyses of several CD GWAS, both of which are discussed in the current chapter. Lastly, the first sequencing studies for CD as well as future challenges are described.

Over the last 5 years, genetic studies have provided major new insights regarding key pathogenic mechanisms underlying Crohn's disease and ulcerative colitis. To date these have mostly been based on genome-wide association studies, but newer genomics technologies are now beginning to complement GWAS findings and add to our understanding of the molecular genetic universe of inflammatory bowel disease [1]. Ultimately the improved understanding of IBD pathogenic mechanisms, including clues regarding environmental factors, can help in the design of improved therapies and development of better preventative strategies in individuals identified as being at risk.

IBD has for many years been recognised to result from a complex interaction of genetic susceptibility with environmental risk factors, producing dysregulation of the mucosal immune system and an inflammatory response targeting the gut flora. Epidemiological studies have highlighted the contribution of smoking and factors which perturb the epithelial barrier and have hinted at the importance of dysbiosis, for example, relating to childhood exposure to antibiotics and the potential importance of enteric infection as a trigger for IBD. Likewise, in the immunological heyday of the 1990s, multiple immune and cytokine pathways were noted to be abnormal in IBD. However, analysis of these entities is difficult—particularly in separating cause from effect in individuals in whom IBD has already developed. An attraction of studying germline genetic variation is the ability to say with certainty what came first.

Family Studies and Genetic Epidemiology

Epidemiological and family studies, in the 1980s and 1990s, provided convincing evidence for a genetic contribution to IBD susceptibility [2–11]. The increased risk can be quantitated by the λ_s ratio, which describes the increased risk to siblings compared to the background population risk. For Crohn's disease this is 17–35 and for UC is 8–15. By comparison, the λ_s for coronary artery disease is ~3, and for type 1 diabetes (T1D) it is ~15, but with ~50 % of this accounted for by the strong MHC association—early indicators that IBD would prove fertile ground for hunting down non-MHC susceptibility genes. Twin and family studies and their importance in the initial identification of the genetic component of IBD are reported and discussed in "Chapter 2."

Early Molecular Genetic Studies

Early approaches to identifying susceptibility genes for IBD hinged mainly on candidate gene studies. The main success in this regard came with the identification of association between UC and specific alleles of the MHC class 2 region. This was initially identified in Japanese cohorts—Asakura et al. reporting *HLA-DR2* to be associated with UC, with Sugimura et al. subsequently identifying the *DRB1*1502* allele as responsible for this [12, 13]. Association between UC and the MHC was subsequently identified in Europeans, with the *DRB1*0103* allele implicated in both severe UC and extra-intestinal manifestations of IBD [14, 15]. Interestingly, the contribution of the MHC to Crohn's disease susceptibility appears to be modest in Europeans and largely confined to the subgroup of patients with exclusively Crohn's colitis [16]. This illustrates how different CD is to classic autoimmune diseases and supports the clinical impression that colon-only CD may actually be closer to UC than to small bowel CD in terms of its aetiology.

Genome-wide linkage studies using large panels of affected sibling pairs dominated the IBD genetic field in the late 1990s (see Fig. 5.1, [17]). Their yield was relatively modest and reflected the lack of statistical power of this technique for detecting the loci of modest effect size that we now know typifies complex disease. *NOD2* represents the exception, which proved this rule, and was identified within a large linkage interval in the peri-centromeric region of chromosome 16 using a combination of positional cloning and positional candidate gene analysis [6, 8]. Much has been written about NOD2 in the last 10 years, and it will be discussed in "Chapter 10". Suffice to say the discovery of NOD2 provided the first major insight into the critical contribution of innate immunity to CD pathogenesis. NOD2 is an intracellular receptor for muramyl dipeptide, a ubiquitous component of bacterial cell walls. Both the original report from Hugot et al. and subsequent detailed resequencing studies have identified the fact that the CD risk variants cluster in the portion of the gene encoding the "leucine rich region" which recognises and binds muramyl dipeptide. Carriage of a single risk variant confers an odds ratio (OR) of 1.5–3, while in homozygotes the OR rises to 17–40—with the association being almost exclusively with small bowel CD. East Asian populations lack the CD-associated coding variants in *NOD2*—perhaps explaining in part the lower prevalence of CD in East Asia and the relative preponderance of UC. Functional analyses have implicated a variety of mechanisms by which NOD2 variants might predispose to CD—including aberrant activation of NF-κB, altered modulation of

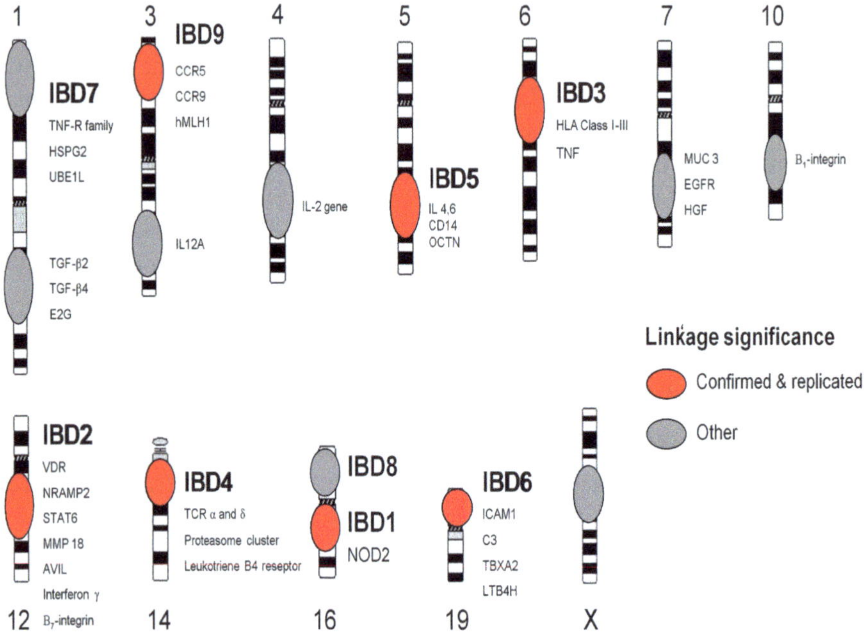

Fig. 5.1 The major focus of genetic research in the mid-1990s was the identification of disease susceptibility loci using GWAS. A large number of markers distributed across the genome are typed in individuals from multiply affected families. Markers that are inherited together with a disease in a family are used to define these regions. Genome-wide scans have identified susceptibility loci on a number of chromosomes. These include *IBD1*, the first and most consistently replicated IBD locus identified on chromosome 16, which is exclusive to CD. This area harbours the *NOD2* gene. Another replicated area of linkage termed *IBD3* is found on chromosome 6. This harbours the HLA region. Modified from Ahmad T, Satsangi J, McGovern D, Bunce M, Jewell DP. Review article: the genetics of inflammatory bowel disease. Alimentary pharmacology & therapeutics. 2001;15(6):731–48. Epub 2001/05/31. With kind permission from Elsevier Limited

TLR signalling, disrupted Paneth cell function with reduced mucosal defensin production and defective autophagy [18]. All highlight the central role played by NOD2 in innate immunity and regulation of the cellular response to bacteria, and it may be precisely because of NOD2's pleiotropic roles in innate immune responses that its mutation exerts such a powerful effect in predisposing to Crohn's disease.

GWAS

Introduction

Genome-wide association study (GWAS) technology allows an unbiased survey of the genome for regions showing association with IBD. By identifying genes, which map within the association intervals, particularly if they are further implicated by,

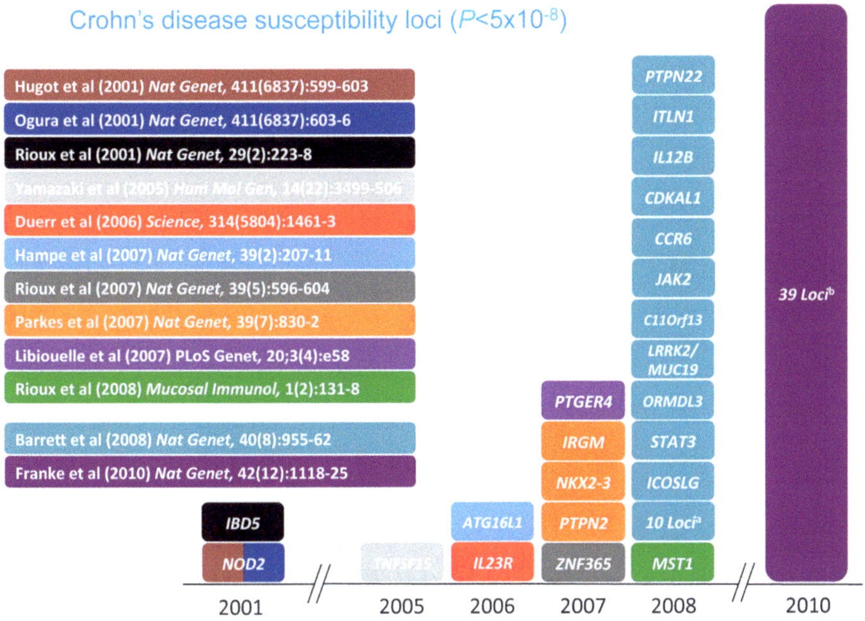

Fig. 5.2 Exponential increase of disease-gene findings for CD

for example, association with coding variants or correlation of genotype with gene expression, inferences can be drawn regarding their primary contribution to IBD pathogenesis. Downstream studies are still required to delineate precise causal variants and explore the allelic spectrum by sequencing and fine mapping, to understand the impact of associated variants on gene expression in relevant cell types and to explore their functional impact on cell biology, immunology and microbial interaction. However, it was GWAS, which have revolutionised our understanding of IBD genetic susceptibility. For more details on the GWAS method, see also "Chapter 4".

GWAS studies have proven successful in IBD on a number of levels—but particularly in terms of the large number of loci identified which show confirmed association with IBD susceptibility, the new pathogenic insights gained and the understanding of the molecular genetic relationship between CD and UC. The index GWAS studies have now been complemented by large GWAS meta-analyses and replication studies from the International IBD Genetics Consortium (IIBDGC). These have included tens of thousands of IBD patients globally and have led to the most recent identification of a total of 163 confirmed IBD susceptibility loci to date [7] (see Fig. 5.2) [19].

Overview of Large-Scale CD Association Studies

Several large-scale association studies (with at least 10,000 SNPs per screen) have been carried out in the last years and enhanced significantly our understanding on the causes of CD. Table 5.1 summarises the study design and most important results

Table 5.1 Overview of large-scale CD association studies

Study	References	Platform	Disease	Ancestry	New loci
Yamazaki et al.	[21]	73 k	CD	Japan	1 (*TNFSF15*)
Duerr et al.	[22]	Illumina 300 k	Ileal CD	North America	1 (*IL23R*)
Hampe et al.	[23]	Non-synonymous SNP 20 k	CD/UC	Germany	1 (*ATG16L1*)
Franke et al.	[60]	Affymetrix [92,387]	CD	Germany	1 (*NELL1*)
Rioux et al.	[24]	Illumina [304,413]	Ileal CD	North America	1 (10q21.1)
Libioulle et al.	[25]	Illumina [302,451]	CD	France/ Belgium	1 (*5p13.1*, upstream of *PTGER4*)
WTCCC	[26]	Affymetrix [469,557]	CD	Europe	9
Parkes et al.	[27]	Affymetrix [469,557]	CD	UK	4
Raelson et al.	[29]	Perlegen [164,279]	CD	Québec founder population, Germany	–
Franke et al.	[41]	Affymetrix [83,360]	CD and SA	Germany	1 (*10p12.2*, shared CD/ SA)
Barrett et al.	[46]	Affymetrix and Illumina [635,547] (imputed)	CD	UK, North America, France/ Belgium	19
Franke et al.	[40]	Affymetrix and Illumina [953,241] (imputed)	CD	Europe	39
McGovern et al.	[32]	Illumina [304,825]	CD	Europe	6
Wang et al.	[44]	Illumina 550 k	CD/UC/T1D	Caucasian	6
Festen et al.	[43]	Affymetrix and Illumina [471,504]	CD and CelD	Europe	7 (shared CD/ Cel)
Ellinghaus et al.	[45]	Affymetrix, Illumina and Perlegen [1,116,213] (imputed)	CD and PS	Europe, North America	1 (SOCS1)
Kenny et al.	[61]	Affymetrix and Illumina [1,060,934] (imputed)	CD	Ashkenazi Jews	5
Kugathasan et al.	[36]	Illumina 550 k	Early-onset IBD	North America, Italy	2 (20q13, 21q22)
Imielinski et al.	[38]	Illumina 550 k	Early-onset IBD	North America, Scotland, Italy	5

of all large association studies published so far. The next sections briefly go into detail by highlighting key studies.

Early Studies

The first ever GWAS was published in 2005, for age-related macular degeneration [20], followed 6 months later by the first GWAS for CD [21]. Here, Yamazaki and colleagues successfully genotyped 72,738 SNPs in 94 Japanese CD patients and 752 healthy control individuals. One thousand eight hundred and eighty-eight significantly associated SNPs were then further tested in a Japanese replication sample. Several SNPs in the *TNFSF15* gene region on chromosome 9q32 were successfully replicated, and a subset of SNPs was then further tested in a study sample from the UK in which the locus was also replicated. The neighbouring *TNFSF8* gene was excluded as a candidate gene by dense SNP fine mapping, which showed that both genes are located on distinct LD blocks. *TNFSF15* encodes the protein TL1A (TNF ligand-related molecule 1A), a tumour necrosis factor (TNF) family member. It is important to note that despite the small sample size (hence low statistical power) and low SNP coverage (more than 500,000 SNPs should be genotyped and then subjected to imputation for adequate genomic coverage according to current standards)—Yamazaki and colleagues identified a novel CD candidate gene that subsequently replicated in both Asian- and Caucasian-descent patients and controls. *TNFSF15* thus represents the first IBD gene to be identified by GWAS, and it is notable that its contribution to disease susceptibility is not restricted to a single ethnic group.

Over a year later, a larger association study was reported for Crohn's disease by an American research group. Duerr and colleagues [22] tested 308,332 SNPs in 547 ileal CD patients (non-Jewish, European ancestry) and 548 healthy controls. Besides the known risk locus *NOD2*, a coding variant in the *IL23R* gene was identified as significantly associated. This coding polymorphism is non-synonymous—that is, it affects the open reading frame of the transcript leading to a change of a single amino acid in the resulting IL23R protein. In the original study, the glutamine variant of the SNP Arg381Gln was found to be significantly less common than the arginine allele, with an allelic frequency of 1.9 % in the non-Jewish patients with ileal CD vs. 7.0 % in non-Jewish controls. The glutamine allele appeared to protect against development of CD in both non-Jewish [OR = 0.26] and Jewish [OR = 0.45] case–control cohorts. Following this, several other groups replicated the Arg381Gln association finding, and additional independent variants were identified at the *IL23R* locus that is associated with both CD and UC.

Also in 2006, a genome-wide candidate SNP association study was published by Hampe et al. [23]. In their study, 7,159 informative non-synonymous SNPs were tested in 735 healthy controls and 368 CD patients. The best-associated 72 SNPs from the screening stage were then tested for association in 380 independent CD trios, 498 CD singleton cases and 1,032 controls. Disease association of rs2241880 in the autophagy-related 16-like 1 gene (*ATG16L1*) was replicated in these samples and confirmed in a UK panel. By haplotype and regression analysis, the authors found that marker rs2241880, a coding SNP (T300A), carries virtually all the disease risk exerted by the *ATG16L1* locus. This study implicated the autophagy pathway in CD pathophysiology for the first time, with the *ATG16L1* association being replicated later by several other groups [24] (see also "Chapter 12").

The "GWAS Era"

Beginning of 2007, Rioux and colleagues published the first high-density GWAS study in CD with >300,000 SNPs under study, these being examined in 998 ileal CD cases and 1007 healthy controls [24]. This study replicated the *ATG16L1* finding—additionally demonstrating the importance of this protein in the autophagy pathway by several in vitro studies. Recent functional studies in ATG16 hypomorphic mice have shown abnormal Paneth cell morphology and elegantly demonstrated the complex interaction between genetic susceptibility, environmental stressors, intact gut flora and the need for a particular (noroviral) trigger to elaborate the full phenotype of intestinal inflammation—thereby perhaps beginning to approach the complexity seen in human IBD. Two other CD GWAS studies were also published in 2007. A French–Belgian team also used the Illumina 300K SNP array [25] and identified a CD-associated region at *5p13.1*. This localises to a 250 kb linkage disequilibrium (LD) block, which maps to a 1.25 Mb gene desert. Despite the lack of protein-coding genes in this interval, Libioulle et al. nevertheless showed that the Crohn's disease-associated alleles in this gene desert correlate with quantitative expression levels of the prostaglandin receptor EP4, encoded by *PTGER4*. This gene resides closest to the associated region is but still 270 kb away from the most associated SNP. This nicely demonstrates that disease-associated variants can affect regulatory regions, which in consequence can influence the expression of distant genes. This and other related studies led to systematic genome-wide SNP-expression correlation analyses, known as eQTL (expression quantitative trait locus) mapping studies. Although the expression patterns of these eQTL studies were not always measured in the disease-relevant tissue(s) of a particular phenotype—in most cases lymphoblastoid cell line resources were exploited—these analyses significantly increased the knowledge on the influence of genetic variation on gene expression in general, besides generating several plausible hypotheses for noncoding disease-associated variants identified by GWAS.

The UK CD GWAS was also published in 2007 as part of the Wellcome Trust Case Control Consortium [26] and set new standards for association studies. The WTCCC included seven common diseases—amongst them CD with ca. 2000 patients—comprising altogether 14,000 patients and 3,000 shared controls. Besides the statistical and technological advance, this so far largest study for CD identified four novel disease loci (*3p21–22*, *5q33.1*) [*IRGM*], *10q24* [*NKX2-3*] and *18p11* [*PTPN2*]; follow-up described in [27] and replicated the *IL23R*, *NOD2*, *5p13.1*, *5q31*, *ATG16L1* and the previously found intergenic region on *10q21.1* [24]. For the novel *3p21* locus, Goyette and colleagues later showed that the R689C variant (rs3197999) in *MST1* is the most likely causative variant at this locus [28]. *MST1* encodes macrophage-stimulating protein (MSP), a protein regulating the innate immune responses to bacterial ligands. R689C is predicted to interfere with MSP binding to its receptor, suggesting a role for this gene in the pathogenesis of IBD. Dense fine mapping of the *3p21* locus was also carried out in the study by Raelson et al. [29]. For the novel WTCCC study locus on *5q33.1*, Parkes and colleagues implicated variants at the *IRGM* gene locus as the strongest associated signals in this region [27]. IRGM belongs to the p47 immunity-related GTPase family. Its mouse homologue, LRG-47 (encoded by Lrgm), critically controls intracellular pathogens by autophagy, and Lrgm–/– mice show markedly increased susceptibility to *Toxoplasma gondii* and *Listeria monocytogenes*. Consistent with this, IRGM induces autophagy—another demonstration of the importance of this pathway in CD aetiology—and thereby controls intracellular *Mycobacterium tuberculosis* in human macrophages. McCarroll et al. identified a 20 kb deletion polymorphism upstream of *IRGM* carried by ~40 % of the Caucasian population, which correlates with expression of *IRGM* and which is in complete LD with the neighbouring GWAS lead SNP rs13361189. Whether this structural variant is causal remains a topic of debate, particularly as Prescott et al. subsequently reported the finding of an insertion–deletion ("indel") polymorphism in the 5′UTR of *IRGM* which disrupts a transcription factor-binding site, and most recently Brest et al. reported that an associated synonymous SNP in the coding sequence of *IRGM* alters a microRNA-binding domain, hence affecting mRNA stability and gene translation [30]. Corroborating the expression data suggesting that the Crohn's disease-associated IRGM variants result in reduced gene expression, Lapaquette et al. reported that knockdown of IRGM by siRNA in human macrophages permitted a substantial increase in the number of adherent *E. coli* able to survive within these macrophages and suggested that this effect was specific for the LF82 serotype which this group have demonstrated plays a key role in triggering Crohn's disease [31]. While the complexity inherent in the multiple other genes and loci associated with Crohn's disease susceptibility must not be ignored, this example nicely illustrates one potentially important pathway all the way from associated gene variant to impact on gene expression and functional impact on innate immunity, allowing a recognised environmental agent to exert its IBD-predisposing effect.

One of the most recent GWAS for CD is the study by McGovern and colleagues who analysed >300,000 SNPs in 896 CD cases and 3,204 healthy controls, all of Caucasian descent [32]. Besides replicating 21 previously known loci, they identified suggestive associations with genes involved in tight junctions/epithelial integrity (*ASHL*, *ARPC1A*), innate immunity (*EXOC2*), dendritic cell biology [*CADM1* (*IGSF4*)], macrophage development (*MMD2*), TGF-β signalling (*MAP3K7IP1*) and *FUT2*. The association at the *FUT2* locus was then further replicated and is of particular interest as the gene product is a physiological trait that regulates gastrointestinal mucosal expression of blood group A and B antigens. About 20 % of Caucasians are so-called nonsecretors who do not express ABO antigens in saliva as a result of being homozygous for the nonsense variant of the *FUT2* W134X SNP (rs601338). No excess of heterozygotes in CD were observed compared to controls, which the authors speculated is in line with the proposed hypothesis that the *FUT2* association is "driven" by an association between nonsecretor status and *FUT2*. The W143X SNP displays evidence of being maintained by strong selective pressure. A large body of evidence suggests that this maintenance may be because of numerous trade-offs surrounding host–microbe interactions. For example, nonsecretors are resistant to infection with the Norwalk (Noro) and respiratory viruses but are more susceptible to duodenal ulcers, rheumatic fever and cholera. Furthermore, the breast milk of secreting mothers provides protection against *Campylobacter jejuni* to their offspring by exploiting the binding affinity of the bacterium to fucosyloligosaccharides. In a follow-up study by Rausch and colleagues [33], it was observed that the *FUT2* genotype explained substantial differences in microbial community composition, diversity and structure in biopsies of the large intestine. They further identified several bacterial species displaying disease-by-genotype associations. These findings indicated that alterations in resident microbial communities may in part be explained by the variety of host susceptibilities surrounding nonsecretor status and that *FUT2* is an important genetic factor influencing host–microbial diversity. Future studies that examine the genotype–microbiome interactions (which seem to play a crucial role in CD aetiology) at genome-wide levels are likely following soon. It has already been demonstrated in mice that a very early interaction between the host and bacteria is necessary to allow a normal response to inflammatory stimuli later in life [34]—supporting the hygiene hypothesis for chronic inflammatory diseases. While the genome seems to partially influence which bacteria will "like" or "dislike" their host, the epigenome seems to be the "mediator" and the "memory" of the various interactions (for more details on host–microbe interactions, see "Chapter 14").

GWAS for Early-Onset CD

Studying early-onset presentations of complex disease is appealing to geneticists because of the expectation that these efforts have a higher chance of identifying novel risk variants. Implicit in this strategy is the assumption that these patients represent a more severe, more genetically influenced group of affected individuals. 15–20 % of IBD patients present in childhood or adolescence with epidemiological and natural history studies clearly demonstrating a rising incidence in this age group. Although early-onset disease is characterised by particular phenotypic features, such as more extensive disease at onset and rapid progression, two recent genome-wide association studies (GWAS) carried out exclusively in this age group have demonstrated marked genetic similarities to adult disease [35].

In the first study, which used a case–control panel that was a subset of that used in the second, Kugathasan et al. performed a GWAS using DNA from 1,011 individuals with paediatric-onset IBD (647 CD and 317 UC) and 4,250 matched controls [36]. They replicated several known loci from non-paediatric association studies (*NOD2*, *IL23R*, *HLA*, *TNFSF15*) and identified two novel disease-associated loci, *20q13* and *21q22*. Although the authors were unable to pinpoint the causal gene in the *20q13* region, they considered the *TNFRSF6B* gene the most compelling candidate based on the critical role of specific polymorphisms within genes involved in the TNF pathway in the pathogenesis of IBD. It is of interest that the protein product for *TNFRSF6B* acts as a decoy receptor (protein called DCR3) in preventing FasL-induced cell death, and a resistance to FasL-dependent apoptosis has previously been shown for T lymphocytes in CD. The authors also observed that the mean serum DCR3 concentration was significantly increased in individuals with IBD carrying the major allelic variants compared to IBD carrying the minor allelic variants. The *21q22* signal resides in a small region of LD that harbours no genes, with the nearest gene being *PSMG1* (proteasome assembly chaperone 1). A Canadian study for early-onset CD (410 patients) later replicated the *20q13* but not the *21q22* finding [37].

In the second GWAS for early-onset IBD, Imielinski and colleagues [38] analysed 550,000 SNPs in 3,426 affected individuals (1,636 CD, 724 UC, 53 unclassified) and 11,963 genetically matched controls, recruited through international collaborations in Europe and North America. The authors identified five new regions associated with early-onset IBD susceptibility, including *16p11* near the cytokine gene *IL27* (rs8049439), *22q12* (rs2412973), *10q22* (rs1250550), *2q37* (rs4676410) and *19q13.11* (rs10500264). The scan also detected associations at 23 of 32 loci previously implicated in adult-onset CD and at 8 of 17 loci implicated in adult-onset UC, highlighting the close pathogenetic relationship between early- and adult-onset IBD.

Essers and colleagues later demonstrated, by analysing 35 common established CD susceptibility loci in early-onset patients (average age of onset of 11.7 years), that paediatric patients do not carry significantly more risk alleles than adult CD patients [39]. This does not exclude the hypothesis though that adult and/or paediatric patients carry significantly different risk variants which are yet unknown. Future large-scale resequencing studies will most likely clarify this hypothesis. As described elsewhere in the text, a small handful of studies have already identified monogenic forms of severe early-onset inflammatory bowel disease. It should be noted here that out of the aforementioned two and five novel loci identified by Kugathasan et al. and Imielinski et al., respectively, none and two were genome-wide significant in the below-mentioned meta-analysis for CD [40], including mostly adult-onset patients, respectively.

Combined Analyses with Other Phenotypes

As predicted from their close clinical relationship, many key susceptibility loci are shared between Crohn's disease and ulcerative colitis. However, a less expected finding prior to the GWAS era was the extent to which overlap would be seen for multiple loci across many immune-mediated diseases (discussed in more detail in "Chapter 7").

Given this overlap, one topic which continues to be the source of significant debate is what statistical thresholds are appropriate for "claiming" association between a given locus and, for example, CD if the same locus has already demonstrated genome-wide significant association with UC or indeed with another immune-mediated disease. Using the conventional genome-wide significance threshold of $p<5\times10^{-8}$ is clearly overly conservative given the markedly high prior probability for association. This issue has been addressed in more detail in the recent report of the analysis of Immunochip data from the IIBDGC [7].

Some of the overlap between immune-mediated diseases reflects known co-morbidities—for example, Crohn's disease and psoriasis, both appearing often in the same patient, share a significant portion of their genetic risk map. Therefore, in some instances, researchers have combined more than one disease in GWAS to identify such shared risk genes or to search for differences. To this end, CD GWAS data sets were combined with GWAS for sarcoidosis [41]; one novel shared locus on *10p12.2*, replicated recently in [42]; celiac disease [43]; *TAGAP* and *PUS10* as novel shared loci, T1D and UC [44]; e.g., T1D risk loci HLA, *PTPN22*, *IL27*, *IL18RAP* and *IL10* are protective for CD and psoriasis [45]; seven shared non-HLA loci plus *SOCS1* on *16p13* as a novel CD risk locus. Ongoing research efforts of the "Immunochip Consortium" are aiming at combining and jointly analysing GWAS data sets for most known autoimmune diseases, an effort that will enlarge the genetic risk map even further.

Meta-Analyses of CD GWAS Studies

While the index GWAS studies were able to identify loci conferring (in complex disease genetic terms) larger effect sizes, they were, in retrospect, underpowered to detect the many more loci that confer an OR of disease of <1.2. Reliable identification of such loci requires analysis of substantially larger sample sets. This becomes possible with subsequent work from the IIBDGC which has conducted two meta-analyses of CD GWAS studies [40, 46] and undertaken a large collaborative experiment using the Immunochip [7]. Each of these studies has identified multiple new CD susceptibility loci of progressively smaller effect size. While critics might argue that these loci exert such a weak effect that their impact on CD pathogenesis is negligible, a more rounded view is that these loci very much help to "join the dots", aiding informatics analyses such as GRAIL in identifying causal genes and helping to define entire pathways where index GWAS studies highlighted just single components. The latter is particularly important where proteins participate in several distinct pathways or where their function has previously been only partially elucidated—such as had been the case for *IRGM*, which some authorities dismissed as a pseudogene until the evidence from CD GWAS corroborated its putative function in autophagy. An additional and potentially important point is that the biological impact may be out of proportion to the strength of the genetic association signal for a particular variant. This is illustrated by the modest association between HMG-CoA reductase gene variants and hypercholesterolaemia, and yet this gene product represents the target of a class of drugs called statins, one of the most effective therapies for reducing cholesterol population-wide.

A detailed discussion of all the CD genes and loci identified in these three IIBDGC studies—amounting to in excess of 100 independent CD susceptibility loci—is beyond the scope of this chapter, and interested readers are referred to the original publications for all details [40, 47]. Below we focus on some of the key themes and key pathways identified.

The Franke et al. meta-analysis published in 2010 comprised GWAS data on 6,333 cases and 15,056 controls and followed up the top association signals in 15,694 cases, 14,026 controls and 414 parent–offspring trios. It identified 30 new susceptibility loci meeting genome-wide significance. Following in silico analyses and manual curation, a number of positional candidate genes were identified as being of interest, including *SMAD3*, *ERAP2*, *IL10*, *IL2RA*, *TYK2*, *FUT2*, *DNMT3A*, *DENND1B*, *BACH2* and *TAGAP*.

These meta-analyses have demonstrated association between CD and an ever-increasing number of loci encoding IL23/Th17 pathway components, such as *TYK2*, *JAK2*, *STAT3*, *ICOSLG* and *CCR6*. Intuitively the IBD-associated variants in *IL23R* might be predicted to exert their effect on adaptive immunity via CD4+ and Th17 pathways. There has been increasing evidence for a major role of Th17 cells in IBD pathogenesis in recent years, significantly spurred by the genetic evidence. However, caution is required before jumping to this conclusion as polymorphisms in *IL23* pathway genes may also impact innate immunity. For example, Buonocore et al. recently reported the accumulation of CD3 negative, IL23-responsive innate lymphoid cells in the colon, these being capable of producing IL17 and IFN-γ and mediating innate colitis in mice [48]. Production of Th17 cytokines by analogous cells in humans appeared higher in colons from IBD cases vs. controls. Additional functional interrogation is required to explore the role of Th17 pathways in mediating mucosal homeostasis and microbial interaction, particularly in light of a recent clinical study demonstrating that anti-IL17 antibody therapy leads to worse clinical outcomes than placebo when trialled in Crohn's disease.

The region encoding interleukin-10 was also found to be associated with CD in the Franke et al. meta-analysis. This locus was originally identified as associated with IBD in a German GWAS in ulcerative colitis, association being documented with noncoding variants upstream of the *IL10* gene [49]. Additional interest in this pathway derived from a study, which identified mutations in the interleukin-10 receptor as causing an extreme form of IBD in infants. The latter study, which was based on analysis of two consanguineous families by exome sequencing, identified homozygous mutations in both the *IL10RA* and *IL10RB* genes as abrogating interleukin-10 signalling and leading to severe intestinal inflammation [50, 51]. IL-10 is known to play an important regulatory role in immune homeostasis, and *IL10* knockout mice represent one of the best animal models of IBD (see also "Chapter 3"). Attempts at manipulating IL-10 signalling for therapeutic benefit in IBD have so far been unsuccessful—but this probably reflects the need for a better understanding of the functional impact of the disease-predisposing variants and improved timing and targeting of any IL-10-based treatment.

Another exome-sequencing study this time in a large panel of IBD cases identified association, with a rare variant in the *CARD9* gene. *CARD9* mediates signalling between pattern recognition receptors such as NOD2 and the pro-inflammatory transcription factor NF-κB, and the genomic interval in which it lies was originally implicated in the Barrett et al. CD meta-analyses. Here, the lead SNP was a common noncoding variant associated with increased disease risk, and in eQTLs data sets, this SNP was found to correlate with strongly increased expression of the *CARD9* gene. The subsequent exome-sequencing study by Rivas et al. neatly mirrored this by identifying a novel low-frequency splice site variant in *CARD9* which impairs the function of the protein and is associated with reduced disease risk [52]. From these findings, we can see that the genetic universe of IBD encompasses the whole allele frequency spectrum from common variants, identified in GWAS studies, low-frequency variants, identified by targeted exome sequencing in large panels, and private mutations identified by whole exome sequencing of extreme cases in families, as exemplified by the *IL10* receptor mutations.

SMAD3 showed association with CD in the Franke et al. meta-analysis. Phosphorylated following TGF-β signalling through its receptor, the SMAD3 protein complexes with SMAD4 and is then translocated to the nucleus to modulate target gene expression and exert broadly immunosuppressive effects. Involvement of TGF-β signalling pathways in CD pathogenesis has recently been corroborated by the findings from Immunochip, which identify association with several other components of the TGF-β pathway including SMAD7, SMURF1 and FURIN [7]. These all play important roles in the TGF-β-mediated induction of Foxp3+ regulatory T cells, with SMAD3 deficiency reciprocally enhancing Th17. TGF-β is also critically involved in epithelial restitution in the gut, where its potent pro-fibrogenic effects may be relevant both to mucosal repair and intestinal stricture formation which is a hallmark of Crohn's disease [53].

Association between CD and variants at the *ERAP2* locus is intriguing, particularly as the CD-associated variant correlates strongly with *ERAP2* gene expression in published eQTL data sets. Regulated by NF-κB, *ERAP2* encodes one of two human endoplasmic reticulum aminopeptidases, which work in concert to trim peptides for presentation on MHC class I and hence critically affect antigen presentation to T cells. Ankylosing spondylitis and psoriasis are also associated with this locus, but with a pattern of associated variants more closely implicating *ERAP1*. Given the close clinical relationship between Crohn's disease and ankylosing spondylitis, and the strong association of HLA-B27 with the latter but not with the former, the divergent association of these closely related molecules is intriguing and will refocus interest on the MHC class I associations in Crohn's disease.

As already indicated above, the NF-κB pathway is another for which multiple components show genetic association with CD susceptibility. NF-κB is a master transcriptional regulator of multiple cytokine genes involved in the inflammatory response and controls epithelial integrity and mucosal immune homeostasis in the presence of gut microflora [54]. Association with the *TNFAIP3* gene locus was first seen in Wang et al. comparative GWAS analysis [44] and is now recognised as a pleiotropic locus associated with multiple other immune-mediated diseases including rheumatoid arthritis, SLE, celiac disease, T1D and psoriasis. This gene encodes the ubiquitin-editing protein A20 protein, a TNF-α-inducible zinc finger protein thought to limit NF-κB-mediated immune responses. Recent data from the

Immunochip experiment have identified a number of other key constituents of the NF-κB pathway as being associated with CD, including *REL* on chromosome 2, *RELA* on chromosome 11 and *NF-κB* itself on chromosome 4.

The first meta-analyses of CD and UC GWAS studies undertaken by IIBDGC treated each form of IBD as a separate phenotype. Even from these analyses, which used overly conservative thresholds for association in view of the higher "priors" for loci meeting genome-wide significance in one phenotype, it was evident that multiple susceptibility loci were shared between the two phenotypes [40, 47]. This applied, for example, to multiple components of the Th17 pathway, namely, CARD9, IL-10 and ICOS ligand as well as many others. Given the phenotypic overlap between CD and UC, such overlap is not surprising. This issue has been evaluated more thoroughly in the analysis of Immunochip data by the IIBDGC, in which it is clear that over 100 of the 163 loci which are associated at genome-wide significance with some form of IBD contribute to both phenotypes [7]—albeit in some instances the association with one form is stronger than with the other. As interesting, and perhaps even more informative with regard to specific pathogenic mechanisms, is the identification of loci, which appear disease-specific. Par excellence this applies to *NOD2* but also to *ATG16L1*, both of which are associated exclusively with CD—indeed *NOD2* variants are actually modestly protective against UC. Evidently the innate immune mechanisms of processing of intracellular bacteria mediated by the protein products of these genes critically influence IBD phenotype. Corroborating this, functional studies of both have implicated a variety of potentially pathogenic mechanisms, including processing of bacterial antigens and disruption of Paneth cells, which are key mediator of innate immunity in the ileum.

It is intriguing that variants in genes linked to epithelial barrier function seem to be specifically associated with UC and not Crohn's disease—the converse of *NOD2* and the autophagy genes. While many complex explanations might exist, these observations correlate nicely with UC being confined to the superficial layers of the colon, while the transmural inflammation of Crohn's disease is caused by defects in cellular innate immunity and bacterial handling in the deeper layers of the lamina propria and beyond.

Future Challenges

To date based on loci meeting genome-wide significance thresholds, approximately 25 % of the heritability of CD has been accounted for. This proportion can be increased significantly based on reanalysis of GWAS data if one accounts for the many hundreds of additional common variants that individually contribute only a tiny fraction of overall variance of disease risk [55]. Nonetheless, accounting for the additional "missing heritability", or at least enough of it to allow substantial progress towards understanding pathogenic mechanisms in CD, is a significant challenge to the IBD genetic community, and various approaches are being adopted.

One of the most important of these is rare variant studies—attempting to identify (possibly highly penetrant) low-frequency and rare variants associated with CD susceptibility. It is already apparent from the relatively limited sequencing efforts conducted to date that rare variants, for example, within GWAS loci, contribute independently to disease risk. The question is whether and to what extent this applies more broadly to loci, which lie outside GWAS intervals and which within multiply affected families make a significant contribution to disease risk. The analytic and logistic challenges are substantial, but high-throughput whole-genome sequencing efforts are underway to identify new variants to take forward into large-scale association studies.

Additional genomics methodologies are also being applied, increasingly at a genome-wide level. This applies to expression analyses and their correlation with germline genetic variation. Such studies are particularly helpful when undertaken in separated cells rather than homogenised tissue or mixed cell populations, due to the now good evidence that gene-regulatory mechanisms are often cell-type specific [56]. A challenge to the IBD community is to ensure that such analyses are undertaken in cell types of potential relevance to CD—including leukocyte subsets and intestinal epithelial cells, but not forgetting less "obvious" candidates such as stromal cells and stem cells.

Genome-wide epigenetic analyses are also increasingly being applied [57–59] to better understand how core regulatory mechanisms such as methylation, histone binding, chromatin remodelling and microRNAs might affect CD susceptibility. The same issues regarding the need to conduct such studies in separated cell types pertain, but efforts are now underway to tackle the logistic challenges involved and understand the interaction between germline variation and environmental factors (infection, gut microbiota, smoking and diet) with such epigenetic mechanisms.

A major aspiration for genetic studies is that they will provide substantial insights regarding environmental triggers and drivers of CD. This is already playing out, and the overlap between susceptibility loci for CD and mycobacterial infection has been strongly highlighted in the Immunochip analysis [7]. Genomic and particularly sequencing technologies are increasingly being used to characterise the microbiota in CD vs. health and seek bacterial and viral triggers for IBD.

As the genetic story unfolds, so interest will refocus on the functional impact of the implicated genetic mechanisms and the extent to which they can be manipulated or subjugated to therapeutic benefit. Preventative strategies may also become relevant at the point that we are better able to identify, by prediction algorithms, those at significant absolute risk of developing disease. Further goals commensurate with clinical translation include detailed pharmacogenetic studies and prognostic modelling to separate patients destined to run a severe disease course from those less likely to have trouble. Clinical translation must remain the ultimate goal for the CD genetic community, and with the exciting progress that has been made over the last 5 years in understanding the pathogenesis of CD, the hope is that CD will be one of the first common diseases that can be cured.

References

1. Vavricka SR, Schoepfer AM, Bansky G, Binek J, Felley C, Geyer M et al (2011) Efficacy and safety of certolizumab pegol in an unselected Crohn's disease population: 26-week data of the FACTS II survey. Inflamm Bowel Dis 17(7):1530–1539
2. Bayless TM, Tokayer AZ, Polito JM II, Quaskey SA, Mellits ED, Harris ML (1996) Crohn's disease: concordance for site and clinical type in affected family members—potential hereditary influences. Gastroenterology 111(3):573–579, Epub 1996/09/01
3. Colombel JF, Grandbastien B, Gower-Rousseau C, Plegat S, Evrard JP, Dupas JL et al (1996) Clinical characteristics of Crohn's disease in 72 families. Gastroenterology 111(3):604–607, Epub 1996/09/01
4. Farmer RG, Michener WM, Mortimer EA (1980) Studies of family history among patients with inflammatory bowel disease. Clin Gastroenterol 9(2):271–277
5. Halfvarson J (2011) Genetics in twins with Crohn's disease: less pronounced than previously believed? Inflamm Bowel Dis 17(1):6–12, Epub 2010/09/18
6. Hugot JP, Chamaillard M, Zouali H, Lesage S, Cezard JP, Belaiche J et al (2001) Association of NOD2 leucine-rich repeat variants with susceptibility to Crohn's disease. Nature 411(6837):599–603, Epub 2001/06/01
7. Jostins L, Ripke S, Weersma RK, Duerr RH, McGovern DP, Hui KY et al (2012) Host-microbe interactions have shaped the genetic architecture of inflammatory bowel disease. Nature 491(7422):119–124, Epub 2012/11/07
8. Ogura Y, Bonen DK, Inohara N, Nicolae DL, Chen FF, Ramos R et al (2001) A frameshift mutation in NOD2 associated with susceptibility to Crohn's disease. Nature 411(6837): 603–606, Epub 2001/06/01
9. Orholm M, Munkholm P, Langholz E, Nielsen OH, Sorensen TI, Binder V (1991) Familial occurrence of inflammatory bowel disease. N Engl J Med 324(2):84–88, Epub 1991/01/10
10. Peeters M, Nevens H, Baert F, Hiele M, de Meyer AM, Vlietinck R et al (1996) Familial aggregation in Crohn's disease: increased age-adjusted risk and concordance in clinical characteristics. Gastroenterology 111(3):597–603, Epub 1996/09/01
11. Satsangi J, Grootscholten C, Holt H, Jewell DP (1996) Clinical patterns of familial inflammatory bowel disease. Gut 38(5):738–741, Epub 1996/05/01
12. Asakura H, Tsuchiya M, Aiso S, Watanabe M, Kobayashi K, Hibi T et al (1982) Association of the human lymphocyte-DR2 antigen with Japanese ulcerative colitis. Gastroenterology 82(3):413–418, Epub 1982/03/01
13. Sugimura K, Asakura H, Mizuki N, Inoue M, Hibi T, Yagita A et al (1993) Analysis of genes within the HLA region affecting susceptibility to ulcerative colitis. Hum Immunol 36(2):112–118, Epub 1993/02/01
14. Orchard TR, Chua CN, Ahmad T, Cheng H, Welsh KI, Jewell DP (2002) Uveitis and erythema nodosum in inflammatory bowel disease: clinical features and the role of HLA genes. Gastroenterology 123(3):714–718, Epub 2002/08/29
15. Satsangi J, Welsh KI, Bunce M, Julier C, Farrant JM, Bell JI et al (1996) Contribution of genes of the major histocompatibility complex to susceptibility and disease phenotype in inflammatory bowel disease. Lancet 347(9010):1212–1217, Epub 1996/05/04
16. Fisher SA, Tremelling M, Anderson CA, Gwilliam R, Bumpstead S, Prescott NJ et al (2008) Genetic determinants of ulcerative colitis include the ECM1 locus and five loci implicated in Crohn's disease. Nat Genet 40(6):710–712, Epub 2008/04/29
17. Ahmad T, Tamboli CP, Jewell D, Colombel JF (2004) Clinical relevance of advances in genetics and pharmacogenetics of IBD. Gastroenterology 126(6):1533–1549, Epub 2004/05/29
18. Schreiber S, Rosenstiel P, Albrecht M, Hampe J, Krawczak M (2005) Genetics of Crohn disease, an archetypal inflammatory barrier disease. Nat Rev Genet 6(5):376–388, Epub 2005/04/30
19. http://www.ibdgenetics.org/
20. Klein RJ, Zeiss C, Chew EY, Tsai JY, Sackler RS, Haynes C et al (2005) Complement factor H polymorphism in age-related macular degeneration. Science 308(5720):385–389, Epub 2005/03/12

21. Yamazaki K, McGovern D, Ragoussis J, Paolucci M, Butler H, Jewell D et al (2005) Single nucleotide polymorphisms in TNFSF15 confer susceptibility to Crohn's disease. Hum Mol Genet 14(22):3499–3506, Epub 2005/10/14
22. Duerr RH, Taylor KD, Brant SR, Rioux JD, Silverberg MS, Daly MJ et al (2006) A genome-wide association study identifies IL23R as an inflammatory bowel disease gene. Science 314(5804):1461–1463, Epub 2006/10/28
23. Hampe J, Franke A, Rosenstiel P, Till A, Teuber M, Huse K et al (2007) A genome-wide association scan of nonsynonymous SNPs identifies a susceptibility variant for Crohn disease in ATG16L1. Nat Genet 39(2):207–211, Epub 2007/01/04
24. Rioux JD, Xavier RJ, Taylor KD, Silverberg MS, Goyette P, Huett A et al (2007) Genome-wide association study identifies new susceptibility loci for Crohn disease and implicates autophagy in disease pathogenesis. Nat Genet 39(5):596–604, Epub 2007/04/17
25. Libioulle C, Louis E, Hansoul S, Sandor C, Farnir F, Franchimont D et al (2007) Novel Crohn disease locus identified by genome-wide association maps to a gene desert on 5p13.1 and modulates expression of PTGER4. PLoS Genet 3(4):e58
26. WTCCC (2007) Genome-wide association study of 14,000 cases of seven common diseases and 3,000 shared controls. Nature 447(7145):661–678, Epub 2007/06/08
27. Parkes M, Barrett JC, Prescott NJ, Tremelling M, Anderson CA, Fisher SA et al (2007) Sequence variants in the autophagy gene IRGM and multiple other replicating loci contribute to Crohn's disease susceptibility. Nat Genet 39(7):830–832, Epub 2007/06/08
28. Goyette P, Lefebvre C, Ng A, Brant SR, Cho JH, Duerr RH et al (2008) Gene-centric association mapping of chromosome 3p implicates MST1 in IBD pathogenesis. Mucosal Immunol 1(2):131–138, Epub 2008/12/17
29. Raelson JV, Little RD, Ruether A, Fournier H, Paquin B, Van Eerdewegh P et al (2007) Genome-wide association study for Crohn's disease in the Quebec Founder Population identifies multiple validated disease loci. Proc Natl Acad Sci U S A 104(37):14747–14752, Epub 2007/09/07
30. Brest P, Lapaquette P, Souidi M, Lebrigand K, Cesaro A, Vouret-Craviari V et al (2011) A synonymous variant in IRGM alters a binding site for miR-196 and causes deregulation of IRGM-dependent xenophagy in Crohn's disease. Nat Genet 43(3):242–245, Epub 2011/02/01
31. Lapaquette P, Bringer MA, Darfeuille-Michaud A (2012) Defects in autophagy favour adherent-invasive Escherichia coli persistence within macrophages leading to increased pro-inflammatory response. Cell Microbiol 14(6):791–807, Epub 2012/02/09
32. McGovern DP, Jones MR, Taylor KD, Marciante K, Yan X, Dubinsky M et al (2010) Fucosyltransferase 2 (FUT2) non-secretor status is associated with Crohn's disease. Hum Mol Genet 19(17):3468–3476, Epub 2010/06/24
33. Rausch P, Rehman A, Künzel S, Häsler R, Ott SJ, Schreiber S et al (2011) Colonic mucosa-associated microbiota is influenced by an interaction of Crohn disease and FUT2 (Secretor) genotype. Proc Natl Acad Sci U S A. 2011 Nov 22;108(47):19030–19035. doi: 10.1073/pnas.1106408108, Epub 2011/11/08
34. Olszak T, An D, Zeissig S, Vera MP, Richter J, Franke A et al (2012) Microbial exposure during early life has persistent effects on natural killer T cell function. Science 336(6080): 489–493, Epub 2012/03/24
35. Henderson P, van Limbergen JE, Wilson DC, Satsangi J, Russell RK (2011) Genetics of childhood-onset inflammatory bowel disease. Inflamm Bowel Dis 17(1):346–361, Epub 2010/09/15
36. Kugathasan S, Baldassano RN, Bradfield JP, Sleiman PM, Imielinski M, Guthery SL et al (2008) Loci on 20q13 and 21q22 are associated with pediatric-onset inflammatory bowel disease. Nat Genet 40(10):1211–1215, Epub 2008/09/02
37. Amre DK, Mack DR, Morgan K, Fujiwara M, Israel D, Deslandres C et al (2009) Investigation of reported associations between the 20q13 and 21q22 loci and pediatric-onset Crohn's disease in Canadian children. Am J Gastroenterol 104(11):2824–2828, Epub 2009/07/23

38. Imielinski M, Baldassano RN, Griffiths A, Russell RK, Annese V, Dubinsky M et al (2009) Common variants at five new loci associated with early-onset inflammatory bowel disease. Nat Genet 41(12):1335–1340, Epub 2009/11/17
39. Essers JB, Lee JJ, Kugathasan S, Stevens CR, Grand RJ, Daly MJ (2009) Established genetic risk factors do not distinguish early and later onset Crohn's disease. Inflamm Bowel Dis 15(10):1508–1514, Epub 2009/03/27
40. Franke A, McGovern DP, Barrett JC, Wang K, Radford-Smith GL, Ahmad T et al (2010) Genome-wide meta-analysis increases to 71 the number of confirmed Crohn's disease susceptibility loci. Nat Genet 42(12):1118–1125, Epub 2010/11/26
41. Franke A, Fischer A, Nothnagel M, Becker C, Grabe N, Till A et al (2008) Genome-wide association analysis in sarcoidosis and Crohn's disease unravels a common susceptibility locus on 10p12.2. Gastroenterology 135(4):1207–1215
42. Cozier YC, Ruiz-Narvaez EA, McKinnon CJ, Berman JS, Rosenberg L, Palmer JR (2012) Fine-mapping in African-American women confirms the importance of the 10p12 locus to sarcoidosis. Genes Immun 13(7):573–578, Epub 2012/09/14
43. Festen EA, Goyette P, Green T, Boucher G, Beauchamp C, Trynka G et al (2011) A meta-analysis of genome-wide association scans identifies IL18RAP, PTPN2, TAGAP, and PUS10 as shared risk loci for Crohn's disease and celiac disease. PLoS Genet 7(1):e1001283, Epub 2011/02/08
44. Wang K, Baldassano R, Zhang H, Qu HQ, Imielinski M, Kugathasan S et al (2010) Comparative genetic analysis of inflammatory bowel disease and type 1 diabetes implicates multiple loci with opposite effects. Hum Mol Genet 19(10):2059–2067, Epub 2010/02/24
45. Ellinghaus D, Ellinghaus E, Nair RP, Stuart PE, Esko T, Metspalu A et al (2012) Combined analysis of genome-wide association studies for Crohn disease and psoriasis identifies seven shared susceptibility loci. Am J Hum Genet 90(4):636–647, Epub 2012/04/10
46. Barrett JC, Hansoul S, Nicolae DL, Cho JH, Duerr RH, Rioux JD et al (2008) Genome-wide association defines more than 30 distinct susceptibility loci for Crohn's disease. Nat Genet 40(8):955–962, Epub 2008/07/01
47. Anderson CA, Boucher G, Lees CW, Franke A, D'Amato M, Taylor KD et al (2011) Meta-analysis identifies 29 additional ulcerative colitis risk loci, increasing the number of confirmed associations to 47. Nat Genet 43(3):246–252, Epub 2011/02/08
48. Buonocore S, Ahern PP, Uhlig HH, Ivanov II, Littman DR, Maloy KJ et al (2010) Innate lymphoid cells drive interleukin-23-dependent innate intestinal pathology. Nature 464(7293): 1371–1375, Epub 2010/04/16
49. Franke A, Balschun T, Karlsen TH, Sventoraityte J, Nikolaus S, Mayr G et al (2008) Sequence variants in IL10, ARPC2 and multiple other loci contribute to ulcerative colitis susceptibility. Nat Genet 40(11):1319–1323, Epub 2008/10/07
50. Glocker EO, Frede N, Perro M, Sebire N, Elawad M, Shah N et al (2010) Infant colitis—it's in the genes. Lancet 376(9748):1272, Epub 2010/10/12
51. Glocker EO, Kotlarz D, Boztug K, Gertz EM, Schaffer AA, Noyan F et al (2009) Inflammatory bowel disease and mutations affecting the interleukin-10 receptor. N Engl J Med 361(21): 2033–2045, Epub 2009/11/06
52. Rivas MA, Beaudoin M, Gardet A, Stevens C, Sharma Y, Zhang CK et al (2011) Deep resequencing of GWAS loci identifies independent rare variants associated with inflammatory bowel disease. Nat Genet 43(11):1066–1073, Epub 2011/10/11
53. Di Sabatino A, Jackson CL, Pickard KM, Buckley M, Rovedatti L, Leakey NA et al (2009) Transforming growth factor beta signalling and matrix metalloproteinases in the mucosa overlying Crohn's disease strictures. Gut 58(6):777–789, Epub 2009/02/10
54. Nenci A, Becker C, Wullaert A, Gareus R, van Loo G, Danese S et al (2007) Epithelial NEMO links innate immunity to chronic intestinal inflammation. Nature 446(7135):557–561, Epub 2007/03/16
55. Visscher PM, Brown MA, McCarthy MI, Yang J (2012) Five years of GWAS discovery. Am J Hum Genet 90(1):7–24, Epub 2012/01/17

56. Fu J, Wolfs MG, Deelen P, Westra HJ, Fehrmann RS, Te Meerman GJ et al (2012) Unraveling the regulatory mechanisms underlying tissue-dependent genetic variation of gene expression. PLoS Genet 8(1):e1002431, Epub 2012/01/26
57. Cooke J, Zhang H, Greger L, Silva AL, Massey D, Dawson C et al (2012) Mucosal genome-wide methylation changes in inflammatory bowel disease. Inflamm Bowel Dis 18(11):2128–2137, Epub 2012/03/16
58. Hasler R, Feng Z, Backdahl L, Spehlmann ME, Franke A, Teschendorff A et al (2012) A functional methylome map of ulcerative colitis. Genome Res 22(11):2130–2137, Epub 2012/07/25
59. Nimmo ER, Prendergast JG, Aldhous MC, Kennedy NA, Henderson P, Drummond HE et al (2012) Genome-wide methylation profiling in Crohn's disease identifies altered epigenetic regulation of key host defense mechanisms including the Th17 pathway. Inflamm Bowel Dis 18(5):889–899, Epub 2011/10/25
60. Franke A, Hampe J, Rosenstiel P, Becker C, Wagner F, Hasler R et al (2007) Systematic association mapping identifies NELL1 as a novel IBD disease gene. PLoS One 2(8):e691, Epub 2007/08/09
61. Kenny EE, Pe'er I, Karban A, Ozelius L, Mitchell AA, Ng SM et al (2012) A genome-wide scan of Ashkenazi Jewish Crohn's disease suggests novel susceptibility loci. PLoS Genet 8(3):e1002559, Epub 2012/03/14

Chapter 6
Genetics of Ulcerative Colitis

Mauro D'Amato, Catherine Labbé, and John D. Rioux

Abstract While it has been known for several decades that susceptibility to ulcerative colitis (UC), one of the two major forms of inflammatory bowel disease (IBD), is in part inherited, it is only recently that considerable progress has been made in identifying some of the key genetic determinants playing a role in predisposition to disease. The advent of genome-wide association studies (GWAS) has provided geneticists with an extremely powerful tool for the identification of causative genes in complex diseases. To date, 163 IBD risk loci have been identified, and functional insight obtained from genes mapping to these associated regions has highlighted the importance of immune function and gut interaction with bacteria, as well as the maintenance of epithelial barrier in UC pathogenesis. The functional characterization of the remaining genes, as well as the identification of rare causative variants in these associated regions, represents a considerable challenge for the future, but promises to unravel novel biological pathways that may be amenable to exploitation for the design of innovative therapeutic strategies in UC.

Ulcerative colitis (UC) was described for the first time in 1859 by British physician Sir Samuel Wilks [1]. One hundred and fifty years of scientific studies and careful observation of its clinical manifestations and yet the pathogenesis of this debilitating inflammatory bowel disease (IBD) remains, for a large part, a mystery. Only in

M. D'Amato, Ph.D. (✉)
Department of Biosciences and Nutrition, NOVUM, Karolinska Institutet,
Hälsovägen 7-9, 14183 Stockholm, Sweden
e-mail: Mauro.damato@ki.se

C. Labbé, M.Sc., Ph.D.
Montreal Heart Institute, Montréal, QC, Canada

Université de Montréal, Montréal, QC, Canada

J.D. Rioux
Université de Montréal and the Montreal Heart Institute, Montréal, QC, Canada

M. D'Amato and J.D. Rioux (eds.), *Molecular Genetics of Inflammatory Bowel Disease*,
DOI 10.1007/978-1-4614-8256-7_6, © Springer Science+Business Media New York 2013

the last decade have tremendous improvements been made in the knowledge of the molecular pathways underlying disease onset and chronic inflammation. These are primarily attributable to large-scale genetic studies that aimed to identify genetic risk factors in UC and led to the discovery of specific genes and associated biological pathways contributing to disease susceptibility.

Early UC genetic studies focussed on the major histocompatibility complex (MHC) located on chromosome 6p; linkage and association to this locus represented the first discoveries of genetic susceptibility to UC [2]. Following this came a series of mostly unsuccessful targeted association studies, unsuccessful owing to limited power to detect significant association due to the number of markers and the size cohorts tested. The technical innovations and discoveries that followed the publication of the complete sequence of the human genome set the stage for the advent of genome-wide association studies (GWAS) [3, 4]. Today, recent GWAS, meta-analyses of GWAS and replication studies testing very large patient cohorts obtained through extensive international collaborations have identified 133 genomic regions associated to UC susceptibility, which cumulatively contribute to explain 7.5 % of disease variance [5–7].

The MHC and UC Susceptibility

In 1967, the MHC region was first identified to be involved in immune-mediated disease when the expression of HLA-B antigens was found to be higher in patients with Hodgkin's lymphoma [8]. Variations in the MHC locus have since then been associated to autoimmune, inflammatory and also infectious disease, confirming this locus to be of paramount importance in immunity. Traditionally, the MHC is split into class I, class II and class III regions, each class containing multiple genes, including human leukocyte antigen (HLA)-encoding genes as well as non-HLA-encoding genes. The sequencing of the 3.44 Mb MHC region identified 224 different genes [9]. Class I (*HLA-A*, *HLA-B*, *HLA-C*) and class II (HLA-DP, HLA-DQ, HLA-DR) genes are involved in antigen presentation and stimulation of T helper cells (T_H cells). Class III region genes encode several proteins with immune functions, including components of the complement system, cytokines and heat shock proteins. The MHC region exhibits striking sequence variability; for example, the *HLA-B* gene is the most polymorphic gene in the human genome, with over 2,000 alleles already identified in different populations [10]. Another particular feature of the MHC region is the presence of haplotypes with very strong and extensive linkage disequilibrium (LD), which often limit the exact localization of causative association signals in classical genetic studies.

Since the 1970s, several linkage and association studies have tried to verify whether specific HLA alleles are implicated in UC susceptibility. More often than not, these early studies produced conflicting results. This can mostly be explained by inadequate coverage of genetic variability in the region, by different typing methodologies and by low statistical power owing to limited cohort sizes. In 1999,

Stokkers et al. published a meta-analysis of several previously reported association studies of HLA alleles, in order to estimate the contribution of the HLA class II alleles to IBD. Specifically, they analyzed 18 association studies of the MHC region in UC [2]. In the pooled results, HLA-DR2, HLA-DR9 and HLA-DRB1*0103 were associated with UC risk, while a protective effect of allele HLA-DR4 was detected. A similar effort published by Fernando et al. a decade later aimed to review 35 years of MHC genetic research and pool the results in an updated meta-analysis [11]. This particular study examined the association of the MHC alleles to six different diseases, and 37 independent studies were considered for UC [2]. This new meta-analysis confirmed the association of UC to HLA-DRB1*0103 and HLA-DRB1*1502 and detected a weak association to the MHC class I related gene A (*MICA*), which had been previously reported [12]. The study identified novel associations to HLA-DR5, HLA- A19 and HLA-A24.

In 2009, the International MHC and Autoimmunity Genetics Network (IMAGEN) published an extensive association study of the MHC region [13]. A total of 1,472 single nucleotide polymorphisms (SNPs) covering common MHC variation were genotyped in over 10,000 DNA samples from cohorts of patients suffering from one of seven immune-related diseases. The cohort for UC consisted of 667 patients from Italy, and three different signals associated to increased UC risk were identified, namely, in the *HLA-DRB1* gene (with HLA-DRB1*1101 possibly responsible for the association signal), the *NOTCH4* gene region and a third locus containing the *BAT8*, *C2*, *RDBP* and *SKIV2L* genes.

As of today, the MHC is still one of the strongest and most consistently replicated association signals in UC. This remains the case even within the context of the more recent genome-wide SNP-based association studies. One of the first in UC was a genome-wide scan of non-synonymous SNPs from Fisher et al. [14]. This study identified a 400-kb haplotype block containing *BTNL2* as well as loci HLA-DQA1, *HLA-DRA*, *HLA-DRB5* and *HLA-DRB1*. The first true GWAS (>100,000 SNPs) of UC to find an association to the MHC was of IBD patients with early-onset disease [15]. Association to the MHC has since been detected in every subsequent GWAS and meta-analyses of GWAS. However, as for many other immune-mediated diseases with association in the MHC, the complexity of the region makes it difficult to pinpoint the actual causal gene(s) or variant(s) for UC.

Limited Success of Early Targeted Associated Studies

Before the GWAS era, association studies to candidate genes outside the MHC were attempted, but they generally resulted in limited success and no locus was consistently replicated. These early association studies were generally performed in small cohorts from different populations and ethnic groups, and the coverage of genetic variability in the queried genes or genomic regions was often low. Two loci that stand out, however, are the *ABCB1* locus and *IL2-IL21* locus as multiple studies have shown consistent results.

The ATP-binding cassette subfamily B member 1 gene (*ABCB1*), also known as the multidrug resistance gene 1 (MDR1), encodes P-glycoprotein 170. This transmembrane protein is an efflux transporter pump that regulates the flow of substances in and out of the cell. It is highly expressed in epithelial cells of the intestinal tract and influences the response to many drugs, including glucocorticoids used it the treatment of IBD [16]. *Abcb1* knockout mice develop severe spontaneous colitis that can be cured with antibiotics, which suggested that intestinal flora is necessary to maintain inflammation in mouse models of IBD [17]. Several candidate gene studies from the last decade have confirmed that genetic variation in the *ABCB1* gene is associated with UC. While a meta-analysis of six targeted association studies published until 2006 concluded that the T allele of *ABCB1* SNP rs1045642 is significantly associated with UC risk [18], a more recent meta-analysis of 16 association studies suggested the association with this allele is only marginal [19]. In addition, this locus was not among the 163 genetic risk factors identified in the most recent GWAS of IBD, suggesting that if truly associated with UC, its effect is likely to be quite modest [6].

Before being described in UC, the locus 4q27 had already been characterized as a well-established immune-related locus, because of its association with celiac disease, type 1 diabetes, Grave's disease, SLE, rheumatoid arthritis and psoriasis [20–24]. This locus contains the cytokine *IL2* and *IL21* genes, which are both functionally interesting candidates. IL21 is involved in the differentiation of T_H17 cells implicated in the pathogenesis of IBD [25], and its expression is increased in intestinal tissue from UC patients compared to controls [26], while *il2* knockout mice develop an IBD-like condition [27]. The first study to implicate the *IL2-IL21* locus in IBD susceptibility was published by Festen et al. [28], who reported the genotyping of four SNPs in a three-stage association study on a total of 3,195 UC cases from North America, the Netherlands and Italy. Association signals at all four SNP sites reached genome-wide significance. The identified locus covers a 480 kb region of high linkage disequilibrium (LD) and contains *IL2*, *IL21* and two other genes, namely, *TENR* and *KIAA1109*, which are less plausible biological candidates to play a role in UC. Soon after, the associations were replicated in German and Spanish studies [29, 30] and later replicated in a large meta-analysis of GWAS [5]. Additionally, the locus is also associated in Chinese populations where LD is less extensive than in Caucasian populations, allowing for the identification of two apparently independent association signals for the *IL2* and *IL21* genes [31].

The GWAS Revolution

The limited success of targeted gene-candidate association studies made it clear that a new strategy would be needed to allow for the identification of genes underlying this complex disease. Success in hypothesis-driven association studies, such as gene-candidate studies, is restricted by the limited knowledge of disease molecular pathogenesis. Early association studies were also plagued by the low statistical power of small cohorts to identify risk genes with modest effect size, which often resulted in false-positive findings. Knowledge of the complete sequence of the human genome, availability of markers covering most of the common variation in the human genome and microarray technology for mass parallel genotyping allowed the introduction of GWAS. Access to large cohorts and stringent statistical methods contributed to the success of GWAS in identifying complex disease genes by international teams of collaborators.

Since 2008, five groups published independent GWAS on UC [14, 15, 32–38]. The findings from these GWAS and from their first comprehensive meta-analysis (Table 6.1) recently led to identification of 47 loci associated with UC susceptibility [5]. This number has recently been brought to 133 (23 UC-specific loci and 110 loci associated with both CD and UC), thanks to a major collaborative effort from the International IBD Genetics Consortium (IIBDGC), who studied genetic information coming from a total of 38,565 IBD patients (17,865 UC and 20,700 CD) and 37,747 controls from 17 participating countries around the world [6].

Historically, the first GWAS in IBD was performed in 2006 for Crohn's disease, when CD-risk variants in the interleukin 23 receptor gene (*IL23R*) were identified [39]. Replication was performed in cohorts of both CD and UC patients, and *IL23R* was found to be a gene common to both IBD subphenotypes [39], with the coding variant Arg381Gln conferring protection against both CD and UC. *IL23R* represents one of the first genes, outside of the MHC, to be successfully associated to UC, and has opened up new avenues for potential therapeutic exploitation (see "Chapters 11 and 15" for more detail).

The first large-scale association study specifically targeting UC was published in 2008 by a UK group [14], who studied 10,886 non-synonymous SNPs and MHC tag SNPs. The screening cohort was composed of 905 UC patients recruited in the UK and 1,465 controls from the 1958 British Birth Cohort [40], and replication was performed in an independent cohort of similar size and origin. This scan resulted in the identification of three UC loci, namely, the MHC region (which confirmed previous findings); the gene encoding the extracellular matrix protein 1 (*ECM1*), expressed in the small and large intestine and involved in NF-κB activation [41]; and the macrophage stimulating gene *MST1*, previously associated to CD and UC [42] and confirmed as a CD locus [43].

The first true GWAS targeting the entire genome in UC was published in 2008 by a German group [37]. The genotyping of 355,262 SNPs (post-quality control) was performed in a screening cohort composed of 1,167 UC cases from Germany and Norway and 777 controls, while three replication cohorts included patients from Germany, the UK and Belgium, for a total of 1,855 UC and 3,091 controls. Again, association signals in the MHC region were detected, together with additional independent loci containing the anti-inflammatory interleukin 10 (*IL10*) gene and the *ARPC2* gene, respectively. This group also reported an analysis of 50 Crohn's disease risk loci in UC [44], which revealed variants in *BSN*, *IL12B*, *NKX2-3*, *PTPN2*, *KIF21B*, *CCNY*, *HERC2*, *STAT3* and 10q21.2 (intergenic) loci to be relevant also to UC.

Table 6.1 Genome-wide association studies of ulcerative colitis

Study	Type	Cohort size (screen)	Ethnic origin	Main findings	New Loci[a]	Replicated loci
[14]	GWAS scan of nsSNPs	905 UC 1,465 HC	UK	ECM1 MST1	3	MHC
[15]	GWAS of early-onset IBD	317 UC 4,250 HC	North America	20q13[b] 21q22[b]	2	IL23R MHC
[37]	GWAS	1,167 UC 777 HC	Germany Norway	IL10 ARP2C	2	MHC
[34]	GWAS	1,052 UC 2,571 HC	North America	1p36 12q15	2	IL23R MHC
[35]	GWAS	2,361 UC 5,417 HC	UK	HNF4A 16q22 LAMB1	3	23
[36]	GWAS	749 UC 2,031 HC	Japan	FCGR2A 13q12 SLC26A3	3	3
[38]	GWAS of early-onset IBD	724 UC 6,158 HC	North America Italy Scotland	Suggestive association to 2q37	1	8
[33]	GWAS	1,043 UC 1,703 HC	Germany	IL17REL[a] 7q22[a]	2	MHC
[32]	Meta-analysis	2,693 UC 6,791 HC	North America Sweden	5p15 2p16 ORMDL3	3	14
[5]	Meta-analysis	6,687 UC 19,718 HC	All cohorts of Caucasian ancestry	16 % of heritability explained	29	18
[6]	Meta-analysis	10,920 UC 15,977 HC	All cohorts of Caucasian ancestry	IBD genetic architecture shaped by host-microbe co-evolution	55	44

UC ulcerative colitis, *HC* healthy controls, *IBD* inflammatory bowel diseases, *GWAS* genome-wide association study

[a]Did not reached genome-wide significance in first screen and was genotyped in replication stage after inclusion criteria were lowered

[b]Significant results from IBD (CD+UC) cohorts

The second GWAS in UC was performed by a North American group [34]. Genotyping for the screening stage of this study was performed on 1,052 UC patients and 2,571 controls from Caucasians of European ancestry, while the replication phase was performed on cohorts from North America (768 UC patients and 721 controls) and Italy (619 UC and 394 controls). Confirming the association with MHC and IL23R, this GWAS also identified two new UC loci: 1p36 (RNF186, OTUD3, PLA2G2E) and 12q15 (*INFG*, *IL26*, *IL22*).

The third UC GWAS was performed by the UK IBD Genetics Consortium [35], who genotyped a total of 4,682 UC patients and 10,235 controls and identified three new loci on 20q13, 16q22 and 7q31. Each of these loci contains candidate genes that are biologically relevant in the context of inflammation. Locus 20q13.12 maps to the 3′ untranslated region of gene hepatocyte nuclear factor 4α (*HNF4A*), which is involved in cell adherence, and rare variants have been associated to maturity-onset diabetes of the young [45, 46]. Locus 16q22 encodes several genes including *CDH1*, the gene encoding glycoprotein E-cadherin that is involved in cell adhesion in the intestinal epithelium and whose expression is reduced in the inflamed colon of UC patients [47]. The strongest candidate gene for locus 7q31 is *LAMB1* encoding laminin β1 subunit. Laminins are present in the lamina propria of the intestines where they help in anchoring the epithelial layer, and expression of laminins is decreased in the colonic mucosa of UC patients [48]. While this study also replicated previously associated UC loci (including 1p36, the MHC, *IL23R*, *MST1* and *NKX2-3*), further data from the UK IBDGC group implicated nine CD loci in UC including *KIF21B*, *IL18RAP*, *IL12B*, *JAK2* and *STAT3* [49].

The fourth UC GWAS (and the first executed in a non-Caucasian population) was performed by a Japanese group [36] on a total of 1,384 cases and 3,057 controls (index and replication cohorts), with three additional UC loci identified: *FCGR2A*, 13q12.13 (intergenic) and *SLC26A3* (7q31). The *FCGR2A* gene encodes a member of the family of immunoglobulin Fc receptor genes that is expressed on platelets, macrophages and neutrophils where it is involved in the phagocytosis of IgG-coated particles [50]. The *SLC26A3* gene is located in locus 7q31, which was previously identified by the UK group, but the association signal appears to somewhat differ in the two populations (stronger for LAMB1 and SLC26A3, respectively, in the UK and Japanese cohorts).

In 2010, the German group carried out an extension of their first GWAS [33], with the goal to increase coverage of the genome-wide variability by using a combination of a new genome-wide SNP genotyping chip (the Affymetrix 6.0) that allowed more SNPs to be studied and of novel imputation approaches. In addition to the known MHC signal, association to new loci, *IL17REL* and 7q22 (near *SMURF1*, *KPNA7*), was identified. *IL17REL* encodes interleukin 17 receptor E-like, a poorly characterized member of the IL17 receptor gene family with no known ligand, which may bind specific IL17 cytokines to drive T_H17 inflammatory responses [51]. In an effort to further characterize the function of these loci in UC pathogenesis, the authors studied expression quantitative trait loci (eQTLs) by looking at expression signatures related to the associated SNPs at both loci. Differential immune gene expression was detected, with *IL17REL* variants modulating the expression of the *IL17RE* receptor, the cytokine *CSF3* and the modulator of T cell response *CD276*, while alleles at SNP rs7809799 in the 7q22 locus modulated the expression of the cytokine *IL1F10*, the transcription regulator of B cell development *FOXP1* and the MHC-associated gene *BTN3A*.

Early-Onset Ulcerative Colitis

In an effort to identify modest effect genes that had not been identified in previous adult-onset studies, a large collaborative GWAS of early-onset IBD was performed [15]. The rationale presented was that it should be easier to identify signals previously missed in (predominantly) adult-onset studies because of the clearer ascertainment of patients (early-onset IBD patients usually present extensive colitis) and stronger family history in early-onset IBD. All patients were diagnosed before turning 19 years old, were of European ancestry and were ascertained in various centres across the USA and one centre in Rome, Italy. The genotyping of the screening cohort (1,011 early-onset IBD patients including 317 UC patients) and replication of association signals led to the identification of two new IBD loci on chromosome 20q13 and 21q22. These loci were found as general IBD loci as they were found to be associated to both CD and UC. The 20q13 locus harbours several candidate genes, among which the authors highlighted *TNFRSF6B* because of the importance of the TNF pathway in IBD pathogenesis [52] and their detection of higher *TNFRSF6B* expression in colonic biopsies in relation to mucosal inflammation. The 21q22 locus is intergenic but the nearest gene codes for the proteasome assembly chaperone 1 (*PSMG1*), which showed a modestly increased expression in IBD patients compared to control.

A follow-up early-onset IBD GWAS was also published by the same group [38], including genotyping of adult-onset loci in their cohorts. While no new UC risk locus was identified in the study, this effort led to the association of two CD loci with UC: *ICOSLG* (21q22) and *ORMDL3* (17q12). ICOSLG is the ligand of ICOS expressed on activated T cells [53], while ORMDL3 is involved in cellular Ca^{2+} homeostasis and appears to be a general inflammation and multi-diseases locus, as it was previously associated to asthma [54].

Meta-analyses: The Power of Many

The next phase in the evolution of association studies in UC was underlined by the activities of the IIBDGC. This consortium comprises several research groups from 16 different countries, who joined forces by sharing data and expertise in order to increase cohort size and improve statistical power to detect modest effect loci in large meta-analysis efforts [55].

The first meta-analysis in UC pooled GWAS data from the previously mentioned North American study, along with new North American and Swedish GWAS [32]. This first UC meta-analysis included a total of 2,693 UC patients and 6,791 controls for the discovery phase, as well as 2,009 UC patients and 1,580 controls for the replication phase. A total of 13 loci achieved genome-wide significance including four new loci: 1q21 (*FCGRA2*—previously identified in the Japanese GWAS), 2p16 (*REL-PUS10*), 17q12 (*ORMDL3*) and 5p15. Locus 2p16 contains the RNA chaperon gene *PUS10* and the gene encoding C-Rel (REL), one of the NF-κB proteins, while for the 5p15 locus, the nearest gene *CEP72* encodes a regulator of microtubule organization during mitosis [56]. Additionally, 14 loci that were previously either formally or suggestively identified were replicated in this study, including *IL23R*, *IL12B*, *TNFRSF15*, IBD5 and *IL10*. Ten traditional CD loci were also identified in this UC meta-analysis including *IRGM*, *STAT3*, *CCL2-CCL7* and 21q21. Following the publication of this first UC meta-analysis, the authors estimated that less than 10 % of UC genetic variance could be explained by the associated loci known at the time [32].

The GWAS meta-analyses of CD and UC that followed had a dramatic impact on our knowledge of IBD genetic risk factors and again are the result of the efforts of the IIBDGC. Specifically, at the beginning of 2011, there were 18 loci significantly associated with UC identified by a few different independent studies. The same year, an IIBDGC meta-analysis of all these studies, including a total of 16,000 UC cases and 32,000 controls, discovered 29 additional loci, bringing the number of known UC loci to 47 and the estimated heritability they explain to 16 %. In order to identify relevant causal gene(s) at each locus, the authors used literature mining (GRAIL), eQTL databases, 1,000 genome sequencing data to search for correlated non-synonymous SNPs and physical proximity within the loci. Only a year later, the IIBDGC expanded the breadth of these analyses reporting one of the largest meta-analyses ever performed, based on 75,000 IBD cases and controls and including data from 15 different GWAS and additional typing on the immunochip, an array specifically designed to capture variation at 200 known risk loci for 12 common autoimmune diseases [6]. In this study, 71 new causative regions were identified, which brought the total tally to 163 independent IBD risk loci, far more than reported for any other complex disease. Of these, 110 appear to be relevant to both Crohn's disease and ulcerative colitis, while 23 show risk effects that are UC specific (the remaining 30 are CD-only loci). Fundamental contribution from this study was also the realization that a large portion of IBD risk loci are shared with other complex immune-mediated diseases (particularly ankylosing spondylitis and psoriasis), primary immunodeficiencies and mycobacterial disease, pointing to an essential role for host factors involved in defence against infection in IBD.

In addition to identifying hundreds of regions associated with increased IBD risk, most recent meta-analyses have also provided key information as to the potential biological pathways involved in UC pathogenesis. Among other methods, functional annotation of the identified UC genes allows to cluster them in pathways and clarify the molecular origin of the disease. In the latest meta-analysis from IIBDGC, where the highest number of UC loci was identified, the most significantly enriched Gene Ontology [57] term was regulation of cytokine production, specifically IFN-γ, IL-12, TNF-α and IL10 signalling. T, B and NK cell activation was the next most significant, and strong enrichment was also seen for response to molecules of bacterial origin and for the JAK-STAT signalling pathway. These pathways add to those already suspected to be relevant from individual gene functions such as, for instance, transcriptional regulation (*PRDM1*, *IRF5* and *NKX2-3*) and intestinal barrier functions (*GNA12*, *HNF4A* and *LAMB1*). A list of prioritized candidate UC-risk genes and their functional properties is reported in Fig. 6.1.

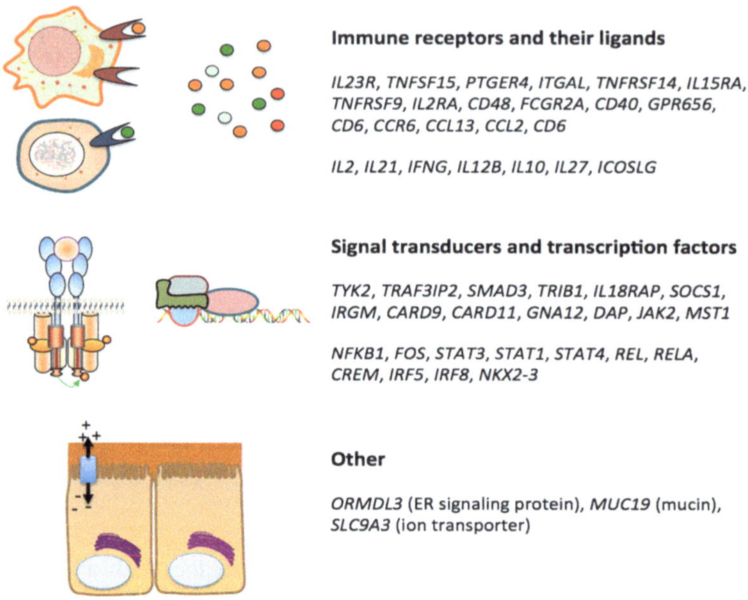

Fig. 6.1 UC risk genes grouped by function. The list of genes and corresponding biological functions is based on the prioritization at their respective risk loci based on the results included in the most recent UC meta-analysis [6]

Many UC Genes Are General IBD Risk Factors

Overlapping phenotype and aggregation of both CD and UC in some families have suggested that part of the genetic risk is common to both diseases. Many genetic studies have tested the association of Crohn's disease loci to UC and vice versa (as mentioned earlier in this chapter). The large meta-analyses of CD and UC, respectively, by Franke et al. and Anderson et al. enabled one of the first systematic analyses of loci shared by these two diseases. Specifically, Anderson et al. identified common loci by comparing the results of these large GWAS studies in CD and UC: common loci were defined when genome-wide significant loci for one disease ($P<5\times10^{-8}$) reached $P<1\times10^{-4}$ in the other disease, and at least 28 shared loci were identified by this means [5]. Among those, many are involved in T cell differentiation (e.g. cytokines *IL21*, *IL10*, *IFNG* and cytokine receptor *IL7R*). Some of them are more specifically associated with the IL23R pathway (*IL23R*, *JAK2*, *STAT3*, *IL12B* and *PTPN2*), which is involved in the maintenance of T_H17 cells and has been involved in several other diseases (see "Chapters 7 and 11"). T_H17 cells are thought to coordinate defence against specific pathogen and mediate inflammation [58], and the original identification of IL23R as an IBD risk factor shattered the paradigm of CD and UC being primarily T_H1- and T_H2-mediated diseases, respectively [39]. Genes involved in TNF signalling are also well represented among common IBD genes (*TNFRSF9*, *TNFRSF14* and *TNFSF15*), and they encode proteins with various immune effects including systemic inflammation and activation of inflammatory transcription factor NF-κB. As mentioned, the latest meta-analysis has brought the number of common CD-UC shared loci to 110 [6], while 23 appear to be the risk loci with UC-specific genetic effects. Interestingly, however, most of these UC-specific genes show the same direction of effect in Crohn's disease, suggesting that nearly all of the biological mechanisms involved in one disease have some role in the other. One intriguing exception is NOD2, which still represents the strongest causal gene in CD but shows significant protective effects in UC, an observation that may reflect biological differences between the two diseases.

Associated Loci Differ Between Ethnic Groups

With the exception of a Japanese study [58], all UC GWAS carried out to date have been performed on Caucasians of European descent. Considering the interethnic differences in genomic architectures that reflect genetic drift, mutations and evolutionary factors, it is likely that some risk variants may be relevant in some populations but not others. One of the ultimate goals of current genetic studies is their exploitation for the design of novel therapeutic approaches, and it is therefore crucial to understand what portion of the genetic risk is detectable and hence also relevant in individual populations. The increasing number of association studies performed in cohorts of individuals of non-European descent, some of these GWAS, allows for some initial information to be evaluated in this context. The risk gene *IL23R*, for example, might not be associated in all ethnic groups, as it was not detected in the Japanese GWAS of UC [36] and in a targeted association study based on a Korean case-control cohort [59]. This may be at least in part due to the fact that some European *IL23R* variants are less common or absent in the Japanese population, like in the case of the protective *IL23R* 381Q allele (1 % compared to 7 % in Caucasians). Evidence for similar interethnic differences exists for other IBD loci, for instance, *NOD2* and *ATG16L1* [60, 61, 62], and large-scale analyses are currently at the core of IIBDGC interethnic studies that aim to shed further light on this important question.

Many Risk Factors Are Still to Be Discovered

Surprisingly, the total number of 133 UC risk loci (a number by far higher than in most other complex diseases) is still estimated to cumulatively account only for a minor portion of disease variance, namely, 7.5 % [6]. It should be noted that several additional loci narrowly miss the genome-wide-significance threshold of significance ($P<5\times10^{-8}$) and thus are not considered to be validated loci; however, this observation suggests that other risk loci are still to be discovered. It has been proposed that additional loci or causative rare variants may not be captured by GWAS platforms that primarily test common variants. Deep re-sequencing of risk loci is a way to identify these rare variants, and next-generation sequencing (NGS) technologies have greatly reduced per sample costs and process time, thus allowing such studies to be feasibly performed on adequate numbers of cases and controls. The first large-scale re-sequencing study performed in IBD was published in October 2011 [63]. Fifty six CD loci were re-sequenced in two DNA pools from 350 CD cases and 350 healthy controls for variants discovery, and follow-up genotyping of 115 low-frequency SNPs (non-synonymous, nonsense or splice-site variants) was performed in 16,054 CD and 12,153 UC patients and 17,575 healthy controls. The study identified a protective splice-site variant in the *CARD9* gene, and additional IBD susceptibility variants were also detected in *IL23R*, *CUL2*, *PTPN22* and *C1orf106*. In total, the authors estimated that 1–2 additional percent of CD heritability (and likely less for UC) is explained by these rare variants. These findings appear to be very modest but they give an important insight into complex disease genetics, where common variants of modest effect and rare variants of higher penetrance coexist in the same genes. Additional UC-focussed re-sequencing efforts will likely allow for the identification of other rare risk variants that may help to both pinpoint exact causative genes (in multiple-gene loci) and improve our understanding of the functional consequences of genetic variation at specific UC risk loci.

The Future of UC Genetics

The last few years have seen tremendous progress and important discoveries in UC genetics; however, most of the known disease heritability is still to be explained. As mentioned, one favoured hypothesis is that the remaining heritability is to be found in new risk loci and in additional rare variants located in known and yet to be identified UC loci. One of the obstacles to these discoveries is the statistical power that will be required, given that current loci have required tens of thousands of samples. Additional independent GWAS and the pooling of these new GWAS into increasingly bigger meta-analyses will certainly lead to the identification of new loci. Re-sequencing studies that aim to identify rare causal variants will be essential companions to these GWAS, and the analysis of news markers such as copy number variations (CNVs) will hopefully provide additional insight into disease heritability. In addition, long-distance regulatory variants need to be more carefully evaluated given that some associated loci are gene deserts, as well as individual SNPs for their potential role in altering epigenetic signatures in DNA methylation, histone acetylation, microRNA binding, etc. Genotyping cohorts of different ethnic background will be helpful in fine-mapping-associated loci. Genetic variation patterns are often different in different populations, and, for example, African populations have more diversity and the linkage disequilibrium blocks are generally shorter than in population of European ancestry, thus allowing for the identification of new or refined association signals.

Finally, there is great need for integrative genomic and system biology approaches where multilayer data are collectively analyzed and exploited to decipher the mysteries of the genome. Together with the avalanche of data coming from several GWAS and their meta-analysis, the considerable challenge of having to understand the role of each and individual genetic risk factor from hundreds of loci has emerged. In this endeavour, large-scale functional studies into the transcriptome and the proteome are now indispensable. The amount of work still ahead in UC genetics seems to be no less than what it was some years ago, but we certainly now have a very solid foundation of knowledge upon which we can build.

References

1. De Dombal FT (1968) Ulcerative colitis: definition, historical background, aetiology, diagnosis, natural history and local complications. Postgrad Med J 44(515):684–692, Epub 1968/09/01
2. Stokkers PC, Reitsma PH, Tytgat GN, van Deventer SJ (1999) HLA-DR and -DQ phenotypes in inflammatory bowel disease: a meta-analysis. Gut 45(3):395–401, Epub 1999/08/14
3. Lander ES, Linton LM, Birren B, Nusbaum C, Zody MC, Baldwin J et al (2001) Initial sequencing and analysis of the human genome. Nature 409(6822):860–921, Epub 2001/03/10
4. Venter JC, Adams MD, Myers EW, Li PW, Mural RJ, Sutton GG et al (2001) The sequence of the human genome. Science 291(5507):1304–1351, Epub 2001/02/22
5. Anderson CA, Boucher G, Lees CW, Franke A, D'Amato M, Taylor KD et al (2011) Meta-analysis identifies 29 additional ulcerative colitis risk loci, increasing the number of confirmed associations to 47. Nat Genet 43(3):246–252, Epub 2011/02/08
6. Jostins L, Ripke S, Weersma RK, Duerr RH, McGovern DP, Hui KY et al (2012) Host-microbe interactions have shaped the genetic architecture of inflammatory bowel disease. Nature 491(7422):119–124, Epub 2012/11/07
7. Franke A, McGovern DP, Barrett JC, Wang K, Radford-Smith GL, Ahmad T et al (2010) Genome-wide meta-analysis increases to 71 the number of confirmed Crohn's disease susceptibility loci. Nat Genet 42(12):1118–1125, Epub 2010/11/26
8. Amiel JL (1967) Study of the leukocyte phenotypes in Hodgkin's disease. In: Curtoni ES, Mattiuz PL, Tosi RM (eds) Histocompatibility testing. Munksgaard, Copenhagen, pp 79–81
9. The_MHC_Consortium (1999) Complete sequence and gene map of a human major histocompatibility complex. The MHC sequencing consortium. Nature 401(6756):921–923, Epub 1999/11/30
10. Robinson J, Waller MJ, Parham P, de Groot N, Bontrop R, Kennedy LJ et al (2003) IMGT/HLA and IMGT/MHC: sequence databases for the study of the major histocompatibility complex. Nucleic Acids Res 31(1):311–314, Epub 2003/01/10
11. Fernando MM, Stevens CR, Walsh EC, De Jager PL, Goyette P, Plenge RM et al (2008) Defining the role of the MHC in autoimmunity: a review and pooled analysis. PLoS Genet 4(4):e1000024, Epub 2008/04/26
12. Orchard TR, Dhar A, Simmons JD, Vaughan R, Welsh KI, Jewell DP (2001) MHC class I chain-like gene A (MICA) and its associations with inflammatory bowel disease and peripheral arthropathy. Clin Exp Immunol 126(3):437–440, Epub 2001/12/12
13. Rioux JD, Goyette P, Vyse TJ, Hammarstrom L, Fernando MM, Green T et al (2009) Mapping of multiple susceptibility variants within the MHC region for 7 immune-mediated diseases. Proc Natl Acad Sci U S A 106(44):18680–18685, Epub 2009/10/23
14. Fisher SA, Tremelling M, Anderson CA, Gwilliam R, Bumpstead S, Prescott NJ et al (2008) Genetic determinants of ulcerative colitis include the ECM1 locus and five loci implicated in Crohn's disease. Nat Genet 40(6):710–712, Epub 2008/04/29

15. Kugathasan S, Baldassano RN, Bradfield JP, Sleiman PM, Imielinski M, Guthery SL et al (2008) Loci on 20q13 and 21q22 are associated with pediatric-onset inflammatory bowel disease. Nat Genet 40(10):1211–1215, Epub 2008/09/02
16. Ho GT, Moodie FM, Satsangi J (2003) Multidrug resistance 1 gene (P-glycoprotein 170): an important determinant in gastrointestinal disease? Gut 52(5):759–766, Epub 2003/04/15
17. Panwala CM, Jones JC, Viney JL (1998) A novel model of inflammatory bowel disease: mice deficient for the multiple drug resistance gene, mdr1a, spontaneously develop colitis. J Immunol 161(10):5733–5744, Epub 1998/11/20
18. Annese V, Valvano MR, Palmieri O, Latiano A, Bossa F, Andriulli A (2006) Multidrug resistance 1 gene in inflammatory bowel disease: a meta-analysis. World J Gastroenterol 12(23):3636–3644, Epub 2006/06/15
19. Zintzaras E (2012) Is there evidence to claim or deny association between variants of the multidrug resistance gene (MDR1 or ABCB1) and inflammatory bowel disease? Inflamm Bowel Dis 18(3):562–572, Epub 2011/09/03
20. van Heel DA, Franke L, Hunt KA, Gwilliam R, Zhernakova A, Inouye M et al (2007) A genome-wide association study for celiac disease identifies risk variants in the region harboring IL2 and IL21. Nat Genet 39(7):827–829, Epub 2007/06/15
21. Todd JA, Walker NM, Cooper JD, Smyth DJ, Downes K, Plagnol V et al (2007) Robust associations of four new chromosome regions from genome-wide analyses of type 1 diabetes. Nat Genet 39(7):857–864, Epub 2007/06/08
22. Zhernakova A, Alizadeh BZ, Bevova M, van Leeuwen MA, Coenen MJ, Franke B et al (2007) Novel association in chromosome 4q27 region with rheumatoid arthritis and confirmation of type 1 diabetes point to a general risk locus for autoimmune diseases. Am J Hum Genet 81(6):1284–1288, Epub 2007/11/14
23. Sawalha AH, Kaufman KM, Kelly JA, Adler AJ, Aberle T, Kilpatrick J et al (2008) Genetic association of interleukin-21 polymorphisms with systemic lupus erythematosus. Ann Rheum Dis 67(4):458–461, Epub 2007/08/28
24. Liu Y, Helms C, Liao W, Zaba LC, Duan S, Gardner J et al (2008) A genome-wide association study of psoriasis and psoriatic arthritis identifies new disease loci. PLoS Genet 4(3):e1000041, Epub 2008/03/29
25. Yang L, Anderson DE, Baecher-Allan C, Hastings WD, Bettelli E, Oukka M et al (2008) IL-21 and TGF-beta are required for differentiation of human T(H)17 cells. Nature 454(7202): 350–352, Epub 2008/05/13
26. Monteleone G, Monteleone I, Fina D, Vavassori P, Del Vecchio Blanco G, Caruso R et al (2005) Interleukin-21 enhances T-helper cell type I signaling and interferon-gamma production in Crohn's disease. Gastroenterology 128(3):687–694
27. Sadlack B, Merz H, Schorle H, Schimpl A, Feller AC, Horak I (1993) Ulcerative colitis-like disease in mice with a disrupted interleukin-2 gene. Cell 75(2):253–261, Epub 1993/10/22
28. Festen EA, Goyette P, Scott R, Annese V, Zhernakova A, Lian J et al (2009) Genetic variants in the region harbouring IL2/IL21 associated with ulcerative colitis. Gut 58(6):799–804, Epub 2009/02/10
29. Glas J, Stallhofer J, Ripke S, Wetzke M, Pfennig S, Klein W et al (2009) Novel genetic risk markers for ulcerative colitis in the IL2/IL21 region are in epistasis with IL23R and suggest a common genetic background for ulcerative colitis and celiac disease. Am J Gastroenterol 104(7):1737–1744, Epub 2009/05/21
30. Marquez A, Orozco G, Martinez A, Palomino-Morales R, Fernandez-Arquero M, Mendoza JL et al (2009) Novel association of the interleukin 2-interleukin 21 region with inflammatory bowel disease. Am J Gastroenterol 104(8):1968–1975, Epub 2009/05/28
31. Shi J, Zhou L, Zhernakova A, Qian J, Zhu F, Sun G et al (2011) Haplotype-based analysis of ulcerative colitis risk loci identifies both IL2 and IL21 as susceptibility genes in Han Chinese. Inflamm Bowel Dis 17(12):2472–2479, Epub 2011/06/08
32. McGovern DP, Gardet A, Torkvist L, Goyette P, Essers J, Taylor KD et al (2010) Genome-wide association identifies multiple ulcerative colitis susceptibility loci. Nat Genet 42(4): 332–337, Epub 2010/03/17

33. Franke A, Balschun T, Sina C, Ellinghaus D, Hasler R, Mayr G et al (2010) Genome-wide association study for ulcerative colitis identifies risk loci at 7q22 and 22q13 (IL17REL). Nat Genet 42(4):292–294, Epub 2010/03/17
34. Silverberg MS, Cho JH, Rioux JD, McGovern DP, Wu J, Annese V et al (2009) Ulcerative colitis-risk loci on chromosomes 1p36 and 12q15 found by genome-wide association study. Nat Genet 41(2):216–220, Epub 2009/01/06
35. Barrett JC, Lee JC, Lees CW, Prescott NJ, Anderson CA, Phillips A et al (2009) Genome-wide association study of ulcerative colitis identifies three new susceptibility loci, including the HNF4A region. Nat Genet 41(12):1330–1334, Epub 2009/11/17
36. Asano K, Matsushita T, Umeno J, Hosono N, Takahashi A, Kawaguchi T et al (2009) A genome-wide association study identifies three new susceptibility loci for ulcerative colitis in the Japanese population. Nat Genet 41(12):1325–1329, Epub 2009/11/17
37. Franke A, Balschun T, Karlsen TH, Sventoraityte J, Nikolaus S, Mayr G et al (2008) Sequence variants in IL10, ARPC2 and multiple other loci contribute to ulcerative colitis susceptibility. Nat Genet 40(11):1319–1323, Epub 2008/10/07
38. Imielinski M, Baldassano RN, Griffiths A, Russell RK, Annese V, Dubinsky M et al (2009) Common variants at five new loci associated with early-onset inflammatory bowel disease. Nat Genet 41(12):1335–1340, Epub 2009/11/17
39. Duerr RH, Taylor KD, Brant SR, Rioux JD, Silverberg MS, Daly MJ et al (2006) A genome-wide association study identifies IL23R as an inflammatory bowel disease gene. Science 314(5804):1461–1463, Epub 2006/10/28
40. Power C, Elliott J (2006) Cohort profile: 1958 British birth cohort (National Child Development Study). Int J Epidemiol 35(1):34–41, Epub 2005/09/13
41. Matsuda A, Suzuki Y, Honda G, Muramatsu S, Matsuzaki O, Nagano Y et al (2003) Large-scale identification and characterization of human genes that activate NF-kappaB and MAPK signaling pathways. Oncogene 22(21):3307–3318, Epub 2003/05/23
42. Goyette P, Lefebvre C, Ng A, Brant SR, Cho JH, Duerr RH et al (2008) Gene-centric association mapping of chromosome 3p implicates MST1 in IBD pathogenesis. Mucosal Immunol 1(2):131–138, Epub 2008/12/17
43. WTCCC (2007) Genome-wide association study of 14,000 cases of seven common diseases and 3,000 shared controls. Nature 447(7145):661–678, Epub 2007/06/08
44. Franke A, Balschun T, Karlsen TH, Hedderich J, May S, Lu T et al (2008) Replication of signals from recent studies of Crohn's disease identifies previously unknown disease loci for ulcerative colitis. Nat Genet 40(6):713–715, Epub 2008/04/29
45. Yamagata K, Furuta H, Oda N, Kaisaki PJ, Menzel S, Cox NJ et al (1996) Mutations in the hepatocyte nuclear factor-4alpha gene in maturity-onset diabetes of the young (MODY1). Nature 384(6608):458–460, Epub 1996/12/05
46. Battle MA, Konopka G, Parviz F, Gaggl AL, Yang C, Sladek FM et al (2006) Hepatocyte nuclear factor 4alpha orchestrates expression of cell adhesion proteins during the epithelial transformation of the developing liver. Proc Natl Acad Sci U S A 103(22):8419–8424, Epub 2006/05/23
47. Karayiannakis AJ, Syrigos KN, Efstathiou J, Valizadeh A, Noda M, Playford RJ et al (1998) Expression of catenins and E-cadherin during epithelial restitution in inflammatory bowel disease. J Pathol 185(4):413–418, Epub 1998/11/26
48. Schmehl K, Florian S, Jacobasch G, Salomon A, Korber J (2000) Deficiency of epithelial basement membrane laminin in ulcerative colitis affected human colonic mucosa. Int J Colorectal Dis 15(1):39–48, Epub 2000/04/15
49. Anderson CA, Massey DC, Barrett JC, Prescott NJ, Tremelling M, Fisher SA, et al (2009) Investigation of Crohn's disease risk loci in ulcerative colitis further defines their molecular relationship. Gastroenterology 136(2):523–529. e3. Epub 2008/12/11
50. Huang ZY, Chien P, Indik ZK, Schreiber AD (2011) Human platelet FcgammaRIIA and phagocytes in immune-complex clearance. Mol Immunol 48(4):691–696, Epub 2010/12/21
51. Wu B, Jin M, Zhang Y, Wei T, Bai Z (2011) Evolution of the IL17 receptor family in chordates: a new subfamily IL17REL. Immunogenetics 63(12):835–845, Epub 2011/07/07

52. van Heel DA, Udalova IA, De Silva AP, McGovern DP, Kinouchi Y, Hull J et al (2002) Inflammatory bowel disease is associated with a TNF polymorphism that affects an interaction between the OCT1 and NF(-kappa)B transcription factors. Hum Mol Genet 11(11): 1281–1289, Epub 2002/05/23
53. Hutloff A, Dittrich AM, Beier KC, Eljaschewitsch B, Kraft R, Anagnostopoulos I et al (1999) ICOS is an inducible T-cell co-stimulator structurally and functionally related to CD28. Nature 397(6716):263–266, Epub 1999/02/04
54. Moffatt MF, Kabesch M, Liang L, Dixon AL, Strachan D, Heath S et al (2007) Genetic variants regulating ORMDL3 expression contribute to the risk of childhood asthma. Nature 448(7152):470–473, Epub 2007/07/06
55. http://www.ibdgenetics.org/
56. Oshimori N, Li X, Ohsugi M, Yamamoto T (2009) Cep72 regulates the localization of key centrosomal proteins and proper bipolar spindle formation. EMBO J 28(14):2066–2076, Epub 2009/06/19
57. http://www.ncbi.nlm.nih.gov/
58. Bettelli E, Korn T, Oukka M, Kuchroo VK (2008) Induction and effector functions of T(H)17 cells. Nature 453(7198):1051–1057
59. Yang SK, Jung Y, Hong M, Kim H, Ye BD, Lee I et al (2011) No association between TNFSF15 and IL23R with ulcerative colitis in Koreans. J Hum Genet 56(3):200–204, Epub 2011/01/14
60. Mahurkar S, Banerjee R, Rani VS, Thakur N, Rao GV, Reddy DN et al (2011) Common variants in NOD2 and IL23R are not associated with inflammatory bowel disease in Indians. J Gastroenterol Hepatol 26(4):694–699, Epub 2010/12/16
61. Yamazaki K et al (2002) Absence of mutation in the NOD2/CARD15 gene among 483 Japanese patients with Crohn's disease. J Hum Genet 47:469–472
62. Yamazaki K et al (2007) Association analysis of genetic variants in IL23R, ATG16L1and 5p13.1 loci with Crohn's disease in Japanese patients. J Hum Genet 52:575–583
63. Rivas MA, Beaudoin M, Gardet A, Stevens C, Sharma Y, Zhang CK et al (2011) Deep resequencing of GWAS loci identifies independent rare variants associated with inflammatory bowel disease. Nat Genet 43(11):1066–1073, Epub 2011/10/11

Chapter 7
Genetic Overlap Between Inflammatory Bowel Disease and Other Diseases

Charlie W. Lees and Alexandra Zhernakova

Abstract The inflammatory bowel diseases (IBDs) Crohn's disease and ulcerative colitis have seen tremendous success in the identification of disease susceptibility alleles, over the past 5 years. Presently 163 susceptibility loci are described in the published literature. Whilst formal fine mapping is largely awaited, initial analysis and refinement of these loci has been carried out using various analytical methods, including eQTL, nsSNP and literature-mining tools (GRAIL). Study of the implicated genes within each loci has yielded multiple novel insights into disease biology. These include the role of innate immunity and killing of intracellular bacteria in Crohn's disease (e.g. NOD2 and autophagy) and the probable pathogenetic role of a defective mucosal barrier in ulcerative colitis. Several pathways, most notably IL23 signalling, have multiple components that are associated with not only both Crohn's disease and ulcerative colitis but also multiple other immune- and nonimmune-mediated diseases. These suggest, for example, that IL23 signalling plays a general role not only in susceptibility to intestinal inflammatory disease but also in immune-mediated disease as a whole. The finely tuned immune balance in individuals with susceptibility alleles is tipped in favour of developing chronic inflammation likely following a requisite and largely undefined environmental trigger/insult. Further intriguing overlaps are described with IBD- and nonimmune-mediated diseases, notably diabetes mellitus where a potential shared aetiology within the handling of the gut microbiota has started to emerge. Of most potential relevance to the aetiology of IBD is the fascinating overlap between Crohn's disease and mycobacterial leprosy, where shared susceptibility genes include *NOD2*, *IL23R*, *TNFSF15* and *RIPK2*. These emerging data

C.W. Lees, M.B.B.S., M.R.C.P., Ph.D. (✉)
Gastrointestinal Unit, Western General Hospital, Crewe Road, Edinburgh, EH4 2XU, UK
e-mail: charlie.lees@ed.ac.uk

A. Zhernakova, M.D., Ph.D.
Genetics Department, University Medical Centre Groningen and University of Groningen, Groningen, The Netherlands

M. D'Amato and J.D. Rioux (eds.), *Molecular Genetics of Inflammatory Bowel Disease*,
DOI 10.1007/978-1-4614-8256-7_7, © Springer Science+Business Media New York 2013

suggest a selective pressure on alleles that on the one hand confer susceptibility to chronic intestinal inflammation whilst offering protection against infectious disease. These and many other insights waiting to be uncovered will shed new light on disease biology across many fields and help to focus drug discovery efforts over many years to come.

Introduction

Genome-wide association studies have transformed the landscape of complex disease genetics since their widespread adoption in the past 5 years. To date, over 1,700 GWAS have been published analysing hundreds of complex diseases and/or traits. Curation of these data is a major challenge. Presently, the National Human Genome Research Institute (NHGRI) catalogue of published GWAS (accessed 22 March 2013) includes 1,556 publications and 9,811 SNPs [1]. The primary aim of these studies is to understand disease biology through gene discovery. GWAS are not, as is often incorrectly stated, 'hypothesis-free'—the null hypothesis is that there is no difference between the allelic structure in cases and appropriately matched healthy controls. But they are 'assumption-free'—i.e. no a priori assumption is made as to which variants predispose/protect from disease 'a' or trait 'b', and hence existing dogmas about the underlying biology of a disease can be challenged.

One of the most intriguing and challenging aspects to consider with GWAS data is the overlap in susceptibility genes between different complex diseases. In so doing we can potentially elucidate shared pathogenetic mechanisms and foster a new wave of cross-fertilisation linking groups of scientists and clinicians, along with their industry partners, who would not have previously considered their 'distinct' diseases, were in the same book let alone on the same page.

Several groups have attempted to look at a section of the GWAS data, e.g. autoimmune disease, and analysed the genetic overlap between all the different diseases studied (some results and challenges of these analyses are described in Box 7.1). In this chapter we are taking the inflammatory bowel diseases (IBDs), Crohn's disease and ulcerative colitis, as our starting point. To date, there are about 100 independent, definitive IBD susceptibility loci described with most recent results bringing the number to 163 during the preparation of this chapter [2, 3]. We have reviewed the literature to ascertain where the overlap exists ('ulcerative colitis OR Crohn's disease OR IBD AND disease x') and manually searched through the NHGRI catalogue of published genome-wide association studies. Before describing the genetic overlap, we will first review several pertinent aspects about the clinical picture of IBD—notably the classical extra-intestinal manifestations described—and second delve into the epidemiological evidence for clustering of other diseases in patients with Crohn's disease and ulcerative colitis.

Clinical Features

Crohn's disease and ulcerative colitis are the two major forms of IBD. They are still considered distinct diseases but share many clinical and pathological features. Crohn's disease can affect any part of the gastrointestinal tract, from mouth to anus, although typically affects regions harbouring the highest concentrations of luminal microbiota—terminal ileum and colon. The inflammatory disease process is transmural and discontinuous with granulomatous inflammation seen histologically. In contrast, ulcerative colitis only affects the colon with continuous mucosal inflammation extending variably from the rectum to involve anything up to the entire colon. Both diseases are thought to arise as a result of an exaggerated mucosal immune response to microbiota in the gut lumen in genetically susceptible individuals. It is presently unknown why cigarette smoking is positively associated with the development of Crohn's disease, whilst being clearly protective for ulcerative colitis.

Both Crohn's disease and ulcerative colitis are associated with important extraintestinal manifestations in a significant proportion of patients. Some of these are probably primarily complications of an uncontrolled systemic inflammatory response, e.g. venous thromboembolic (VTE) disease, whilst most are somehow related to the underlying immune-mediated process. These poorly understood phenomena include those affecting joints (enteropathic arthropathies, seronegative arthritides, sacro-ileitis), skin (erythema nodosum, pyoderma gangrenosum), hepato-biliary system (autoimmune hepatitis, primary sclerosing cholangitis), blood (lymphoma, VTE) and kidneys (renal oxalate stones). Clearly some of these processes are separate diseases in their own right, whilst others may represent complications of immunosuppressive therapy and/or the underlying disease (e.g. lymphoma). We will therefore now consider the epidemiologic evidence for clustering of other disease with Crohn's disease and ulcerative colitis.

Epidemiologic Evidence for Disease Clustering with IBD

To begin our interpretation of the genetic overlap between IBD and other diseases, it is first pertinent to consider the evidence for clustering of other diseases. The presently available epidemiologic data indicate that there is probably an increased incidence of the following diseases in patients suffering from Crohn's disease or ulcerative colitis: asthma [4, 5], ankylosing spondylitis [6–8], atopic dermatitis [9, 10], psoriasis [4, 5, 11], multiple sclerosis [12] and primary sclerosing cholangitis [13]. Of the other diseases where there is a genetic overlap with IBD, no clinical clustering is apparent (or has yet to be clearly demonstrated) for systemic lupus erythematosus [14], sarcoidosis [15], coeliac disease [16–18], type 1 diabetes mellitus [4–6], and type 2 diabetes mellitus.

Genetic Overlap

Following the publication of the two large meta-GWAS in Crohn's disease [19] and ulcerative colitis [20], the tally of IBD susceptibility genes was about 100. Of these, roughly one third were associated with Crohn's disease specifically, one third with ulcerative colitis specifically and one third with both (i.e. with the development of IBD as a whole). An examination of these various susceptibility genes led to the following observations: (1) Crohn's disease results from defective processing of intracellular bacteria (defects in innate immunity including autophagy), (2) ulcerative colitis results from a defective mucosal barrier and (3) defects in both the IL23/Th17 and IL10 immune pathways predispose generally to the development of idiopathic intestinal inflammation [3].

A fascinating related exercise, and the primary purpose of this chapter, centres on the study of the genetic overlap between IBD susceptibility genes and those for other diseases, and what novel insights this potentially yields into (a) disease biology and (b) common targets for therapeutic intervention.

For those immune-mediated diseases where there is both clinical and genetic overlap, we might postulate that common susceptibility alleles predispose to chronic inflammation in different end organs dependent on some, probably unknown, environmental trigger: for example, the gut microbiota in IBD or an airway allergen in asthma. Presently, however, there is little direct evidence to support this supposition. Even more intriguing are the diseases where there is no clinical clustering and no apparent (or at least no known) common pathogenetic mechanism. The most striking example here is the genetic overlap between IBD and diabetes mellitus, where there is a glut of common susceptibility genes (see Fig. 7.1). Already, however, recently emerging data would point to a potential role for common defects in the handling/processing of the gut microbiota by the mucosal immune system as central to disease pathogenesis of type 1 diabetes mellitus [21].

Clearly this is a rapidly evolving field of scientific enquiry, with vast amounts of novel data being released into the public domain almost daily. Therefore, rather than attempt to describe the entire spectrum of genetic overlap between IBD and other diseases, we will focus in on a couple of key areas. Box 7.1 describes the genetic overlap between immune-mediated diseases as a whole, including the many roles that the IL23/Th17 signalling pathway plays in chronic inflammation. We will then describe and discuss the overlap between (1) Crohn's disease, ulcerative colitis and coeliac disease and (2) Crohn's disease and infectious disease, before considering the important future challenges with (3) data analysis, (4) fine mapping of disease loci and (5) implications for biotech/big pharma and the challenges of future drug design.

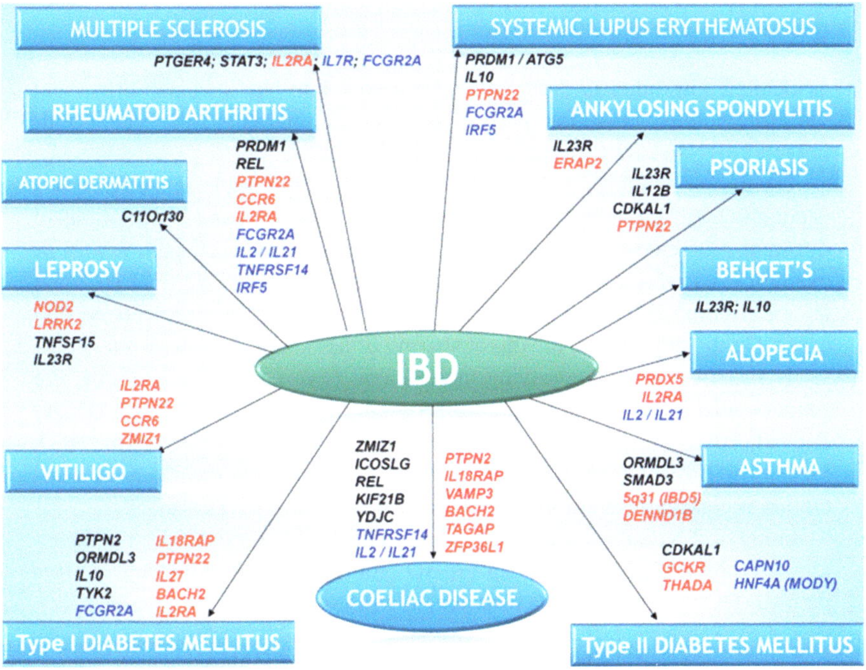

Fig. 7.1 Disease overlap between IBD, coeliac disease and other immune- and nonimmune-mediated diseases. To ensure clarity of present data as well as presentation, only those genes that have attained genome-wide levels of significance in all diseases are depicted here. IBD loci (CD and UC) are depicted in *black*, CD only loci in *red* and UC only loci in *blue*. From Lees CW, Barrett JC, Parkes M, Satsangi J. New IBD genetics: common pathways with other diseases. Gut 2011;60:1739–53 (reprinted with permission from BMJ Publishing Group Ltd)

Box 7.1 Genetic Overlap Among Autoimmune Diseases as a Whole

The emerging observation from GWAS findings is extensive sharing of genetic loci associated to various immune-related diseases. With every new published study, the number of disease-specific loci for any disease goes down. Meta-analysis and cross-studies of diseases suggest that up to 30 % of genetic risk factors are directly shared between different immune-related diseases (e.g. coeliac disease and T1DM and Crohn's disease and arthritis) [22–25]. From 39 genome-wide significant loci reported for coeliac disease, only ten are not reported to be associated with other immune-related diseases [26].

Moreover, many disease-associated genes are known to be closely related in the shared immunological pathways. For example, *IL12A* gene is associated to coeliac disease and multiple sclerosis, whereas *IL12B*—the other subunit of IL12 molecule—is associated with Crohn's disease and psoriasis. It is now evident that many loci which we call 'disease specific' may also be

(continued)

Box 7.1 (continued)

associated with other immune-related diseases, but with a weaker effect and therefore undetected so far. This possibility is supported by the recent findings of cross-disease meta-analyses.

The puzzle of shared and specific associations in immune-related diseases is complicated by two opposite observations. On the one hand, associations to different molecules often point to the same immunological pathways. For example, shared and specific associations of *CTLA4*, *PTPN2*, *PTPN22* and *ICOSLG* point to a shared role for *T* cell activation in immune-related diseases. On the other hand, the direct sharing of associated loci does not necessarily mean the same underlying mechanisms play a role. For example, the functional R620W variant in *PTPN22* confers susceptibility to SLE, T1D and RA but protection to Crohn's disease. Similar picture is observed in the HLA locus, which is shared by all immune-related diseases, but the nature of association is clearly different in the various diseases.

In Fig. 7.2 we selected IL23 signalling as one of the most shared pathway identified by GWAS and relevant in IBD biology. Multiple molecules in this pathway are associated with a wide variety of immune-mediated diseases [27].

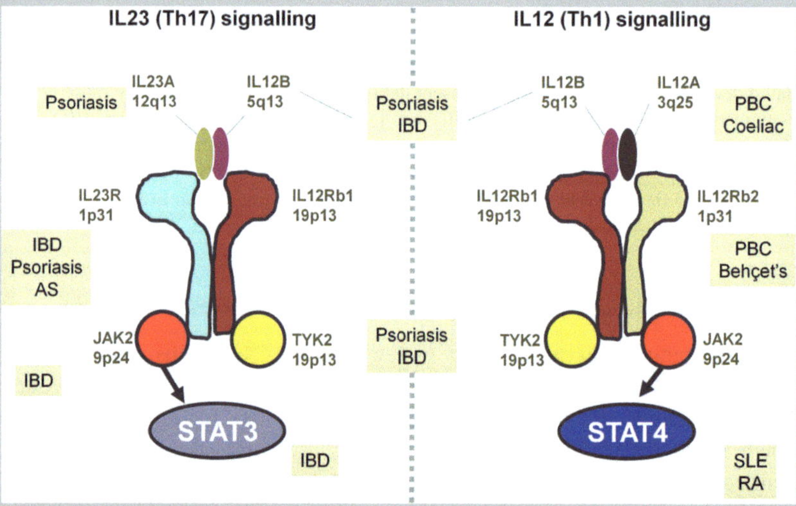

Fig. 7.2 IL23 (*left*) and IL12 (*right*) signalling depicting confirmed disease susceptibility loci. *IBD* inflammatory bowel disease, *RA* rheumatoid arthritis, *AS* ankylosing spondylitis, *SLE* systemic lupus erythematosus, *PBC* primary biliary cirrhosis

Ulcerative Colitis, Crohn's Disease and Coeliac Disease

Coeliac disease is another common gastrointestinal disease, which affects at least 1 % of the Western population [28]. It is a multifactorial disease caused by many different genetic factors that act in combination with non-genetic causes. One of the most important environmental trigger is dietary gluten, a storage protein of wheat, barley and rye. In genetically predisposed individuals gluten peptide triggers a cascade of immune responses and leads to the destruction of the intestinal epithelium and mucosa and to the lymphocytic infiltration of the small bowel [29]. Clinical manifestation of this disease includes both classic forms, such as diarrhoea, abdominal distension and short stature, as well as extra-intestinal, atypical presentation, including anaemia, neurological symptoms, dermatitis herpetiformis and osteoporosis. Coeliac disease is a life-long condition and the only known treatment is to completely exclude gluten from the diet.

Coeliac disease is one of the best-understood immune-related diseases to date. Firstly, the environmental triggering factor (gluten) [30, 31] is known. Secondly, the major genetic risk factor—presence of HLA DQ2 or HLA DQ8 molecules—is present in almost all patients. The mechanism of HLA association to coeliac disease is well established [30, 31]. The HLADQ molecules function as cell surface receptors for exogenous peptide antigens (gluten in the case of coeliac disease) on antigen presenting cells, presenting them to T helper cells. Practically, only individuals positive for DQ2 or DQ8 develop coeliac disease. Thirdly, similar to majority of immune-related diseases, there is involvement of non-HLA disease susceptibility genes [26]. Many of these genes are shared with other autoimmune diseases.

The history of genetic studies in coeliac disease goes back to 1974, when the association to HLA was described for the first time [32]. Since then multiple linkage and candidate gene studies have been performed without major success. With the era of GWAS, the situation changed dramatically, and in the result of two GWAS, several replication efforts and fine mapping of immune-related loci, 57 independent non-HLA genetic variants, located in 39 loci, showed genome-wide significant association to coeliac disease ($<5 \times 10^{-8}$) [33, 34].

Studies on the co-occurrence of coeliac disease and IBD suggest that the frequency of both Crohn's disease and ulcerative colitis is elevated in coeliac disease patients [35–37]. Leeds et al. reported that the risk of developing IBD was tenfold higher in a population of coeliac disease patients than in the general population. However, the prevalence of coeliac disease in patients with IBD does not appear to be increased [16]. Similar prevalence of coeliac disease in IBD patients and in healthy individuals is expected from the similar frequency of HLA DQ2/DQ8, which is required for developing coeliac disease. Increased frequency of IBD in coeliac patients may suggest shared genetics and pathogenesis. Indeed, from 39 non-HLA loci associated to coeliac disease, 14 are reported to be associated to IBD (CD or UC or both) in the GWAS catalogue [1] (see Fig. 7.3). Six loci are shared in coeliac disease and Crohn's disease (*IL18RAP*, *BACH2*, *TAGAP*, *ZMIZ1*, *ZFP36L1*), four are shared with ulcerative colitis (*MMEL1/TNFRSF14*, *TNFAIP3*, *IL2/IL21*

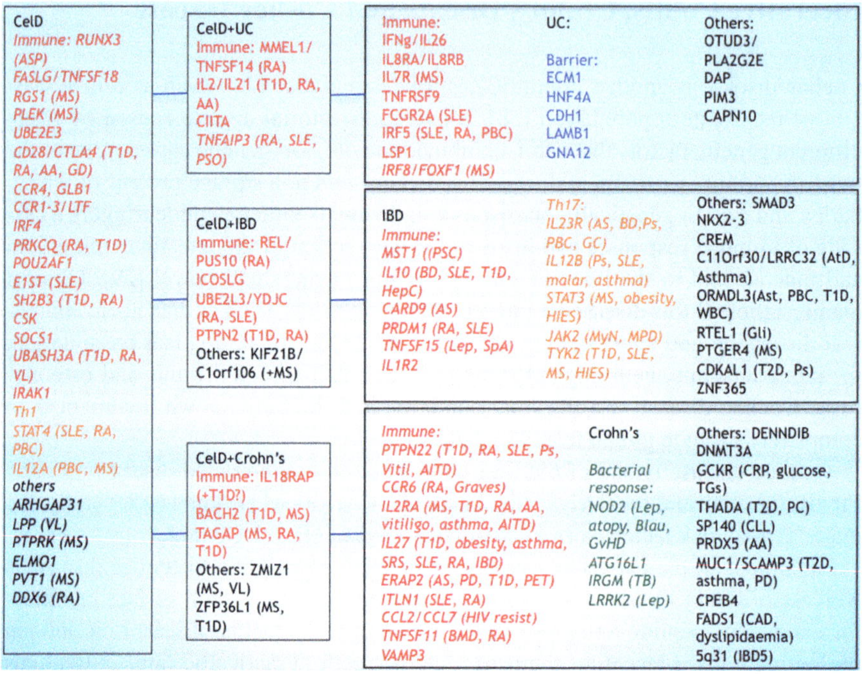

Fig. 7.3 Shared and non-shared genes in coeliac disease and IBD. Majority of loci associated to both coeliac disease and IBD contain genes associated to immune response, in particular genes involved in Th17 signalling. The cluster of genes involved in response to bacteria is specifically associated to Crohn's disease, whereas in ulcerative colitis a group of specific genes involved in intestinal barrier is associated. *Red*—immune mediated, *black*—others, *green*—innate, *orange*—Th17, *blue*—epithelial barrier

and *CIITA*) and five are coeliac-IBD loci—*REL/PUS10*, *ICOSLG*, *PTPN2*, *UBE2L3/YDJC* and *KIF21B/C1orf106*. For most of the shared loci, association is not limited to these diseases, but is also reported in other autoimmune and inflammatory conditions (see Fig. 7.4). Shared coeliac-IBD loci mostly point to the immune-mediated genes, whereas autophagy pathway, associated to Crohn's disease (including *NOD2*, *ATG16L1*, *IRGM* and *LRRK2* genes), and epithelial barrier genes, implicated in UC (*ECM1*, *HNF4A*, *CDH1*, *LAMB1* and *GNA12*), remain specific for these diseases and did not show association to coeliac disease (see Fig. 7.5).

Crohn's Disease and Infectious Diseases

When in 1913 Thomas Dalziel, an Edinburgh-trained surgeon working in Glasgow, first described chronic interstitial enteritis in a seminal paper in the British Medical Journal, he was struck by the pathophysiological similarities to Johne's disease [38]. This is a chronic granulomatous disease affecting the intestines of cattle; the

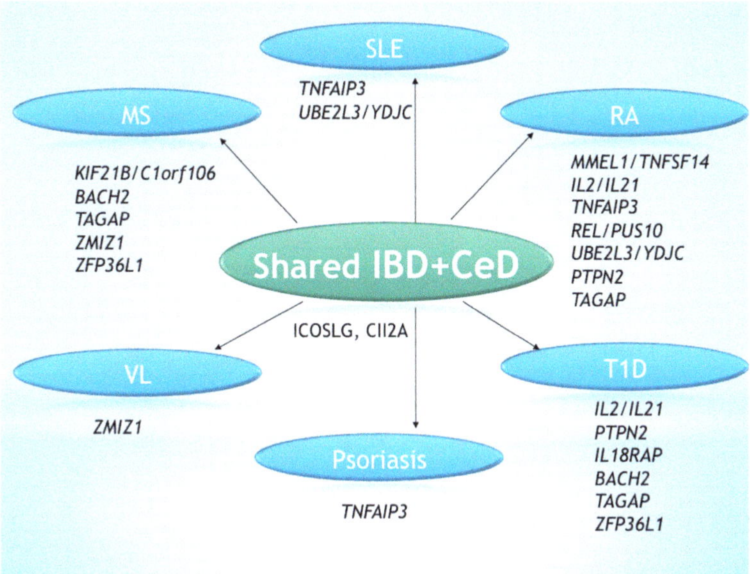

Fig. 7.4 Genetic overlap of coeliac disease and IBD with other immune-mediated diseases. Overlapping associated loci are indicated under the disease names. *T1D* type 1 diabetes, *VL* vitiligo, *RA* rheumatoid arthritis, *MS* multiple sclerosis, *SLE* systemic lupus erythematosus, *CeD* coeliac disease, *IBD* inflammatory bowel disease (either Crohn's disease or ulcerative colitis)

causative organism is known to be *Mycobacterium avium paratuberculosis* (MAP). The IBD clinical and research community retains a nagging doubt that a single infectious aetiological agent remains to be discovered, akin to Barry Marshall's description of the role Helicobacter pylori plays in peptic ulcer disease. MAP has been described as a putative agent, but the balance of evidence in the literature strongly argues against this, not least the definitive Australian anti-mycobacterial therapy trial which showed no difference in outcomes between long-term antibiotic therapy and placebo.

However, the role of microorganisms in the aetiology of Crohn's disease has come back into stark focus in the past 5 years. This is largely because we now have the requisite sequencing and analytical tools to study the gut microbiota. Indeed, the generally accepted hypothesis for the pathogenesis of IBD is of an exaggerated mucosa immune response to normal gut microbiota (as opposed to a single pathogenic organism) in genetically susceptible hosts. However, further interest has been derived directly from gene discovery. The *NOD2* association with ileal Crohn's disease, first described in 2001 [39, 40], initially directed intense research activity at the innate immune system. But, subsequent to the discovery of *ATG16L1* and *IRGM* as Crohn's disease susceptibility genes in 2008 [41, 42], attention shifted to understanding how autophagy contributes to disease pathogenesis. In a series of elegant function experiments, independent groups then demonstrated how these autophagy genes function in co-operation with *NOD2* to process intracellular

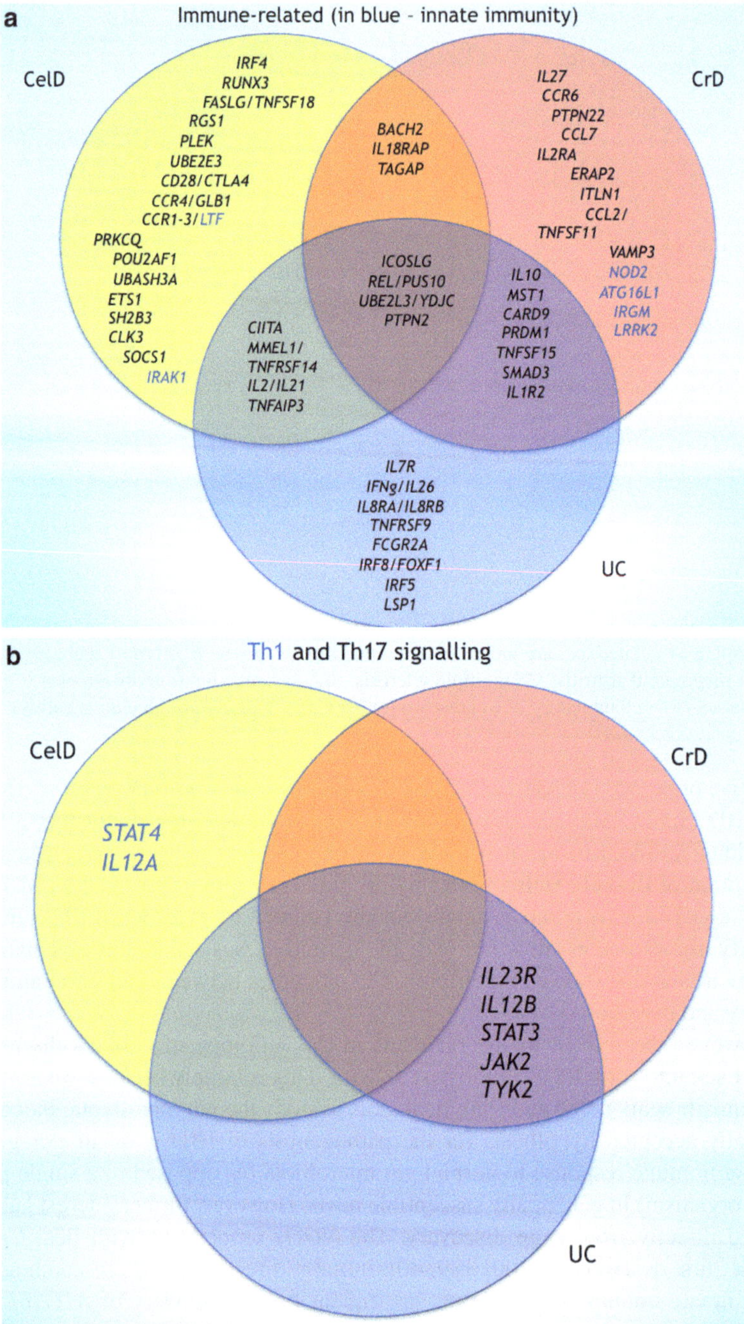

Fig. 7.5 (**a**) Major pathways involved in IBD and CelD and shared genes within the pathways. Immune-related genes. (**b**) Major pathways involved in IBD and CelD and shared genes within the pathways. Th1 (in *blue*) and Th17 (in *black*) pathway. (**c**) Major pathways involved in IBD and CelD and shared genes within the pathways. Epithelial barrier (in *blue*) and others (in *black*) pathway

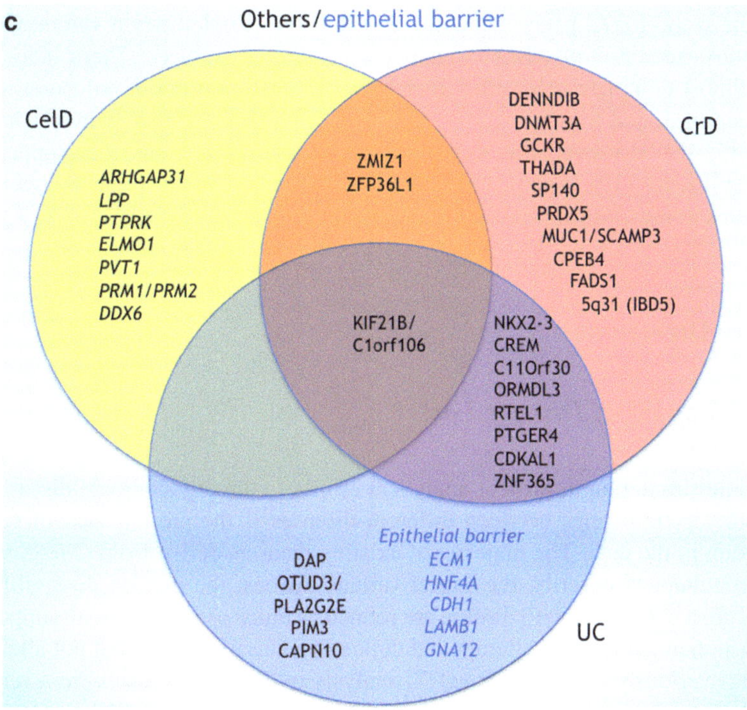

Fig. 7.5 (continued)

bacteria. Defects in any of these innate immunity genes ultimately result in defective killing (and hence persistence) of intracellular bacteria. The parallel description of the role *IRGM* plays in killing mycobacterial tuberculosis [43] turned some commentators back to those earlier thoughts of Dalziel.

This discussion took an expected twist, again from a large-scale GWAS. But this time the GWAS analysed the inherited contribution to mycobacterial leprosy. Studying patients with leprosy in China [44], Zhang and colleagues found that several Crohn's disease susceptibility genes were also associated with the development of leprosy, notably *NOD2*, *IL23R* and *RIPK2*. This is a complicated, emerging story as it is not yet clear from present data how the disease susceptibility alleles in these genes compare directly between Crohn's and leprosy. The allelic frequency and haplotypic structure vary considerably between the Caucasian populations with Crohn's disease and the HAN Chinese with leprosy (as it does between healthy controls in these two distinct populations). Whilst we await the precise detailing of this relationship, an intriguing hypothesis begins to emerge: the selective advantage conferred on these alleles by protecting individuals from the ravages of leprosy, have in the modern, Westernised 'hygenic' world predisposed people exposed to the right environmental/microbiological triggers to developing Crohn's disease.

Whilst much of this argument remains speculative at present, there are several other supporting strands of evidence from the recent gene discovery. One such example is the *FUT2* gene. About 20 % of healthy people carry a variant G428A that blocks secretion of ABO antigens onto epithelial surfaces. These individuals have almost complete protection against symptomatic norovirus infection as has been demonstrated through studies of various outbreaks of this 'winter vomiting bug' [45, 46]. However, this same genotype predisposes to the development of Crohn's disease [19]. Again we see a variant that potentially protects against an infectious disease whilst predisposing to the development of Crohn's disease.

Future Directions

Fine Mapping of Disease Loci

One major limitation to present analytical efforts to directly compare disease susceptibility genes/alleles between different diseases is the present lack of genetic resolution in the data. The majority of existing datasets derive from GWAS which do not attempt to describe the causal variant. At best the linkage disequilibrium blocks most associated with disease are refined as much as possible with supporting evidence then garnered to attempt and define the causative gene (but not allele(s)). This has variously involved (1) eQTL analysis in different datasets from relevant tissue/cell types, (2) literature-mining tools (e.g. GRAIL), (3) nsSNP analysis and (4) refinement based purely on genomic location.

Current experimental approaches to fine-map disease susceptibility loci include deep re-sequencing efforts, custom-designed gene chip arrays (immunochip) and genome-wide sequencing. Pertinent to IBD genetics are the deep re-sequencing efforts led by Mark Daly [47] and the IIBDGC immunochip experiment [48]. This chip array contains over 200 k SNPs chosen by experts in twelve immune-mediated diseases. For Crohn's disease and ulcerative colitis, the SNP selection was chosen to allow (1) deep replication of the meta-GWAS analysis and (2) fine mapping of confirmed disease susceptibility loci. All publically available SNPs (and some derived from pilot re-sequencing experiments) [34] were included. These data in themselves present a massive analytical challenge. Once solved for IBD, comparison with other diseases will require similar resolution in their genetic data. Similar analysis in coeliac disease has recently been published; analysis in other diseases is ongoing. The next step in these studies will be combination of datasets for the more specific identification of shared loci and alleles. Only then may a true picture of the genetic architecture and interrelatedness of complex genetic disease emerge. In the meantime we will continue to take the lead from the fascinating back stories starting to emerge from this present genetic revolution.

Correlation of associated variants with gene expression is another powerful tool for fine mapping of the associated variants on the gene level. Most of the signals associated to immune-related diseases localise in the plausible regulatory regions in proximity to or within the 5′ and 3′ ends of transcription. This suggests that the associated variants can influence the disease mechanism via deregulation of gene expression. Indeed, about 50 % of genetic variants associated to coeliac disease influence expression of the genes located in the proximity of the associated SNP cis-eQTL (expression quantitative trait locus mapping) [33]. Overall, genome-wide studies of gene expression linked with GWAS data have shown that approximately 40 % of disease-associated SNPs affect expression levels of genes in cis [49]. One example of using the eQTL as fine mapping is association of rs917997, located on LD block which includes *IL1RL1*, *IL18R1*, *IL18RAP* and *SLC9A4* genes. All four genes are attractive functional candidates for coeliac disease and IBD; however, strong effect of rs917997 on the expression level of *IL18RAP* allows prioritising the *IL18RAP* among other genes in the block [33].

Common Targets for Therapeutic Intervention

There is an emerging trend to personalise the way patients are treated based on clinical, (pharmaco) genetic, biochemical and serological markers. This field of precision medicine is still very much in its infancy. It aims to direct the use of a particular therapeutic agent to those individuals who are (a) predicted to gain most benefit from the drug 'a' at timepoint 'b' in disease course, (b) predicted to respond to drug 'a' and (c) predicted to have a favourable side effect profile to drug 'a'. This is of potential concern for big pharma who are keen to maximise the financial return to offset huge research and development costs (and of course to keep shareholders on side). Combining insights from the present line of enquiry (i.e. genetic/function overlap between different diseases) with 'intelligent' drug design (targeting only those individuals with e.g. defective autophagy, plus incorporation of pharmacogenetic readouts) may be the most fruitful way out of this catch.

Conclusions

Novel insights into disease biology and therapeutics have arisen from the study of the genetic overlap between the IBDs and other immune- and nonimmune-mediated complex diseases. The rapid phase of gene discovery now requires a period of consolidation for curation and refinement of susceptibility loci. Fine-mapping strategies currently being adopted include the immunochip effort, deep re-sequencing and whole-genome sequencing. We should not expect associated variants to be deleterious mutations, because individuals with intestinal inflammatory diseases born and develop normally. These are rather regulatory variants which all together and in combination with environmental factors lead to fine-tuning our immune and inflammatory response and can shift a balance towards a particular disease. Meanwhile, does our present lack of genetic resolution limit application of the present data? Probably not; we should be in a position to apply our knowledge for treatment modalities before knowing the particular associated variant. An understanding of the different shared and specific molecular pathways underlying diseases may already provide new targets for therapeutic interventions (e.g. IL23 pathway in Crohn's disease, ulcerative colitis, psoriasis and ankylosing spondylitis).

References

1. A catalog of published genome-wide association studies. Division of Genomic Medicine. http://www.genome.gov/gwastudies/index.cfm?pageid=26525384#searchForm
2. Jostins L, Ripke S, Weersma RK, Duerr RH, McGovern DP, Hui KY et al (2012) Host-microbe interactions have shaped the genetic architecture of inflammatory bowel disease. Nature 491(7422):119–124, Epub 2012/11/07
3. Lees CW, Barrett JC, Parkes M, Satsangi J (2011) New IBD genetics: common pathways with other diseases. Gut 60(12):1739–1753, Epub 2011/02/09
4. Bernstein CN, Wajda A, Blanchard JF (2005) The clustering of other chronic inflammatory diseases in inflammatory bowel disease: a population-based study. Gastroenterology 129(3):827–836, Epub 2005/09/07
5. Weng X, Liu L, Barcellos LF, Allison JE, Herrinton LJ (2007) Clustering of inflammatory bowel disease with immune mediated diseases among members of a northern California-managed care organization. Am J Gastroenterol 102(7):1429–1435, Epub 2007/04/18
6. Cohen R, Robinson D Jr, Paramore C, Fraeman K, Renahan K, Bala M (2008) Autoimmune disease concomitance among inflammatory bowel disease patients in the United States, 2001–2002. Inflamm Bowel Dis 14(6):738–743, Epub 2008/02/27
7. Palm O, Moum B, Ongre A, Gran JT (2002) Prevalence of ankylosing spondylitis and other spondyloarthropathies among patients with inflammatory bowel disease: a population study (the IBSEN study). J Rheumatol 29(3):511–515, Epub 2002/03/23
8. Wright V, Watkinson G (1965) Sacro-iliitis and ulcerative colitis. Br Med J 2(5463):675–680, Epub 1965/09/18
9. Hammer B, Ashurst P, Naish J (1968) Diseases associated with ulcerative colitis and Crohn's disease. Gut 9(1):17–21, Epub 1968/02/01
10. Pugh SM, Rhodes J, Mayberry JF, Roberts DL, Heatley RV, Newcombe RG (1979) Atopic disease in ulcerative colitis and Crohn's disease. Clin Allergy 9(3):221–223, Epub 1979/05/01
11. Yates VM, Watkinson G, Kelman A (1982) Further evidence for an association between psoriasis, Crohn's disease and ulcerative colitis. Br J Dermatol 106(3):323–330, Epub 1982/03/01
12. Gupta G, Gelfand JM, Lewis JD (2005) Increased risk for demyelinating diseases in patients with inflammatory bowel disease. Gastroenterology 129(3):819–826, Epub 2005/09/07
13. Olsson R, Danielsson A, Jarnerot G, Lindstrom E, Loof L, Rolny P et al (1991) Prevalence of primary sclerosing cholangitis in patients with ulcerative colitis. Gastroenterology 100 (5 Pt 1):1319–1323
14. Nitzan O, Elias M, Saliba WR (2006) Systemic lupus erythematosus and inflammatory bowel disease. Eur J Intern Med 17(5):313–318
15. Black H, Mendoza M, Murin S (2007) Thoracic manifestations of inflammatory bowel disease. Chest 131(2):524–532
16. Casella G, D'Inca R, Oliva L, Daperno M, Saladino V, Zoli G et al (2010) Prevalence of celiac disease in inflammatory bowel diseases: an IG-IBD multicentre study. Dig Liver Dis 42(3): 175–178
17. Gao Y, Linet MS, Gridley G, Mellemkjaer L, Hemminki K, Goldin LR et al (2008) Shared susceptibility for celiac disease and inflammatory bowel disease? Scand J Gastroenterol 43(10):1279–1280

18. Leeds JS, Horoldt BS, Sidhu R, Hopper AD, Robinson K, Toulson B et al (2007) Is there an association between coeliac disease and inflammatory bowel diseases? A study of relative prevalence in comparison with population controls. Scand J Gastroenterol 42(10):1214–1220
19. Franke A, McGovern DP, Barrett JC, Wang K, Radford-Smith GL, Ahmad T et al (2010) Genome-wide meta-analysis increases to 71 the number of confirmed Crohn's disease susceptibility loci. Nat Genet 42(12):1118–1125
20. Anderson CA, Boucher G, Lees CW, Franke A, D'Amato M, Taylor KD et al (2011) Meta-analysis identifies 29 additional ulcerative colitis risk loci, increasing the number of confirmed associations to 47. Nat Genet 43(3):246–252, Epub 2011/02/08
21. Wen L, Ley RE, Volchkov PY, Stranges PB, Avanesyan L, Stonebraker AC et al (2008) Innate immunity and intestinal microbiota in the development of type 1 diabetes. Nature 455(7216): 1109–1113
22. Coenen MJ, Trynka G, Heskamp S, Franke B, van Diemen CC, Smolonska J et al (2009) Common and different genetic background for rheumatoid arthritis and coeliac disease. Hum Mol Genet 18(21):4195–4203
23. Hinks A, Martin P, Flynn E, Eyre S, Packham J, Barton A et al (2010) Investigation of type 1 diabetes and coeliac disease susceptibility loci for association with juvenile idiopathic arthritis. Ann Rheum Dis 69(12):2169–2172
24. Smyth DJ, Plagnol V, Walker NM, Cooper JD, Downes K, Yang JH et al (2008) Shared and distinct genetic variants in type 1 diabetes and celiac disease. N Engl J Med 359(26): 2767–2777, Epub 2008/12/17
25. Zhernakova A, Stahl EA, Trynka G, Raychaudhuri S, Festen EA, Franke L et al (2011) Meta-analysis of genome-wide association studies in celiac disease and rheumatoid arthritis identifies fourteen non-HLA shared loci. PLoS Genet 7(2):e1002004
26. Kumar V, Wijmenga C, Withoff S (2012) From genome-wide association studies to disease mechanisms: celiac disease as a model for autoimmune diseases. Semin Immunopathol 34(4): 567–580
27. Ahern PP, Schiering C, Buonocore S, McGeachy MJ, Cua DJ, Maloy KJ et al (2010) Interleukin-23 drives intestinal inflammation through direct activity on T cells. Immunity 33(2):279–288, Epub 2010/08/25
28. Sollid LM (2002) Coeliac disease: dissecting a complex inflammatory disorder. Nat Rev Immunol 2(9):647–655
29. Jabri B, Sollid LM (2009) Tissue-mediated control of immunopathology in coeliac disease. Nat Rev Immunol 9(12):858–870
30. Kim CY, Quarsten H, Bergseng E, Khosla C, Sollid LM (2004) Structural basis for HLA-DQ2-mediated presentation of gluten epitopes in celiac disease. Proc Natl Acad Sci U S A 101(12): 4175–4179
31. Tjon JM, van Bergen J, Koning F (2010) Celiac disease: how complicated can it get? Immunogenetics 62(10):641–651
32. Mulder DJ (1974) Letter: HL-A antigens and coeliac disease. Lancet 2(7882):727
33. Dubois PC, Trynka G, Franke L, Hunt KA, Romanos J, Curtotti A et al (2010) Multiple common variants for celiac disease influencing immune gene expression. Nat Genet 42(4): 295–302
34. Trynka G, Hunt KA, Bockett NA, Romanos J, Mistry V, Szperl A et al (2011) Dense genotyping identifies and localizes multiple common and rare variant association signals in celiac disease. Nat Genet 43(12):1193–1201
35. Cottone M, Marrone C, Casa A, Oliva L, Orlando A, Calabrese E et al (2003) Familial occurrence of inflammatory bowel disease in celiac disease. Inflamm Bowel Dis 9(5):321–323, Epub 2003/10/14
36. Tursi A, Giorgetti GM, Brandimarte G, Elisei W (2005) High prevalence of celiac disease among patients affected by Crohn's disease. Inflamm Bowel Dis 11(7):662–666
37. Yang A, Chen Y, Scherl E, Neugut AI, Bhagat G, Green PH (2005) Inflammatory bowel disease in patients with celiac disease. Inflamm Bowel Dis 11(6):528–532

38. K. D (1913) Classic articles in colonic and rectal surgery. BMJ
39. Hugot JP, Chamaillard M, Zouali H, Lesage S, Cezard JP, Belaiche J et al (2001) Association of NOD2 leucine-rich repeat variants with susceptibility to Crohn's disease. Nature 411(6837): 599–603
40. Ogura Y, Bonen DK, Inohara N, Nicolae DL, Chen FF, Ramos R et al (2001) A frameshift mutation in NOD2 associated with susceptibility to Crohn's disease. Nature 411(6837): 603–606
41. Hampe J, Franke A, Rosenstiel P, Till A, Teuber M, Huse K et al (2007) A genome-wide association scan of nonsynonymous SNPs identifies a susceptibility variant for Crohn disease in ATG16L1. Nat Genet 39(2):207–211
42. Parkes M, Barrett JC, Prescott NJ, Tremelling M, Anderson CA, Fisher SA et al (2007) Sequence variants in the autophagy gene IRGM and multiple other replicating loci contribute to Crohn's disease susceptibility. Nat Genet 39(7):830–832, Epub 2007/06/08
43. Singh SB, Davis AS, Taylor GA, Deretic V (2006) Human IRGM induces autophagy to eliminate intracellular mycobacteria. Science 313(5792):1438–1441
44. Han JW, Zheng HF, Cui Y, Sun LD, Ye DQ, Hu Z et al (2009) Genome-wide association study in a Chinese Han population identifies nine new susceptibility loci for systemic lupus erythematosus. Nat Genet 41(11):1234–1237
45. Carlsson B, Kindberg E, Buesa J, Rydell GE, Lidon MF, Montava R et al (2009) The G428A nonsense mutation in FUT2 provides strong but not absolute protection against symptomatic GII.4 Norovirus infection. PLoS One 4(5):e5593
46. Lindesmith L, Moe C, Marionneau S, Ruvoen N, Jiang X, Lindblad L et al (2003) Human susceptibility and resistance to Norwalk virus infection. Nat Med 9(5):548–553
47. Rivas MA, Beaudoin M, Gardet A, Stevens C, Sharma Y, Zhang CK et al (2011) Deep resequencing of GWAS loci identifies independent rare variants associated with inflammatory bowel disease. Nat Genet 43(11):1066–1073, Epub 2011/10/11
48. http://www.ibdgenetics.org/
49. Fehrmann RS, Jansen RC, Veldink JH, Westra HJ, Arends D, Bonder MJ et al (2011) Trans-eQTLs reveal that independent genetic variants associated with a complex phenotype converge on intermediate genes, with a major role for the HLA. PLoS Genet 7(8):e1002197

Chapter 8
Molecular Profiling of IBD Subtypes/Response to Therapy

Isabelle Cleynen and Séverine Vermeire

Abstract There is an unmet need for molecular markers to define and predict disease outcome as reported clinical features are not accurate enough and as more therapeutic classes become available. It is the ultimate goal of physicians treating patients with IBD, to have a molecular (serum, DNA, tissue-based) profile of the patient which allows the most appropriate management: What is the most likely course of disease? Which is the most appropriate therapy with highest chances of success? Which should be the intensity of follow-up? Pharmacogenetic research in IBD has also had only modest success so far. One reason is that identifying molecular markers which influence the response to a drug is more difficult than, for instance, the study of genetic markers that influence toxicity. In addition, treatment response in a heterogenic disease like IBD is influenced by many factors such as disease duration, behavior, and severity. Only TPMT testing prior to start of thiopurine analogues has shown clinical applicability, but does not replace blood monitoring during treatment. What is needed in pharmacogenetic and other predictive studies to advance the field are patients treated with standardized doses of the drug and fixed endpoints and criteria for response. The setting of a clinical trial may well be the preferred method for this, and attempts to collect DNA from these patients should be enforced. Recently, the International IBD Genetics Consortium (IIBDGC) has prioritized a number of projects on prediction of side effects of therapy (anti-TNF and demyelinization, 5-ASA and nephrotoxicity, azathioprine-induced pancreatitis and leucopenia) and has collected an unprecedented number of DNA samples worldwide.

I. Cleynen, Ph.D. • S. Vermeire, M.D., Ph.D. (✉)
Department of Clinical and Experimental Medicine, TARGID, KU Leuven, Herestraat 49-3000, Leuven, Belgium

Division of Gastroenterology, University Hospitals Leuven, Herestraat 49-3000, Leuven, Belgium
e-mail: Severine.Vermeire@uzleuven.be

M. D'Amato and J.D. Rioux (eds.), *Molecular Genetics of Inflammatory Bowel Disease*, DOI 10.1007/978-1-4614-8256-7_8, © Springer Science+Business Media New York 2013

Introduction

Crohn's disease (CD) and ulcerative colitis (UC), collectively called inflammatory bowel diseases (IBD), are disorders of multifactorial etiology and present as a multitude of phenotypes, clinical behaviors, and severity. The heterogeneity of the disease has important implications towards clinical management: patients with a more severe disease course might benefit from early introduction of immunomodulators and/or biologicals, while patients with a favorable disease prognosis could be spared from intense treatment and possible side effects.

Once patients are identified as requiring early aggressive therapy, a new need arises: the need to predict the response to the therapeutic strategy. Given the potential toxicity, loss of response and antibody development, and high cost of treatment with biologicals, reliable prediction of response may enable better selection of candidates for treatment with biologicals, thus decreasing morbidity in patients with a low likelihood of response and enhancing cost-effective use.

Patient Classification

An accurate classification of a complex disease, like IBD, has several benefits with respect to patient counseling, assessing risk for disease progression, and particularly with respect to choosing the most appropriate therapy for an individual patient. Tailoring therapy to the individual patient should be a natural behavior when managing IBD patients. This already starts at diagnosis since important differences in initial disease presentation exist, and furthermore, during the disease course, not all patients have the same disease behavior. In addition, other factors including specific contraindications, anticipation of side effects, and special circumstances (traveling, pregnancy, cancer, long-lasting remission) will influence not only the choice of the drugs but also their dosing and the choice between mono- or combination therapy.

Also for basic scientists, classifications allowing better understanding the pathophysiology of the different manifestations of CD and UC are welcomed.

The Classic Clinical Classification: From Rome to Vienna, Montreal, and Paris

Over the years, three separate phenotypic classifications have been reported, all resulting from international working groups. Even before the first Rome classification in 1988, Farmer et al. had reported on an anatomical classification of CD into ileal, colonic, and ileocolonic disease [1]. Subsequently, the team from Mount Sinai in New York showed the importance of distinguishing the different behaviors of the disease [2]. All three international working groups (Rome in 1988, Vienna in 1998,

Table 8.1 Vienna and Montreal classification for Crohn's disease

	Vienna	Montreal
Age at diagnosis	A1 below 40 years	A1 below 16 years
	A2 above 40 years	A2 between 17 and 40
		A3 above 40 years
Location	L1 ileal	L1 ileal
	L2 colonic	L2 colonic
	L3 ileocolonic	L3 ileocolonic
	L4 upper	L4 isolated upper disease[a]
Behavior	B1 non-stricturing, non-penetrating	B1 non-stricturing, non-penetrating
	B2 stricturing	B2 stricturing
	B3 penetrating	B3 penetrating
		p perianal disease modifier[b]

[a]L4 is a modifier that can be added to L1–L3 when concomitant upper GI disease is present
[b]"p" is added to B1–B3 when concomitant perianal disease is present

and Montreal in 2005) agreed that not only the anatomic location and the clinical behavior were important variables to include in the classification of CD patients but also the age at onset of disease. The classification of Montreal modified the Vienna Crohn's disease classification, mainly by acknowledging that perianal disease should be classified separately from internal fistulizing behavior and by further refining age at onset [3]: early onset of disease is now categorized separately for patients diagnosed below 16 years, whereas A2 and A3 account, respectively, for ages of diagnosis 17–40 and above 40 years (Table 8.1). This was an important modification to add, as several (genetic) studies have linked the young subgroup to specific genetic markers. The decision to separate perianal fistulizing disease from fistulizing disease as a whole was taken from observations that perianal fistulizing disease is not necessarily associated with intestinal fistulizing disease. Also, perianal disease in more recent studies was shown to be a marker of bad outcome (Table 8.1).

In 2011, pediatric IBD experts proposed a modification of the Montreal classification to allow the more dynamic changes in children including failure to thrive. This results in the Paris classification which reported as most important modifications further division of age at diagnosis into A1a (0 to <10 years), A1b (10 to <17 years), A2 (17–40 years), and A3 (>40 years); second, distinction of disease above the distal ileum as L4a (proximal to ligament of Treitz) and L4b (ligament of Treitz to above distal ileum); third, allowing both stenosing and penetrating disease to be classified in the same patient (B2B3); and last, to report on the presence or absence of growth failure (G(1) vs. G(0), respectively) [4].

For UC, the first classification was proposed in Montreal. It was concluded that age at onset, disease extent, and severity were important parameters for subgrouping of patients (Tables 8.2 and 8.3). Also severity of disease was a parameter which was included in the classification of UC and was again driven mainly by the fact that this dictates the choice of treatment (oral vs. intravenous, steroids or not) (Table 8.3).

Table 8.2 Montreal classification of extent of UC

Extent		Anatomy
E1	Ulcerative proctitis	Involvement limited to the rectum, i.e., proximal extent of inflammation is distal to the rectosigmoid junction
E2	Left-sided ulcerative colitis (distal UC)	Involvement limited to a proportion of the colorectum distal to the splenic flexure
E3	Extensive ulcerative colitis (pancolitis)	Involvement extends proximal to the splenic flexure

Table 8.3 Montreal classification of severity of UC

Severity		Definition
S0	Clinical remission	Asymptomatic
S1	Mild ulcerative colitis	≤4 stools per day (with or without blood), absence of systemic disease, normal inflammatory markers (ESR)
S2	Moderate ulcerative colitis	>4 stools per day, but with minimal signs of systemic toxicity
S3	Severe ulcerative colitis	≥6 bloody stools daily, pulse rate ≥90 bpm, temperature ≥37.5 °C, hemoglobin <10.5 g/dL, and ESR ≥30 mm per hour

Towards a New Clinical Classification: Prediction of a "Complicated Disease Course"

In 2002, Cosnes et al. demonstrated in a large tertiary referral cohort that the majority of CD patients evolve over time from a noncomplicated (non-stricturing, non-penetrating) disease behavior to a more complicated disease with stenosis and/or fistulas [5]. The group of Liege also confirmed that—in contrast to the disease location—the clinical behavior changes as disease duration becomes longer [6]. Together with the introduction of anti-TNF agents over a decade ago, this has added a new discussion to the field: can we predict disease progression and thus patient selection for these therapies? The primary goal of all treatments for patients is remission, no steroids, and mucosal healing, and this goal should be achieved in a timely manner. Recent clinical trials in CD clearly illustrate that treatment with biological agents in combination with azathioprine early in the disease course results in higher steroid-free remission rates and higher mucosal healing rates. However, not all patients are good candidates for such a strategy as immunomodulators and biological agents are not free from adverse events, including opportunistic infections and lymphomas. Therefore, predictors of complicated disease behavior are needed to help select those patients who will benefit most from early aggressive therapy.

Predicting the course of the disease has been challenging. Most attempts to identify prognostic markers have focused on clinical parameters.

Beaugerie et al. defined a young age at diagnosis (<40 years), the presence of perianal disease, and the requirement for steroids at diagnosis as risk factors for a subsequent 5-year disabling course of CD [7]. This observation was partially confirmed by Loly et al. who identified ileocolonic involvement as another predictive factor of a disabling disease course [8]. Using another definition, these authors also demonstrated that stricturing behavior and weight loss at diagnosis were both predictive of a severe disease course. In two independent studies using population-based cohorts from New Zealand and Olmsted County, perianal involvement at diagnosis was confirmed not only as predictive of complicated disease behavior but also of more rapid progression of CD [9, 10]. Other clinical features which point towards a more severe disease with a potential bad outcome include extensive small bowel involvement, deep ulcers at endoscopy, and the presence of growth failure in children [11].

Although a number of clinical risk factors for complicated disease have been described that would be useful in the daily decision making of physicians regarding whether or not to treat "more aggressively," they lack specificity and hence are not useful for prediction of disease course in clinical practice [7, 8]. Another problem of these studies is that definitions for complicated, severe, and disabling CD vary between them, for example, from the need for two courses of steroids to formation of ileostomy.

Value of Serology in Classifying Patients

Given the limitations of clinical factors in identifying patients at risk of a complicated disease course, serological and genetic markers have been explored as alternative predictors. In the past 20 years, several serologic markers have been associated with IBD and CD in particular. These markers are antibodies directed against bacterial antigens, including anti-Saccharomyces cerevisiae antibodies (ASCA), anti-laminarobioside carbohydrate antibodies (ALCA), anti-laminarin carbohydrate antibodies (anti-L), anti-chitobioside carbohydrate antibodies (ACCA), anti-chitin carbohydrate antibody (anti-C), anti-mannobioside carbohydrate antibodies (AMCA), anti-outer membrane protein C of Escherichia coli antibodies (anti-OmpC), anti-Pseudomonas fluorescens antibodies (anti-I2), and anti-flagellin antibodies (CBir1). In UC a strong antibody response is found against neutrophil antigens (pANCA) [12–14]. Although these serologic markers were originally investigated as a diagnostic tool, their clinical value may turn out to be their association with a "complicated" disease course (development of strictures, fistulas, abscesses, and necessity for resections). However, a conditio sine qua non before pronouncing on the predictive value of these markers is the stability of antibody responses over time, and this has not always been the case. The observed increased antibody responses in CD patients with longer disease duration in a large Belgian cohort highlight the importance of prospective longitudinal studies [15]. Up to now, one relatively small pediatric longitudinal study has reported a relation between the presence and titer of antibodies (ASCA, anti-OmpC, anti-I2, anti-CBir1), at a time when no fistula or strictures were clinically apparent, and the subsequent development of complicated disease behavior during a median follow-up of 18 months [16]. Similar conclusions were drawn by Rieder et al. in an adult population [17].

Molecular Profiling of IBD Subtypes

Genetic Profiling

Compared to serologic factors genetic factors are more appealing for risk stratification since they are present long before the onset of the disease and before any environmental factor plays a role. They remain stable over time and are unaffected by disease flares. Moreover, as several studies have correlated microbial seroreactivity with genetic mutations in pattern recognition receptors [18, 19], genetic markers might well prove superior to antibodies for use in daily clinical practice.

Many studies have suggested that *NOD2* variants are associated with a shorter time to onset of stricturing disease as well as the need for surgery [20–22]. Specifically, a large study by Seiderer et al. demonstrated a strong risk increase for 1,007 fs homozygotes ($n=19$) for ileal stenoses and risk for surgery [23]. The same group confirmed these findings in a prospective study [24]. Recently, a meta-analysis was performed to get an accurate estimate of the prognostic power of *NOD2* in complicated CD [25]. They found that while the predictive power associated with a single *NOD2* mutation is weak, the presence of two *NOD2* mutations had 98 % specificity for complicated disease. The sensitivity however remained poor. It should be noted that the meta-analysis was limited by the fact that most studies included did not differentiate between 1,007 fs heterozygotes and 1,007 fs homozygotes.

With the identification of several other susceptibility genes through genome-wide association studies (GWA studies), the research on genetic markers and their role in disease progression has recently gained more attention. One analysis demonstrated a significant increase in frequency of the G-allele of *ATG16L1* in patients with ileal disease with or without colonic involvement (61.7 %, $n=652$) compared to those with pure colonic disease (52.2 %, $n=253$, $p=0.00025$) [26]. In a large Dutch cohort, Weersma et al. not only described a higher odds ratio for CD susceptibility with an increasing number of risk alleles in some of the known susceptibility variants (*NOD2*, *IBD5*, *DLG5*, *ATG16L1*, *IL23R*) but also a more severe disease, a greater need for surgery, and a younger age at onset [27]. Henckaerts et al. studied the influence of known CD-associated susceptibility loci on changes in disease behavior. They reported a higher risk of stricturing behavior in patients who were homozygous for the rs1363670 G-allele in AK097548, located near *IL12B* [28]. The GG genotype was also associated with a shorter time to development of strictures, especially in patients with ileal involvement. In the same cohort, male patients carrying at least one rs12704036 T-allele in a gene desert on chromosome 7q36 had the shortest time to development of non-perianal fistula, while presence of a C-allele at the *CDKAL1* rs6908425 SNP and absence of *NOD2* variants were both independently associated with development of perianal fistula, particularly in smokers with colonic involvement.

Although these retrospective cohort studies point towards a role for CD-associated polymorphisms in disease progression, they often lack confirmation in other (prospective) cohorts. In fact, the only gene reproducibly associated with a more severe disease course is *NOD2*. So why were these studies not more successful? The SNPs tested mostly are known susceptibility markers for IBD when compared to healthy controls and might thus be more generally applicable to *all* IBD patients and less useful to differentiate patients. It is also hypothesized that CD genetics consists of disease susceptibility genes/loci on the one side and disease modifying genes/loci on the other. Work from the International IBD Genetics Consortium (IIBDGC) underscores this idea: it was shown that reanalysis of GWA study data in function of disease behavior (mild vs. aggressive disease) identified a number of SNPs that specifically predispose to a more aggressive disease course in CD. The same SNPs however were not associated with the disease overall in the original GWA study [29]. So far, ca 160 loci have been associated with IBD (including the results of the GWA studies and Immunochip meta-analysis) [30]. For most of the identified loci, the causal variant(s) still need(s) to be determined. Another problem the genotype-subphenotype analyses are faced with is that the definition of subphenotypes is often subjective and different in different studies.

It should be considered that genetic markers alone will probably never fully explain or predict evolution of disease, because of the incomplete penetrance of phenotypes, the rather low absolute risk in the general population, and the role of environmental factors in shaping the disease (gene-environment interactions).

We recently tried to reclassify the disease purely on the basis of genetic variants [31]. A number of subgroups could be identified in CD patients, which differed from healthy controls and which could also not be detected by the classical phenotypes of disease anatomy, behavior, or age at onset. Validation of these findings in the large dataset of the IIBDGC is ongoing.

Transcriptional Profiling

In oncology, gene expression profiling has been successfully used to identify transcriptional signatures that predict several aspects of disease behavior, including risk of metastasis and response to chemotherapy [32–34]. These gene expression-based biomarkers have also been translated into clinical practice. In autoimmune and inflammatory conditions, such techniques have generally not detected signatures with equivalent prognostic utility. Typically, the tissues examined (PBMC, mucosal biopsies) are heterogeneous and, hence, any transcriptional variation detected will predominantly reflect differences in the cellular composition between samples. Researchers from the University of Cambridge recently reported a transcriptional signature in separated CD8+ *T* cells which predicted prognosis in CD and UC patients [35]. Interestingly, the same signature was previously also found to predict disease prognosis in systemic lupus erythematosus and in ANCA-associated vasculitis patients [36]. There was a higher incidence of relapsing disease in the subgroup of patients who had an elevated expression of genes involved in antigen-dependent T-cell responses, including signaling initiated by both IL-7 and TCR ligation. The fact that this signature could be found across several autoimmune and immune-mediated diseases suggests that, although being distinct autoimmune and inflammatory conditions, the course of these diseases may be influenced by common pathways. These subgroups, which can be identified by measuring expression of only three genes, raise the prospect of individualized therapy and suggest new potential therapeutic targets in autoimmunity.

Response to Therapy

As in almost all human diseases necessitating medical therapy, a variable response is also observed for most drugs used in IBD. Between 20 and 30 % of patients are refractory to any given medication despite optimal dose and duration. Besides the response, side effects and toxicity are also variable. The need for prediction of response to therapy is as pressing as the need to predict the disease course and will become even more important as more therapeutic classes (anti-TNF, cell adhesion molecule inhibitors) are becoming available.

Clinical Predictors

Several clinical predictors have been suggested. Concurrent use of immunosuppressive medication, younger age, nonsmoking, shorter duration of disease, and colonic location of disease are confirmed clinical predictors of response to anti-TNF (mostly infliximab) in independent cohorts. Clinical predictors alone, however, do not adequately predict the response to infliximab in many patients. Moreover, the mechanism by which they influence response is poorly understood.

Serological Predictors

Supplementary to the clinical assessment, serological markers could be useful to follow up the response to therapy.

Serological Antibodies

Similar as for prediction of disease course, serological antibodies have been suggested as predictor for response to therapy. In the case of prediction of response to anti-TNF therapy, several studies have been performed. No clear association between patterns of serological markers and response to anti-TNF therapy could be shown, with some studies showing an association while others did not [12, 37, 38].

Inflammatory Proteins

The most commonly used acute phase protein is C-reactive protein (CRP). CRP is a pentameric molecule produced and released by hepatocytes on triggering by the cytokines interleukin (IL)-6, IL-1, and TNFα. CRP plays an important role as non-invasive inflammatory marker in patients with IBD and especially in patients with CD. However, its upregulation is heterogeneous; where a strong CRP response has been observed in CD, this response is only modest in UC [39–42]. The Leuven group investigated whether CRP is helpful in optimizing therapy with infliximab in the individual patient with moderate to severe CD. In patients on infliximab maintenance therapy, they showed that more patients with high baseline levels of CRP responded to infliximab than patients with normal levels (90.8 % vs. 82.6 %). Early normalization of CRP levels in turn correlated with sustained long-term response, and CRP levels remained significantly higher among patients who lost their response to infliximab, compared with those with a sustained response. CRP thus appears a good marker of disease activity in patients treated with infliximab, with increased levels of CRP indicating mucosal inflammation and a likelihood of clinical relapse.

Molecular Predictors

Genetics

The success of genetic markers in predicting outcome to CD or UC therapy has been limited, in contrast to other fields such as oncology, where molecular markers have demonstrated clinical utility in predicting response to chemotherapy. For example, the response to cetuximab, a monoclonal antibody to epidermal growth factor receptor in metastatic colorectal cancer is influenced by the KRAS mutation status, as the benefit of cetuximab seems limited to patients with KRAS wild-type tumors [43]. Likewise, germ line mutations may also correlate with clinical outcome to chemotherapy. A sub-analysis of a large phase III study with bevacizumab (Avastin), a humanized monoclonal antibody to vascular endothelial growth factor (VEGF), in metastatic pancreatic cancer, showed that overall survival and progression-free survival were influenced by SNPs in the tyrosine kinase domain of the VEGF receptor-1 [44]. GWAS technology has been successfully applied to identify genetic risk variants for flucloxacillin-induced hepatotoxicity. With only 51 cases of flucloxacillin-induced liver injury and 282 controls, a clear association was seen for the major histocompatibility complex (MHC) region at rs2395029, a marker in complete linkage disequilibrium with *HLA-B*5701*. Likewise, a GWAS study in 85 subjects with statin-induced myopathy and 90 controls identified common variants in *SLCO1B1* that were strongly associated with an increased risk of statin-induced myopathy [45, 46].

In oncology, and in (auto-)immune disease, the response to treatment can be influenced by polymorphisms in specific drug-metabolizing enzymes (affecting active drug concentrations [47]) and drug sensitivity proteins (drug receptor genetic variants) and by heterogeneity in the patient's genetic background.

Most research on genetic predictive factors for drug response in IBD has been conducted on azathioprine, corticosteroids, and infliximab.

Azathioprine

The only class of drugs where genetic testing for response/adverse event prediction is useful and recommended are the thiopurine analogues. Azathioprine is metabolized by the enzyme thiopurine methyl transferase (TPMT), and the activity of this enzyme is under genetic control [48]. Genetic variants in the *TPMT* gene result in lower TPMT enzyme activity, and this is associated with an increased risk for hematopoietic toxicity. In practice, genotyping the most common *TPMT* variants or measuring TPMT enzyme activity can be done. Both techniques have advantages, and which technique is used is in part dependent on the availability. TPMT enzyme activity is measured in red blood cells with a radiochemical or high-performance liquid chromatography assay. The results can be influenced by blood transfusions, but also other drugs may interfere with the TPMT enzyme activity (e.g., diuretics, 5-aminosalicylic acid). TPMT enzyme activity will identify patients with high TPMT activity that metabolize 6-mercaptopurine (6-MP) to 6-methyl-MP and therefore may be resistant to treatment with thiopurine drugs.

Genotyping is easier, but genotypes do not fully correlate with the enzyme activity, especially in the case of wild-type (some patients will have reduced TPMT activity) or heterozygous (some will have a normal TPMT activity) individuals [49–52]. Therefore, measuring TPMT enzyme activity will give a more accurate picture. In patients, in whom *TPMT* genetic testing is performed, azathioprine or 6-MP can be initiated at normal doses (2.5 and 1.5 mg/kg, respectively) in the case of a wild-type genotype or normal TPMT enzyme activity. When TPMT activity is intermediate, or when patients are heterozygous for the common *TPMT* variants, a dose reduction of 50 % is recommended. Finally, low or absent TPMT activity and/or compound heterozygous/homozygous mutant patients should not be initiated on azathioprine or 6-MP, given the high risk of myelotoxicity. Socioeconomic analyses show that TPMT phenotyping or genotyping is cost-effective [53]. However, also in patients with a normal TPMT activity, hematologic toxicity can develop. Monitoring of blood counts and liver transaminases thus remains necessary in all patients, as long as they are taking this drug.

Corticosteroids

Corticosteroids (CS) are potent inhibitors of T-cell activation and cytokine secretion and mediate their anti-inflammatory effect through binding the intracellular glucocorticoid (GC) receptor (GRα). They are effective as induction therapy in moderate to severe active UC and CD. The long-term outcome data look less promising, however, as European as well as North American studies showed that 25–30 % of patients became steroid dependent and 20 % steroid resistant within 1 year [54, 55].

Several mechanisms have been proposed for the resistance to CS. Overexpression of MDR1 (multidrug resistance) and subsequent elevated P-glycoprotein-mediated efflux of the drug was the first [56]. The *MDR1* gene encodes the drug efflux pump P-glycoprotein-170 (Pgp-170) and is expressed on the surface of lymphocytes and intestinal epithelial cells. *MDR1* maps to the IBD susceptibility locus on chromosome 7, and *mdr1* knockout mice spontaneously develop colitis, making it an excellent functional and positional candidate gene for susceptibility to IBD. Indeed, associations between *MDR1* C3435T and UC and *MDR1* G2677C/T and IBD have been described [57, 58]. In addition to the *MDR1* C3435T association, Potocnik et al. [59] reported an association between particular SNPs in introns 13 and 16 of the *MDR1* gene and CS-refractory CD and UC.

Several other associations with SNPs in the *TNF* (tumor necrosis factor) gene and the macrophage *MIF* (migration inhibitory factor) gene and CS dependency or sensitivity have also been reported [60, 61].

What is evident from reviewing the literature on this topic is that trials in patient cohorts treated with fixed doses of CS and followed with well-defined response criteria are needed. In this respect, a very nice recent study used available RCTs and was able using GWAS to identify and confirm novel pharmacogenetic determinants in the *GLCCI1* gene predicting response to inhaled glucocorticoids [62].

Anti-TNF Therapy

The use of monoclonal antibodies to TNF has greatly improved the quality of life of patients suffering from IBD, but this therapy is expensive and not free from side effects. Also, 10–30 % of the patients do not respond to anti-TNF therapy (primary nonresponders). If early (and sustained) response could be accurately predicted, management could be optimized.

Several hypotheses have been tested: (1) early studies looking at genetic variants with respect to anti-TNF outcome focused logically on the TNF and TNF receptor pathway. Specific mutations in these genes were studied but results were either negative, or positive results could not be confirmed in larger cohorts [38, 63, 64]. (2) Since nuclear factor-κB signaling and TNFα levels are lower in cells carrying a CD-associated *NOD2* variant, it was also hypothesized that patients carrying a mutation in the *NOD2* gene respond differently to treatment with a TNF-blocking agent.

However, in three independent cohorts of patients, including the ACCENT study cohort, no significant associations were found [65, 66]. (3) One of the mechanisms of action of infliximab is induction of apoptosis of monocytes and T-lymphocytes [67, 68]. It was therefore postulated that failure to induce apoptosis would be associated with lack of efficacy. We analyzed the effect of polymorphisms in apoptotic genes on outcome [69]. In luminal CD, there was a response rate of 74.7 % in patients with the Fas ligand *FasL* -843 CC/CT genotype compared with a response rate of 38.1 % in patients with the TT genotype (OR = 0.11 [0.08–0.56], $p < 0.01$). Patients with the *caspase-9* 93 TT genotype all responded, in contrast with 66.7 % of patients with the CC and CT genotypes (OR = 1.50 [1.34–1.68], $p = 0.04$). The poorer outcome in patients with the *FasL* -843 CC/CT genotype and carrying *caspase-9* -93C could be overcome by concomitant use of azathioprine. It is of interest that a similar effect was observed for the *Fas* -670 GG genotype. (4) Building on the hypothesis that infliximab leads to complement activation and antibody-dependent cellular cytotoxicity (ADCC), polymorphisms in the Fcγ receptor 3A gene (*FCGR3A*) have been studied. Fcγ receptors bind the Fc portion of the immunoglobulin and have a well-defined role in triggering activation of innate effector cells and adaptive immune responses. The *FCGR3A* 158 V allotype displays a higher affinity for IgG1 and an increased ADCC and influences the therapeutic response to rituximab, an anti-CD20 IgG1 used in the treatment of non-Hodgkin lymphomas [70]. Similarly, Louis et al. [71] showed an association between *FCGR3A* 158 V/F and biological (50 % decrease of CRP) and possibly clinical response to infliximab in CD. This association could not be confirmed in a subset of 344 patients from the large and well-defined ACCENT 1 cohort of 573 patients [72]. However, there was again a trend towards a greater decrease in CRP after infliximab in V/V homozygotes as compared with V/F heterozygotes and F/F homozygotes (−79.4 %, −76.5 %, and −64.3 %, respectively, at week 6, $p = 0.085$; one-tailed $p = 0.043$). We extended the analysis to also include variants in *FCGR3B*. In a cohort of 719 CD patients treated with infliximab, we showed that the V allele variant for *FCGR3A* 158 V/F was associated with higher biological response to IFX (CRP back to normal levels). Being NA1 positive for *FCGR3B* was associated with decreased short-term biological response [73]. As *FCGR3A* 158 V/F, the *FCGR3B* NA1 polymorphism also results in a higher affinity for IgG in neutrophils, where FCGR3B is exclusively expressed. Although these findings might not have immediate clinical impact, they may enhance the understanding of the complex mechanisms of action of anti-TNF agents in CD.

Dubinsky et al. performed a GWA study to look for associations of novel "pharmacogenetic" genome-wide identified loci with primary nonresponse to anti-TNFα in pediatric IBD patients. Three pharmacogenetic GWAS loci were significant in a final predictive model: TACR1 which is a receptor for substance P, a known pro-inflammatory molecule; PHACTR3 (phosphatase and actin regulator 3) which is associated with the nuclear scaffold in proliferating cells; and FAM19A4 which is thought to be structurally related to MIP1α and functions as a chemokine.

Besides predicting response to anti-TNF, our group has investigated if the psoriasiform eruptions, which occur in about 20 % of anti-TNF-treated patients, would have a genetic predisposition [74]. We showed associations between occurrence of these skin-related adverse events and the *IL23R* and *IL12B* genes. This finding is noteworthy given that these genes have been implicated in the susceptibility to psoriasis.

Finally, for drugs such as methotrexate, cyclosporin, and tacrolimus, which are used less often in IBD, no or very few studies on genetic predictive factors have been conducted.

Transcriptomics

Another suggestion to explain the resistance to CSs (also see above) is altered functions of the GR and an excessive synthesis of pro-inflammatory cytokines. For example, steroid-resistant asthma patients do not respond to high doses of inhaled CS. In these patients, reduced peripheral T-lymphocyte GR-binding affinity and increased expression of GRβ (a truncated splice variant of the normal isoform GRα) are observed. GRβ is unable to activate steroid-responsive genes. Honda et al. [75] reported GRβ mRNA expression in 83 % of the patients with steroid-resistant UC compared to only 9 % in steroid-responsive patients and 10 % in healthy controls and chronic active CD patients. These results were confirmed in a recent study from Japan, where the authors looked at the frequency of GRα- and GRβ-positive cells in colonic biopsies of GC-sensitive ($n=6$) and GC-resistant ($n=8$) UC patients [76]. They also found that there were significantly more GRβ-positive cells in the CS-resistant group than in the CS-sensitive and the control groups. Whereas GRα mRNA was expressed in all UC patients, GRβ mRNA was expressed in only one patient in the CS-sensitive group and in seven patients in the CS-resistant group. Interestingly, the Foxp3+ cell count was also significantly higher in the CS-sensitive group. This points towards a possible predictive value of specific expression levels in the response to therapy.

Instead of looking for specific candidate genes, hypothesis-free mucosal expression studies looking at differences between responsive and resistant patients may pave the way. Microarray technology has been applied to advance our understanding on the reasons for nonresponse to anti-TNF agents. To identify predictive gene profiles, Arijs et al. studied mucosal gene expression in infliximab-naive CD and UC patients [77, 78]. By comparing pretreatment colonic mucosal expression profiles of responders with nonresponders, mucosal gene signatures predictive for (non-)response in UC and Crohn's colitis (CDc) were identified. In UC, these markers separated responders from nonresponders with 95 % sensitivity and 85 % specificity. A great overlap of the predictive genes identified in UC and the ones identified in CDc was observed, suggesting that the mechanism of resistance to infliximab therapy is the same for both diseases. These predictive genes showed a lower expression at baseline in responders vs. nonresponders. The proteins encoded by these genes were predominantly involved in immune signaling, indicating that there may be a larger immune burden at baseline in nonresponders than in responders. *IL-13Rα2* (interleukin receptor alpha 2) was one of the key biomarkers to predict nonresponse of infliximab. This molecule was also found to drive fibrosis [79].

Conclusion

Whereas older classifications have focused on anatomy, disease behavior, and age at onset, more recently the need has arisen to also subdivide patients based on their natural disease course. In this respect, clinical predictors of poor outcome in CD have been identified as young age (<40 years) at diagnosis, extensive small bowel disease, perianal fistulizing disease, and needing corticosteroids at diagnosis. Patients with these features will have a significant risk of complicated disease behavior and will benefit from early aggressive therapy. The search for molecular markers to define and predict disease outcome further has moved forward rapidly with the help of modern technologies. However, we are not ready to implement this part of molecular research into clinical practice yet. As physicians treating patients with IBD, the ultimate goal is to have at our disposal a molecular (serum, DNA, tissue-based) profile of the patient which allows the most appropriate management: What is the most likely course of disease? Which is the most appropriate therapy with highest chances of success? Which should be the intensity of follow-up?

Pharmacogenetic research in IBD has witnessed only modest success. One reason is that identifying molecular markers which influence the response to a drug is more difficult than, for instance, the study of genetic markers that influence toxicity. Side effects are usually easy to define and identify, in contrast to efficacy scores, which are often less well defined. In addition and as already explained, treatment response in a heterogenic disease like IBD is influenced by many factors such as disease duration, behavior, and severity. At the moment, only TPMT testing prior to start of thiopurine analogues has shown clinical applicability, but does not replace blood monitoring during treatment. Other reported genetic associations for the different therapeutic classes in IBD have not (yet) shown consistent or robust results. Studies looking at mucosal gene expression profiles could reveal novel pathways of nonresponse. What is needed in pharmacogenetic and other predictive studies to advance the field are patients treated with standardized doses of the drug and fixed endpoints and criteria for response. The setting of a clinical trial may well be the preferred method for this, and attempts to collect DNA from these patients should be enforced.

Recently, the IIBDGC has prioritized a number of projects on prediction of side effects of therapy (anti-TNF and demyelinization, 5-ASA and nephrotoxicity, azathioprine-induced pancreatitis and leucopenia) and is collecting DNA samples from these patients worldwide.

References

1. Farmer RG, Hawk WA, Turnbull RB Jr (1975) Clinical patterns in Crohn's disease: a statistical study of 615 cases. Gastroenterology 68(4 Pt 1):627–635
2. Greenstein AJ, Lachman P, Sachar DB, Springhorn J, Heimann T, Janowitz HD et al (1988) Perforating and non-perforating indications for repeated operations in Crohn's disease: evidence for two clinical forms. Gut 29(5):588–592, Epub 1988/05/01

3. Silverberg MS, Satsangi J, Ahmad T, Arnott ID, Bernstein CN, Brant SR et al (2005) Toward an integrated clinical, molecular and serological classification of inflammatory bowel disease: Report of a Working Party of the 2005 Montreal World Congress of Gastroenterology. Can J Gastroenterol 19(suppl A):5–36
4. Levine A, Griffiths A, Markowitz J, Wilson DC, Turner D, Russell RK et al (2011) Pediatric modification of the Montreal classification for inflammatory bowel disease: the Paris classification. Inflamm Bowel Dis 17(6):1314–1321, Epub 2011/05/12
5. Cosnes J, Cattan S, Blain A, Beaugerie L, Carbonnel F, Parc R et al (2002) Long-term evolution of disease behavior of Crohn's disease. Inflamm Bowel Dis 8(4):244–250, Epub 2002/07/20
6. Louis E, Collard A, Oger AF, Degroote E, Aboul Nasr El Yafi FA, Belaiche J (2001) Behaviour of Crohn's disease according to the Vienna classification: changing pattern over the course of the disease. Gut 49(6):777–782, Epub 2001/11/16
7. Beaugerie L, Seksik P, Nion-Larmurier I, Gendre JP, Cosnes J (2006) Predictors of Crohn's disease. Gastroenterology 130(3):650–656, Epub 2006/03/15
8. Loly C, Belaiche J, Louis E (2008) Predictors of severe Crohn's disease. Scand J Gastroenterol 43(8):948–954, Epub 2008/12/17
9. Tarrant KM, Barclay ML, Frampton CM, Gearry RB (2008) Perianal disease predicts changes in Crohn's disease phenotype-results of a population-based study of inflammatory bowel disease phenotype. Am J Gastroenterol 103(12):3082–3093, Epub 2008/12/18
10. Thia KT, Sandborn WJ, Harmsen WS, Zinsmeister AR, Loftus EV Jr (2010) Risk factors associated with progression to intestinal complications of Crohn's disease in a population-based cohort. Gastroenterology 139(4):1147–1155, Epub 2010/07/20
11. Allez M, Lemann M, Bonnet J, Cattan P, Jian R, Modigliani R (2002) Long term outcome of patients with active Crohn's disease exhibiting extensive and deep ulcerations at colonoscopy. Am J Gastroenterol 97(4):947–953, Epub 2002/05/11
12. Arnott ID, Landers CJ, Nimmo EJ, Drummond HE, Smith BK, Targan SR et al (2004) Sero-reactivity to microbial components in Crohn's disease is associated with disease severity and progression, but not NOD2/CARD15 genotype. Am J Gastroenterol 99(12):2376–2384, Epub 2004/12/02
13. Mow WS, Vasiliauskas EA, Lin YC, Fleshner PR, Papadakis KA, Taylor KD et al (2004) Association of antibody responses to microbial antigens and complications of small bowel Crohn's disease. Gastroenterology 126(2):414–424
14. Targan SR, Landers CJ, Yang H, Lodes MJ, Cong Y, Papadakis KA et al (2005) Antibodies to CBir1 flagellin define a unique response that is associated independently with complicated Crohn's disease. Gastroenterology 128(7):2020–2028, Epub 2005/06/09
15. Ferrante M, Henckaerts L, Joossens M, Pierik M, Joossens S, Dotan N et al (2007) New serological markers in inflammatory bowel disease are associated with complicated disease behaviour. Gut 56(10):1394–1403
16. Dubinsky MC, Lin YC, Dutridge D, Picornell Y, Landers CJ, Farrior S et al (2006) Serum immune responses predict rapid disease progression among children with Crohn's disease: immune responses predict disease progression. Am J Gastroenterol 101(2):360–367, Epub 2006/02/04
17. Rieder F, Schleder S, Wolf A, Dirmeier A, Strauch U, Obermeier F et al (2010) Serum anti-glycan antibodies predict complicated Crohn's disease behavior: a cohort study. Inflamm Bowel Dis 16(8):1367–1375, Epub 2009/12/22
18. Henckaerts L, Pierik M, Joossens M, Ferrante M, Rutgeerts P, Vermeire S (2007) Mutations in pattern recognition receptor genes modulate seroreactivity to microbial antigens in patients with inflammatory bowel disease. Gut 56(11):1536–1542, Epub 2007/06/28
19. Devlin SM, Yang H, Ippoliti A, Taylor KD, Landers CJ, Su X et al (2007) NOD2 variants and antibody response to microbial antigens in Crohn's disease patients and their unaffected relatives. Gastroenterology 132(2):576–586, Epub 2007/01/30

20. Abreu MT, Taylor KD, Lin YC, Hang T, Gaiennie J, Landers CJ et al (2002) Mutations in NOD2 are associated with fibrostenosing disease in patients with Crohn's disease. Gastroenterology 123(3):679–688, Epub 2002/08/29
21. Ahmad T, Armuzzi A, Bunce M, Mulcahy-Hawes K, Marshall SE, Orchard TR et al (2002) The molecular classification of the clinical manifestations of Crohn's disease. Gastroenterology 122(4):854–866
22. Economou M, Trikalinos TA, Loizou KT, Tsianos EV, Ioannidis JP (2004) Differential effects of NOD2 variants on Crohn's disease risk and phenotype in diverse populations: a metaanalysis. Am J Gastroenterol 99(12):2393–2404, Epub 2004/12/02
23. Seiderer J, Schnitzler F, Brand S, Staudinger T, Pfennig S, Herrmann K et al (2006) Homozygosity for the CARD15 frameshift mutation 1007fs is predictive of early onset of Crohn's disease with ileal stenosis, entero-enteral fistulas, and frequent need for surgical intervention with high risk of re-stenosis. Scand J Gastroenterol 41(12):1421–1432, Epub 2006/11/15
24. Seiderer J, Brand S, Herrmann KA, Schnitzler F, Hatz R, Crispin A et al (2006) Predictive value of the CARD15 variant 1007fs for the diagnosis of intestinal stenoses and the need for surgery in Crohn's disease in clinical practice: results of a prospective study. Inflamm Bowel Dis 12(12):1114–1121
25. Adler J, Rangwalla SC, Dwamena BA, Higgins PD (2011) The prognostic power of the NOD2 genotype for complicated Crohn's disease: a meta-analysis. Am J Gastroenterol 106(4):699–712, Epub 2011/02/24
26. Prescott NJ, Fisher SA, Franke A, Hampe J, Onnie CM, Soars D et al (2007) A nonsynonymous SNP in ATG16L1 predisposes to ileal Crohn's disease and is independent of CARD15 and IBD5. Gastroenterology 132(5):1665–1671, Epub 2007/05/09
27. Weersma RK, Stokkers PC, van Bodegraven AA, van Hogezand RA, Verspaget HW, de Jong DJ et al (2009) Molecular prediction of disease risk and severity in a large Dutch Crohn's disease cohort. Gut 58(3):388–395
28. Henckaerts L, Van Steen K, Verstreken I, Cleynen I, Franke A, Schreiber S et al (2009) Genetic risk profiling and prediction of disease course in Crohn's disease patients. Clin Gastroenterol Hepatol 7(9):972–980
29. Lee J, Anderson C, Wesley E, Ahmad T, Edwards C, Parkes M, et al (2010) Identification of a polymorphism that predisposes to longitudinal disease behaviour in Crohn's disease and may have prognostic utility [abstract]. JCC 4(1):S04:20
30. Jostins L, Ripke S, Weersma RK, Duerr RH, McGovern DP, Hui KY et al (2012) Host-microbe interactions have shaped the genetic architecture of inflammatory bowel disease. Nature 491(7422):119–124, Epub 2012/11/07
31. Cleynen I, Mahachie John JM, Henckaerts L, Van Moerkercke W, Rutgeerts P, Van Steen K et al (2010) Molecular reclassification of Crohn's disease by cluster analysis of genetic variants. PLoS One 5(9):e12952
32. Chang JC, Wooten EC, Tsimelzon A, Hilsenbeck SG, Gutierrez MC, Elledge R et al (2003) Gene expression profiling for the prediction of therapeutic response to docetaxel in patients with breast cancer. Lancet 362(9381):362–369, Epub 2003/08/09
33. Watanabe T, Kobunai T, Sakamoto E, Yamamoto Y, Konishi T, Horiuchi A et al (2009) Gene expression signature for recurrence in stage III colorectal cancers. Cancer 115(2):283–292, Epub 2008/11/19
34. Watanabe T, Kobunai T, Tanaka T, Ishihara S, Matsuda K, Nagawa H (2009) Gene expression signature and the prediction of lymph node metastasis in colorectal cancer by DNA microarray. Dis Colon Rectum 52(12):1941–1948, Epub 2009/11/26
35. Lee JC, Lyons PA, McKinney EF, Sowerby JM, Carr EJ, Bredin F et al (2011) Gene expression profiling of CD8+ T cells predicts prognosis in patients with Crohn disease and ulcerative colitis. J Clin Invest 121(10):4170–4179, Epub 2011/09/29
36. McKinney EF, Lyons PA, Carr EJ, Hollis JL, Jayne DR, Willcocks LC et al (2010) A CD8+ T cell transcription signature predicts prognosis in autoimmune disease. Nat Med 16(5):586–591, Epub 2010/04/20

37. Esters N, Pierik M, van Steen K, Vermeire S, Claessens G, Joossens S et al (2004) Transmission of CARD15 (NOD2) variants within families of patients with inflammatory bowel disease. Am J Gastroenterol 99(2):299–305, Epub 2004/03/30
38. Taylor KD, Plevy SE, Yang H, Landers CJ, Barry MJ, Rotter JI et al (2001) ANCA pattern and LTA haplotype relationship to clinical responses to anti-TNF antibody treatment in Crohn's disease. Gastroenterology 120(6):1347–1355, Epub 2001/04/21
39. Kushner I (1990) C-reactive protein and the acute-phase response. Hosp Pract (Off Ed) 25(3A):13, 6, 21–28. Epub 1990/03/30
40. Tall AR (2004) C-reactive protein reassessed. N Engl J Med 350(14):1450–1452, Epub 2004/04/09
41. Vermeire S, Van Assche G, Rutgeerts P (2005) The role of C-reactive protein as an inflammatory marker in gastrointestinal diseases. Nat Clin Pract Gastroenterol Hepatol 2(12):580–586, Epub 2005/12/06
42. Vermeire S, Van Assche G, Rutgeerts P (2006) Laboratory markers in IBD: useful, magic, or unnecessary toys? Gut 55(3):426–431, Epub 2006/02/14
43. Van Cutsem E, Kohne CH, Hitre E, Zaluski J, Chang Chien CR, Makhson A et al (2009) Cetuximab and chemotherapy as initial treatment for metastatic colorectal cancer. N Engl J Med 360(14):1408–1417, Epub 2009/04/03
44. Lambrechts D, Delmar P, Buysschaert I, Claes B, Yesilyurt B, Verslype C, et al (2009) VEGFR-1 polymorphisms as potential predictors of clinical outcome in bevacizumab-treated patients with metastatic pancreatic cancer. Joint ECCO-ESMO Multidisciplinary Congress. 20–24 Sept 2009: Abstract No. 16LBA
45. Daly AK, Donaldson PT, Bhatnagar P, Shen Y, Pe'er I, Floratos A et al (2009) HLA-B*5701 genotype is a major determinant of drug-induced liver injury due to flucloxacillin. Nat Genet 41(7):816–819, Epub 2009/06/02
46. Link E, Parish S, Armitage J, Bowman L, Heath S, Matsuda F et al (2008) SLCO1B1 variants and statin-induced myopathy—a genomewide study. N Engl J Med 359(8):789–799, Epub 2008/07/25
47. Evans WE, McLeod HL (2003) Pharmacogenomics–drug disposition, drug targets, and side effects. N Engl J Med 348(6):538–549, Epub 2003/02/07
48. Weinshilboum RM, Sladek SL (1980) Mercaptopurine pharmacogenetics: monogenic inheritance of erythrocyte thiopurine methyltransferase activity. Am J Hum Genet 32(5):651–662, Epub 1980/09/01
49. Colombel JF, Ferrari N, Debuysere H, Marteau P, Gendre JP, Bonaz B et al (2000) Genotypic analysis of thiopurine S-methyltransferase in patients with Crohn's disease and severe myelosuppression during azathioprine therapy. Gastroenterology 118(6):1025–1030, Epub 2000/06/02
50. Cuffari C, Hunt S, Bayless T (2001) Utilisation of erythrocyte 6-thioguanine metabolite levels to optimise azathioprine therapy in patients with inflammatory bowel disease. Gut 48(5):642–646, Epub 2001/04/17
51. Dubinsky MC, Yang H, Hassard PV, Seidman EG, Kam LY, Abreu MT et al (2002) 6-MP metabolite profiles provide a biochemical explanation for 6-MP resistance in patients with inflammatory bowel disease. Gastroenterology 122(4):904–915, Epub 2002/03/23
52. Schwab M, Schaeffeler E, Marx C, Zanger U, Aulitzky W, Eichelbaum M (2001) Shortcoming in the diagnosis of TPMT deficiency in a patient with Crohn's disease using phenotyping only. Gastroenterology 121(2):498–499, Epub 2001/08/07
53. Winter J, Walker A, Shapiro D, Gaffney D, Spooner RJ, Mills PR (2004) Cost-effectiveness of thiopurine methyltransferase genotype screening in patients about to commence azathioprine therapy for treatment of inflammatory bowel disease. Aliment Pharmacol Ther 20(6):593–599, Epub 2004/09/09
54. Faubion WA Jr, Loftus EV Jr, Harmsen WS, Zinsmeister AR, Sandborn WJ (2001) The natural history of corticosteroid therapy for inflammatory bowel disease: a population-based study. Gastroenterology 121(2):255–260, Epub 2001/08/07
55. Munkholm P, Langholz E, Davidsen M, Binder V (1994) Frequency of glucocorticoid resistance and dependency in Crohn's disease. Gut 35(3):360–362, Epub 1994/03/01

56. Farrell RJ, Murphy A, Long A, Donnelly S, Cherikuri A, O'Toole D et al (2000) High multidrug resistance (P-glycoprotein 170) expression in inflammatory bowel disease patients who fail medical therapy. Gastroenterology 118(2):279–288, Epub 2000/01/29
57. Brant SR, Panhuysen CI, Nicolae D, Reddy DM, Bonen DK, Karaliukas R et al (2003) MDR1 Ala893 polymorphism is associated with inflammatory bowel disease. Am J Hum Genet 73(6):1282–1292, Epub 2003/11/12
58. Ho GT, Nimmo ER, Tenesa A, Fennell J, Drummond H, Mowat C et al (2005) Allelic variations of the multidrug resistance gene determine susceptibility and disease behavior in ulcerative colitis. Gastroenterology 128(2):288–296, Epub 2005/02/03
59. Potocnik U, Ferkolj I, Glavac D, Dean M (2004) Polymorphisms in multidrug resistance 1 (MDR1) gene are associated with refractory Crohn disease and ulcerative colitis. Genes Immun 5(7):530–539, Epub 2004/10/27
60. Cucchiara S, Latiano A, Palmieri O, Canani RB, D'Inca R, Guariso G et al (2007) Polymorphisms of tumor necrosis factor-alpha but not MDR1 influence response to medical therapy in pediatric-onset inflammatory bowel disease. J Pediatr Gastroenterol Nutr 44(2): 171–179, Epub 2007/01/27
61. Griga T, Wilkens C, Wirkus N, Epplen J, Schmiegel W, Klein W (2007) A polymorphism in the macrophage migration inhibitory factor gene is involved in the genetic predisposition of Crohn's disease and associated with cumulative steroid doses. Hepatogastroenterology 54(75):784–786, Epub 2007/06/27
62. Tantisira KG, Lasky-Su J, Harada M, Murphy A, Litonjua AA, Himes BE et al (2011) Genomewide association between GLCCI1 and response to glucocorticoid therapy in asthma. N Engl J Med 365(13):1173–1183, Epub 2011/10/14
63. Mascheretti S, Hampe J, Kuhbacher T, Herfarth H, Krawczak M, Folsch UR et al (2002) Pharmacogenetic investigation of the TNF/TNF-receptor system in patients with chronic active Crohn's disease treated with infliximab. Pharmacogenomics J 2(2):127–136, Epub 2002/06/07
64. Pierik M, Vermeire S, Steen KV, Joossens S, Claessens G, Vlietinck R et al (2004) Tumour necrosis factor-alpha receptor 1 and 2 polymorphisms in inflammatory bowel disease and their association with response to infliximab. Aliment Pharmacol Ther 20(3):303–310, Epub 2004/07/28
65. Mascheretti S, Hampe J, Croucher PJ, Nikolaus S, Andus T, Schubert S et al (2002) Response to infliximab treatment in Crohn's disease is not associated with mutations in the CARD15 (NOD2) gene: an analysis in 534 patients from two multicenter, prospective GCP-level trials. Pharmacogenetics 12(7):509–515, Epub 2002/10/03
66. Vermeire S, Louis E, Rutgeerts P, De Vos M, Van Gossum A, Belaiche J et al (2002) NOD2/CARD15 does not influence response to infliximab in Crohn's disease. Gastroenterology 123(1):106–111, Epub 2002/07/10
67. Shen C, Maerten P, Geboes K, Van Assche G, Rutgeerts P, Ceuppens JL (2005) Infliximab induces apoptosis of monocytes and T lymphocytes in a human-mouse chimeric model. Clin Immunol 115(3):250–259, Epub 2005/05/17
68. ten Hove T, van Montfrans C, Peppelenbosch MP, van Deventer SJ (2002) Infliximab treatment induces apoptosis of lamina propria T lymphocytes in Crohn's disease. Gut 50(2):206–211
69. Hlavaty T, Pierik M, Henckaerts L, Ferrante M, Joossens S, van Schuerbeek N et al (2005) Polymorphisms in apoptosis genes predict response to infliximab therapy in luminal and fistulizing Crohn's disease. Aliment Pharmacol Ther 22(7):613–626, Epub 2005/09/27
70. Cartron G, Dacheux L, Salles G, Solal-Celigny P, Bardos P, Colombat P et al (2002) Therapeutic activity of humanized anti-CD20 monoclonal antibody and polymorphism in IgG Fc receptor FcgammaRIIIa gene. Blood 99(3):754–758, Epub 2002/01/25
71. Louis E, El Ghoul Z, Vermeire S, Dall'Ozzo S, Rutgeerts P, Paintaud G et al (2004) Association between polymorphism in IgG Fc receptor IIIa coding gene and biological response to infliximab in Crohn's disease. Aliment Pharmacol Ther 19(5):511–519, Epub 2004/02/28

72. Louis EJ, Watier HE, Schreiber S, Hampe J, Taillard F, Olson A et al (2006) Polymorphism in IgG Fc receptor gene FCGR3A and response to infliximab in Crohn's disease: a subanalysis of the ACCENT I study. Pharmacogenet Genomics 16(12):911–914, Epub 2006/11/17
73. Arias MT, Cleynen I, Claes K, Drobne D, Gils A, Juergens M et al (2011) The role of FC Gamma Receptor (FCGR) mutations in the response to infliximab in inflammatory bowel disease. Gastroenterology 140(5):S-1
74. Cleynen I, Van Moerkercke W, Juergens M, Ballet V, Drobne D, Vandecandelaere P et al (2010) Anti-TNF induced cutaneous lesions in IBD patients: characterization and search for predisposing factors. Gut 59(suppl 3):A1
75. Honda M, Orii F, Ayabe T, Imai S, Ashida T, Obara T et al (2000) Expression of glucocorticoid receptor beta in lymphocytes of patients with glucocorticoid-resistant ulcerative colitis. Gastroenterology 118(5):859–866, Epub 2000/04/28
76. Fujishima S, Takeda H, Kawata S, Yamakawa M (2009) The relationship between the expression of the glucocorticoid receptor in biopsied colonic mucosa and the glucocorticoid responsiveness of ulcerative colitis patients. Clin Immunol 133(2):208–217, Epub 2009/08/04
77. Arijs I, Li K, Toedter G, Quintens R, Van Lommel L, Van Steen K et al (2009) Mucosal gene signatures to predict response to infliximab in patients with ulcerative colitis. Gut 58(12): 1612–1619, Epub 2009/08/25
78. Arijs I, Quintens R, Van Lommel L, Van Steen K, De Hertogh G, Lemaire K et al (2010) Predictive value of epithelial gene expression profiles for response to infliximab in Crohn's disease. Inflamm Bowel Dis 16(12):2090–2098, Epub 2010/09/18
79. Strober W, Kitani A, Fichtner-Feigl S, Fuss IJ (2009) The signaling function of the IL-13Ralpha2 receptor in the development of gastrointestinal fibrosis and cancer surveillance. Curr Mol Med 9(6):740–750, Epub 2009/08/20

Chapter 9
Epigenetics of Inflammatory Bowel Disease

Robert Häsler, Stefan Schreiber, Stephan Beck, and Philip Rosenstiel

Abstract The term epigenome refers to the tissue- and cell-type-specific collection of DNA methylation, histone modifications, and chromatin accessibility and the set of coding and noncoding RNA molecules (Bernstein et al., Cell 125:315–326, 2006) that are dynamically modulated throughout the lifetime of an individual. Epigenetic modifications are critical for regular developmental processes in the intestine, but variation in the epigenome has also been associated with the development of intestinal diseases, including inflammatory bowel disease (Vavricka et al., Inflammatory Bowel Diseases 17:1530–1539, 2011). We hypothesize that plasticity of the epigenome in different cellular compartments links genetic susceptibility and environmental influences and may determine "decision points" in the progression towards disease onset (i.e., manifestation) and/or progression of IBD. This chapter reviews selected aspects of IBD research with the aim to link the current knowledge of genetic, epigenetic, and functional studies into an integrated view on the role of epigenetic variation in intestinal inflammation.

R. Häsler, Ph.D. • P. Rosenstiel, M.D. (✉)
Institute of Clinical Molecular Biology, Christian-Albrechts-University of Kiel, 24105, Kiel, Germany
e-mail: p.rosenstiel@mucosa.de

S. Schreiber, M.D.
Department of General Internal Medicine, University Hospital Schleswig-Holstein, Campus Kiel, Kiel, Germany

Institute of Clinical Molecular Biology, Christian-Albrechts-University of Kiel, 24105, Kiel, Germany

S. Beck, M.Sc., Ph.D. (✉)
UCL Cancer Institute, 72 Huntley Street, Paul O'Gorman Building, London, WC1E 6BT, UK
e-mail: s.beck@ucl.ac.uk

M. D'Amato and J.D. Rioux (eds.), *Molecular Genetics of Inflammatory Bowel Disease*,
DOI 10.1007/978-1-4614-8256-7_9, © Springer Science+Business Media New York 2013

Introduction

The term epigenome refers to the tissue- and cell-type-specific collection of DNA methylation, histone modifications, and chromatin accessibility and the set of coding and noncoding RNA molecules [1] that are dynamically modulated throughout the lifetime of an individual. Epigenetic modifications are critical for regular developmental processes in the intestine, but variation in the epigenome has also been associated with the development of intestinal diseases, including inflammatory bowel disease (IBD). We hypothesize that plasticity of the epigenome in different cellular compartments links genetic susceptibility and environmental influences and may determine "decision points" in the progression towards disease onset (i.e., manifestation) and/or progression of IBD. This chapter reviews selected aspects of IBD research with the aim to link the current knowledge of genetic, epigenetic, and functional studies into an integrated view on the role of epigenetic variation in intestinal inflammation.

Epigenetics: Background, Technology, and Potential for Common Disease Research

Epigenetics can be viewed as paradigm for phenotypic plasticity and was introduced as a separate field to complement genetics by Conrad Waddington in the early 1940s when studying how the genotype relates to different phenotypes. Although the underlying mechanisms were unknown at the time, Waddington envisioned the existence of an "epigenotype" to explain the phenotypic plasticity observed during normal development [2]. Since then, many of the mechanisms have been worked out in the context of a wide range of biological processes, such as X-chromosome inactivation in female mammals [3], parent-of-origin-specific gene expression (imprinting) [4], and developmental [5] and cellular [6] reprogramming to name but a few. Furthermore, altered epigenetic mechanisms have been linked to cancer as early as 1983 [7] and more recently also to other common diseases [8, 9]. Based on these findings, our perception of epigenetics has changed over the years and was recently redefined as "structural adaptation of chromosomal regions so as to register, signal or perpetuate altered activity states" [10].

Great progress has also been made in elucidating the types of epigenetic marks that register, signal, and perpetuate the activity states and the enzymes that read, write, and erase these marks which, in concert with other modifiers, bring about the structural adaptation of chromosomal regions. There is ongoing debate on what constitutes a true epigenetic mark but also agreement that all marks under consideration at least modulate the epigenome and hence are here referred to as chromatin or epigenome modulators of which there are three main categories. The best studied is DNA methylation in the context of CpG dinucleotides. Low methylation at promoters and high methylation at gene bodies are usually associated with gene expression and, conversely, high promoter and low gene body methylation are associated with gene silencing [11]. This simple on/off concept has recently become more complex following the discovery of non-CpG methylation and other cytosine modifications [12]. Based on current knowledge, genomic cytosine bases can exist in at least six states (unmethylated, C5-methylated, N3-methylated, C5-hydroxymethylated, C5-formylated, C5-carboxylated) and more modifications may exist and at other bases as well. On the protein level, histone tails are the target for an ever-increasing number of posttranslational modifications (that form the second category), including acetylation, methylation, phosphorylation, ribosylation, ubiquitylation, sumoylation, citrullination, and some even more exotic modifications [13]. With respect to function, they can loosely be grouped into activating, repressive, or bivalent modifications. The latter define a combination of activating (e.g., H3K4Me3) and repressive (H3K427Me3) modifications that have been shown to mark poised chromatin which is typical for developmental genes [1]. The third and final category comprises all the remaining modulators, including the enzymes that lay down the modifications (the "writers"), the proteins that recognize them (the "readers"), and the enzymes that remove them (the "erasers") as well as nucleosomes, chromatin-remodeling complexes, and noncoding RNAs. Collectively, these chromatin modifiers provide function to the genome and define the epigenome.

The main bottleneck that has hampered epigenetic analysis of common diseases in the past has been technology. While genome-wide association studies (GWAS) [14] using single-nucleotide polymorphisms (SNPs) uncovered well over 1,000 new disease loci across all investigated human diseases with a tally of over 160 loci in IBD and significantly advanced the genetic analysis [15], no comparable technology was available for epigenetic analysis. This has changed with the emergence of genome-wide methods [16, 17] for the analysis of DNA methylation which is the most informative and accessible epigenetic modification in a clinical context. The currently most promising platform with respect to accuracy, coverage, throughput, and cost is the Illumina 450k Infinium BeadChip which is essentially the epigenetic equivalent of the 500k SNP chip that proved highly successful for GWAS. An obvious next step was to adapt GWAS to epigenetic analysis to enable epigenome-wide association studies (EWAS). Although both analyses have much in common, EWAS also presents new challenges. As the epigenotype is cell-type specific, special care must be taken to select the correct study material. In other words, blood-derived DNA (which is suitable for all GWAS) is not necessarily suitable for all EWAS. Another problem is a phenomenon known as reverse causality. While GWAS associations are usually linked to the underlying causal variation by linkage disequilibrium, EWAS associations can also be the consequence (rather than the cause) of the phenotype under investigation. This problem can be overcome by inclusion of prospectively sampled individuals in the study design as demonstrated in the first EWAS for type 1 diabetes [18]. As the genotype and epigenotype are inherently linked, the need to distinguish genetic from epigenetic effects adds further complexity but can be addressed, e.g., by using monozygotic twins that are disease discordant for the discovery phase [19]. These advances have paved the way to apply epigenetic analysis to common diseases, and the first wave of EWAS is now well underway, including for inflammatory bowel diseases.

Clinical Relevance of Epigenetic Events in Inflammatory Bowel Diseases

Inflammatory bowel diseases are complex disorders, which are known to be strongly influenced by the genetic background [20]. The high familiar concordance observed in IBD initially introduced this concept [21]. Further studies identifying various disease-associated variants supported this hypothesis [22]. Several identified variants additionally provide insight into potential disease relevant molecular mechanisms. A prominent example is *NOD2*, which was the first disease gene identified for Crohn's disease [23, 24] and which is functionally linked to bacterial recognition. Variants of *IL23R* [25] and *IL12B* [26] are associated with both Crohn's disease and ulcerative colitis and are involved in immune system activation.

As for many complex disorders, the identified genetic variants cannot explain the entire disease risk: In Crohn's disease, currently 140 variants are known to be disease associated, and similarly, in ulcerative colitis the number of currently identified variants is 133 [27]. In this context, one has to keep in mind that the probability of accumulating all the variants at once in one single genome is extremely low, especially since many of those variants have very low frequencies. Consequently, the disease risk explained by genetics for a given genome is of purely theoretical nature. Twin concordance rates, which are higher than the currently explained disease risk [28], indicate that several variants are not identified yet. The resulting gap is generally referred as missing heritability [29]. However, the space beyond this gap is even less explored.

By definition, complex disorders are influenced not only by the genetic background. Environmental factors, such as nutrition, toxin exposure, or the intestinal microbiota—to name but few—are being discussed as potential contributors to disease risk and manifestation. Similarly, a high family concordance rate does not necessarily have to be attributed to the genetic background exclusively. Shared environment, nutrients, or toxins could also explain part of the family concordance. Finally, all these factors may interact leading to additional events of pathophysiological relevance [30].

One integral part of this disease risk which cannot be explained by the genetic background exclusively is epigenetic modifications. Traditionally, epigenetic events are defined as heritable modifications in DNA expression without changing the DNA sequence in itself [31]. Besides DNA methylation, histone modification and nucleosome positioning are integrated in this definition. More recent definitions include micro-RNAs as regulators of gene activity in the absence of DNA sequence variation [8]. In complex disorders, a combination of heritable as well as de novo events is being considered potentially disease relevant.

Epigenetic Events in Complex Diseases: Heading the Way to Inflammatory Disorders

Several scenarios, most of them with an oncological background, are known where epigenetic modifications lead to disease manifestation. A popular example is the global hypomethylation often observed in cancer cells [32]. In the same line, it has been shown that hypomethylation of several genes (e.g., 16^{INK4a}–*p14*^{ARF}(*CDKN2A*) and *MGMT*) can be a causal event in early tumorigenesis [33]. Following the expanded definition of epigenetics, miRNAs which are widely downregulated in human tumors [34] as a result of hypomethylated miRNA promoters may play an important epigenetic role in cancers [35]. Beside the large number of studies in cancer, various other diseases have been the target of epigenetic research, showing the pathophysiological relevance of epigenetic modifications and their interactions to environmental factors [9, 36–39]. Interestingly, only very few studies address epigenetic events in inflammatory diseases, where a regulatory network of signal-specific and gene-specific functions is required controlling appropriate responses [40]. Initial studies have shown a link between the hypomethylation of Toll-like receptor 2 (TLR2) and increased proinflammatory response to bacterial peptidoglycan in cystic fibrosis [41]. Bacterial infection as an environmental factor was shown to have impact on the epigenetic status of the genome [42], while a recent study presented a functional map of the psoriasis epigenome [43], illustrating how this potentially could be linked to the transcriptome. Similar transcriptional control is provided by micro-RNAs, who are believed to target up to 30 % of all genes [44]. In concordance with DNA methylation, micro-RNAs have been shown to have significant impact on diseases, including inflammatory disorders [45, 46].

Disease-Associated DNA Methylation in Inflammatory Bowel Disease

Taken together, this illustrates the potential impact epigenetic modifications may have on disease risk, manifestation, and progression in Crohn's disease and ulcerative colitis. In fact, several studies have addressed this issue. First approaches in 1996 showed that DNA hypomethylation is a common pattern observed in the rectal mucosa of ulcerative colitis patients [47]. Interestingly, this effect was observed in patients with long-standing ulcerative colitis, supporting the hypothesis that epigenetic modifications in a given tissue are increasing over time. Epigenetic maturation and its potential impact on the onset of disease which is in early adulthood have been studied in mouse models indicating that mucosal epigenetic maturation continues after early adulthood in mouse, which could play a role in age-associated increase in colitis susceptibility [48].

Most studies in this field focused on the methylation of individual inflammation or immune-process-associated target genes. IFNγ methylation was investigated in various cell types present in the human gut, concluding that its methylation status is relevant for the modulation of cytokine secretion in the mucosa [49]. This subject was followed up in 2011, where IFNγ methylation levels correlated with immune response to microbial components and expression of IFNγ in ulcerative colitis patients, suggesting a categorization of patients based on this response [50]. Quantification of DNA methylation of the promoter region of interferon regulatory factor 5 (*IRF5*) aimed to create a link between epigenetics and genetics, since A 5-bp insertion-deletion (indel) polymorphism in the promoter of *IRF5* has been associated with inflammatory bowel diseases [51]: However, the results implicate that epigenetic dysregulation of the *IRF5* promoter is unlikely to be associated with IBD [52].

Recently, evolving technology enabled assessment of disease-associated methylation in tissues derived from patients inflammatory bowel disease on a broader scale: Quantification of CpG methylation in a set of 1,505 CpG sites corresponding to 807 genes identified seven sites being differentially methylated between healthy and disease individuals [53]. This was expanded to a genome-wide level in Crohn's disease, where 50 methylation sites were identified to be epigenetically modified, including several genes involved in immune activation such as *MAPK13*, *FASLG*, *PRF1*, *S100A13*, *RIPK3*, and *IL21R* [54]. We have recently published a first epigenome-wide DNA methylation analysis (EWAS) combining 27k Illumina, MedIP-Chip and expression arrays from intestinal biopsies of twins discordant for UC. The integrated analysis identified 61 epigenetic disease loci, which were validated in a larger case-control cohort of unrelated individuals [62].

One of the major drawbacks in current approaches investigating the pathophysiological impact of epigenetic modifications is the lack of tools to specifically validate single CpG modifications in a model system. Currently, only demethylation agents, such as azacitidine and decitabine, which have been used in the treatment of myelodysplastic syndrome, are available [55, 56]. By inhibiting methyltransferases, these agents work genome wide. Consequently, it is unclear to which extent the observed cellular effects can be attributed to primary modifications of the methylation of target genes, or to secondary effects, or to interactions of all these.

Regulatory miRNA Networks in the Pathophysiology of Inflammatory Bowel Diseases

In contrast to DNA methylation, epigenetic research in the field of micro-RNAs (miRNAs) has access to such target-specific tools: Sequences, complementary to the micro-RNAs, so-called anti-miRs (or antagomirs), can be used to modulate endogenous miRNA levels. In addition, reporter gene assays represent a powerful tool to validate miRNA findings in model systems. After their discovery in 1992 [57], miRNAs have been found in all eukaryotes, and recent genome-wide computational screens for miRNA targets in humans predict that at least 10 % [58] to 30 % [44] of all genes are regulated by iRNAs. Several studies indicate that miRNAs play an important role in inflammatory scenarios [59–61], including the hypothesis that miRNAs are required to control and balance a specific inflammatory response [62]. Several miRNAs were identified to play a potential pathophysiological role in inflammatory bowel diseases, especially when addressing the disease subtypes specifically: In Crohn's disease miRNAs were associated with ileal and colonic manifestations [63], suggesting that the specificity of miRNA patterns may help to identify disease subtypes. In ulcerative colitis, variants in a noncoding region were shown to alter miRNA functionality [64], providing an explanation on how these variants could exhibit their functional effect. Similarly, a variant in the *IL23R* gene, which is associated with IBD, has been reported to result in inhibition of miRNA binding to this allele, altering the control of this gene which finally may lead to sustained IL23R signaling, promoting the chronicity of IBD [65]. This was followed by recent approaches creating genome-wide maps of circulating miRNAs in ulcerative colitis, supporting the hypothesis that many previously identified variants located in noncoding regions might contribute to disease susceptibility by altering miRNA sequences [66]. Interestingly, some abnormally expressed miRNA could be linked to inactive colonic mucosa of patients with IBD [67], suggesting that not only an active inflammation results in dysregulation of miRNAs.

In summary, the results of studies targeting DNA methylation as well as miRNAs in inflammatory bowel disease represent not only a set of independent disease-associated mechanisms but also create a link between variants in noncoding regions and effects on pathophysiologically relevant target genes. Finally, epigenetics might help to answer the question whether we not only inherit the genetic background of our ancestors but also the footprints of their lifestyle.

Dialogue Between Epigenetics, Environmental Influences, and the Intestinal Microbiome in Inflammatory Bowel Disease

Due to the increasing prevalence of IBD in industrialized societies, the question arises which environmental factors lead to changing manifestation of disease, as this observation cannot be attributed to changes in genetic background of the respective populations [20]. While many factors have been discussed, the most drastic lifestyle changes within the last century are likely related to childhood infection rates (due to vaccination and antibiotics), increased hygiene in general and nutritional habits. It has been shown in epidemiological studies that improvement of hygienic conditions (such as warm water or water toilets) is positively correlated with incidence rates for Crohn's disease [68, 69]. Likewise in Europe there is a striking north–south and west–east gradient of IBD prevalence, and immune-mediated diseases in general are much more common in larger cities than in rural areas and are related to the presence of bacterial antigens [70–72]. Of course, it could be speculated that all these observations are influenced by mere confounding and the true factors are yet to be identified. Still, several striking hypotheses have been raised by genetic studies as well as functional underpinnings that point to a crucial role of the balance of intestinal host–microbiome interactions, and it is tempting to speculate that this long-term influencing factor actually is a major determinant of epigenetic profiles along the entire gastrointestinal tract. For this hypothesis several facts about this type of stable host–microbe interaction are important. Large international efforts have been made to systematically profile the properties and functional repertoires of human microbial communities [73, 74]. These studies have clearly shown a huge diversity of microbial species that is specific to the body region as well as to the individual (microbial "fingerprint"). Even after drastic life history events, e.g., intestinal infections or courses of combination antibiotic therapies, intestinal microbial consortia display evidence of resilience, i.e., after a certain time the specific consortia return to their previous diversity that is similar to the one before the event. It has been proposed that only few stable states of the human intestinal core microbiome exist, the so-called enterotypes [75]. These metagenomic states, representing differences in core metabolic activities and pathways, could be caused by the genetic (and epigenetic) makeup of the host, but on the other hand the enterotypes together may also imprint on the long-term epigenetic (and thus functional) profiles in the different cellular compartments of the intestinal mucosa [76]. Exciting data point to long-term influences of dietary modifications on microbial communities that in turn cause functional changes in the human host. This principle was first described in animal models of obesity, where microbial communities that were transplanted from obese individuals led to increased energy harvest and weight gain in lean individuals [77]. This principle of microbiota-transmissible susceptibility has now been expanded to a number of immune-mediated diseases including IBD. In a genetic model of amino acid malnutrition resulting in dysbiotic microbial communities and increased susceptibility to colitis, it has been shown that long-term dietary supplement with chemically modified tryptophan resulted in changes of antimicrobial peptide profiles and decreased inflammatory responses [78]. Interestingly, the inflammatory phenotype could be transmitted to germfree wild-type animals by stool transplantation pointing to a crucial role of the microbiome in exerting this long-term effect. Further, the state of the intestinal microbiome has been shown to imprint on long-term functional properties of natural killer cells that result in different outcomes after experimental induction of colitis [79]. This effect was only restricted to a defined "vulnerable" period in the immunological life history and linked to changes in DNA methylation patterns making it likely that changes in the cell-type-specific epigenomes may modulate inflammatory responses in the long term. Along this line, in a larger cohort of monozygotic twins, stable correlations between the presence of distinct bacterial species and certain host transcripts or transcript profiles have been shown [80]. In IBD twins, this stable correlation is lost, which points to a gradual loss of epigenetic control of this two-way interaction.

It will thus be interesting to link the more classical view of nutrigenomics that is regularly defined as the investigation of how food components impact on phenotype–genotype interactions [81] with the "other" dimension of our intestinal genome, the microbiome, and related epigenetic marks. The advent of large-scale sequencing now allows for time and cost-efficient investigation of different sequence spaces, including the many epigenomes of the intestinal tract and their potential functional consequences (see Fig. 9.1). For the first time, the hypothesis that epigenetic modifications are the missing connection between genetic predisposition, environmental influences, and disease manifestation can be tested and put into a functional and clinical perspective. Several consortia have been launched within the framework of the International Human Epigenome Consortium (IHEC) including the BLUEPRINT [82] and DEEP networks that exactly address these questions in the different cellular compartments of the intestinal mucosa.

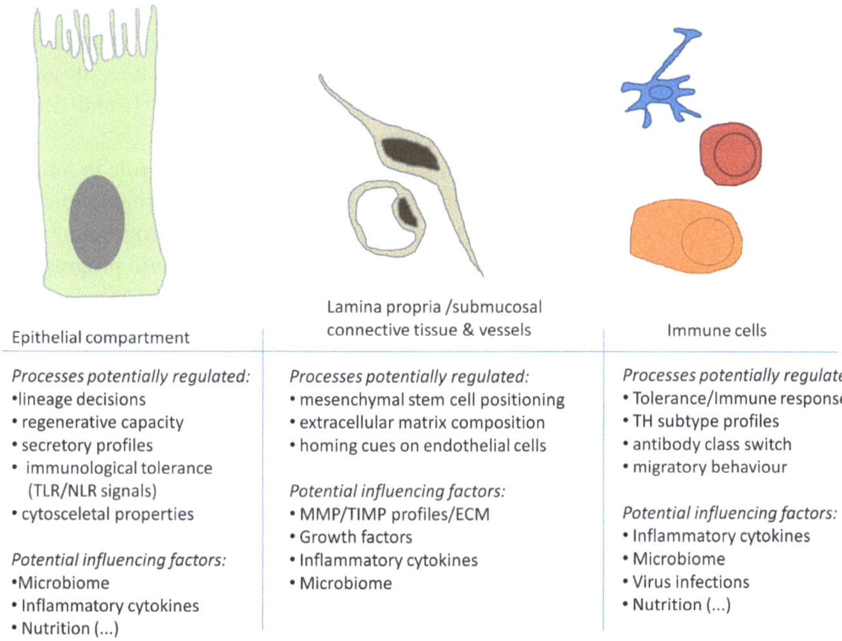

Fig. 9.1 Scheme of cellular compartments of the intestinal mucosa that are potentially influenced by epigenetic alterations in IBD

Is Epigenetics the Missing Link Between Inflammation and Cancer?

Tissue damage, wound healing, and continuously increased cell proliferation are only a few mechanisms of inflammation, which are believed to contribute to the initiation and development of cancer [83]. However, many elements of this link are still not understood. Since tumor tissue is often found to be globally hypomethylated and locally hypermethylated [32] which is believed to inactivate tumor suppressor genes [84], a key question is how inflammation can promote such changes in methylation. A general principle discussed in this context is inflammation-mediated cytosine damage: DNA damage caused by inflammatory agents such as reactive oxygen species can lead to inappropriate methylation, finally resulting in the development of cancer [85]. One pioneering study in this field demonstrated that epigenetic modification of the promoter of E-cadherin is associated with ulcerative colitis in patients undergoing colectomy [86], hypothesizing about its role in the progression from chronic inflammation to cancer. Interestingly, assessment of the methylation of 11 genes comparing 48 ulcerative colitis-associated cancers, 21 ulcerative colitis-associated dysplasias and 69 sporadic colorectal cancers could not show that epigenetic modifications lead to more aggressive clinical courses [87]. In contrast to that, other studies linked altered methylation in several target genes to predisposition, manifestation, or progression of colorectal cancer in patients with ulcerative colitis as well as in model systems: Studies on the alternate reading frame p14 (ARF) [88], WNT signaling pathway genes [89], DNA mismatch repair genes [90], genes coding for the tumor suppressors ESR1 and N33 [91], and death-associated protein kinase DAPK [92] supported this concept.

Similarly, several miRNAs have been shown to potentially play a role in inflammatory bowel-associated neoplastic transformation: miRNA-31 dysregulation was presented as a candidate in the context of chronic inflammation progressing into tumor. This micro-RNA has been also shown to increase with disease progression in IBD patients [93]. Neurotensin, which promotes inflammation and colon cancer by activating neurotensin-1 receptor, has been shown to stimulate the expression of miR-21 and miR-155, suggesting a functional link. Furthermore, tissue levels of both micro-RNAs correlated with tumor stage in human colon tumor samples [94]. Recent studies have identified several miRNAs being regulated during the progression from dysplasia to cancer in patients with IBD (miR-122, miR-181a, miR-146b-5p, let-7e, miR-17, miR-143) [95].

As the potential development of colorectal cancer is one of the most serious complications for patients with IBD, the need of a more detailed understanding how inflammation can progress is evident. Epigenetic modifications, especially DNA methylation, may represent an inflammatory memory in the intestinal mucosa. However, as many of the studies provide mostly an exemplary view on a selected group of patients, drawing conclusions on the validity of the results for larger cohorts should be undertaken carefully. Further functional studies, documenting both the functional background and the clinical validity, will be required to further progress in this field.

Outlook: Epigenetic Strategies in Diagnostics and Treatment of Inflammatory Bowel Disease

Biomarkers to classify diseases or disease subtypes have been always a major goal in epigenetic research. However, providing validated biomarkers of adequate diagnostic value is a challenging and considerably expensive endeavor. Most publications are descriptive and use the term biomarker in the context of an observed molecular pattern, without the final validation which could demonstrate the validity of this observation in a clinical setting. In fact, most current approaches cannot afford taking their findings into very large cohorts. Such cohorts could be only created in joint efforts between academia and industry [96]. As epigenetic modifications in IBD have been linked to cancer, several therapeutical avenues from the field of cancer drugs could be potentially of interest in the therapy of IBD. However, it has been recently questioned whether premature claims on the effectiveness of some drugs are the result of peer pressure rather than the result of validated clinical research [97].

Independent of these shortcomings, several studies aimed to utilize the specificity of observed epigenetic patterns. Easily accessible biomaterials, such as peripheral blood, are of particular interest in this context: In children, where noninvasive methods are often favored, a study has presented 11 CD-associated serum miRNAs potentially suitable for diagnostic purposes [98]. Similarly, differentiating active ulcerative colitis from Crohn's disease was possible using a defined set of miRNAs from peripheral blood in adults [99], further supporting the hypothesis of specific patterns. A recent study confirmed this concept with a different and reduced set of miRNAs in circulating blood [100], which suggests that the number of specific signals is substantially larger than the number of biomarkers currently published.

A major issue of the upcoming large-scale epigenomic studies in IBD will be the elucidation of cellular specificity of such epigenetic events. As epigenetic profiles are highly cell-type specific [16, 17, 101], most of the previous studies aiming to develop clinical biomarkers or novel therapeutic principles suffer from the fact that sum signals (i.e., the entire mucosa or whole peripheral blood) were investigated. Even reproducible changes could thus reflect secondary differences in cellular composition rather than pathophysiologically relevant differences in epigenetic profiles. If epigenetic marks are to be translated to clinical therapies, the molecular chain of events has to be detected and linked to defined cell populations. It is evident that epigenetic variation may have broad consequences on cellular phenotypes in all functional compartments of the intestinal mucosa (see Fig. 9.1). It is thus important to experimentally understand the impact of certain marks in a functional context, e.g., how are lineage decisions influenced by epigenetic alterations in intestinal epithelial stem cells? What is the impact of epigenetic modifications on tolerance to microbial stimuli in professional migratory immune cells? Is there a trans-generational effect of inflammatory effects that can be attributed to epigenetic principles? If these aims can be reached, it is likely that we can start looking for epigenetic marks (that could possibly be linked to microbiome changes) even prior to clinical manifestation of disease. It will be a challenge to identify therapeutic principles that specifically target single epigenetic modifications; so far compounds like HDAC inhibitors completely lack target specificity, but still have been found efficacious in defined inflammatory indications like systemic sclerosis. Patterns of epigenetic marks represent a dynamic picture into etiology. The ultimate goal of EWAS is to merge high-resolution information on epigenetic variation such as differential DNA methylation or miRNA levels with functional consequences on mRNA regulation and clinical phenotypes into a molecular risk map that will contribute to a clearer understanding of the etiology of IBD. This map will help bridging the gap between unexplained disease susceptibility and disease manifestation and may lead to the identification of novel diagnostic and therapeutic targets. Broadening the scope of such studies to longitudinal studies that follow high-risk populations (e.g., IBD kindred cohorts) into manifestation may even result in biomarkers for identifying susceptible individuals prior to disease manifestation. Applying targeted preventive measures (e.g. modification of the intestinal microbiome) in such high-risk individuals would—for the first time—aim for a causative intervention, which in the end may only be possible before the onset of clinically overt disease.

Acknowledgements This work was supported by Nationales Genomforschungsnetz (NGFN) grants by the BMBF, the EU grant gEUVADIS, the Deutsche Forschungsgemeinschaft (DFG) under contract numbers SFB877, RO2994/5-1, and the Cluster of Excellence Inflammation at Interfaces. SB was supported by a Royal Society Wolfson Research Merit Award (WM100023) and grants from the Wellcome Trust (084071) and EU-FP7 BLUEPRINT (282510). We apologize to those researchers whose important contributions to the field we were unable to include.

References

1. Bernstein BE, Mikkelsen TS, Xie X, Kamal M, Huebert DJ, Cuff J et al (2006) A bivalent chromatin structure marks key developmental genes in embryonic stem cells. Cell 125(2): 315–326, Epub 2006/04/25
2. Waddington CH (2012) The epigenotype. 1942. Int J Epidemiol 41(1):10–13
3. Gendrel AV, Heard E (2011) Fifty years of X-inactivation research. Development 138(23):5049–5055, Epub 2011/11/10
4. Ferguson-Smith AC (2011) Genomic imprinting: the emergence of an epigenetic paradigm. Nat Rev Genet 12(8):565–575, Epub 2011/07/19
5. Reik W (2007) Stability and flexibility of epigenetic gene regulation in mammalian development. Nature 447(7143):425–432, Epub 2007/05/25
6. Cherry AB, Daley GQ (2012) Reprogramming cellular identity for regenerative medicine. Cell 148(6):1110–1122, Epub 2012/03/20
7. Feinberg AP, Vogelstein B (1983) Hypomethylation distinguishes genes of some human cancers from their normal counterparts. Nature 301(5895):89–92, Epub 1983/01/06
8. Feinberg AP (2007) Phenotypic plasticity and the epigenetics of human disease. Nature 447(7143):433–440, Epub 2007/05/25
9. Petronis A (2010) Epigenetics as a unifying principle in the aetiology of complex traits and diseases. Nature 465(7299):721–727, Epub 2010/06/11
10. Bird A (2007) Perceptions of epigenetics. Nature 447(7143):396–398, Epub 2007/05/25
11. Suzuki MM, Bird A (2008) DNA methylation landscapes: provocative insights from epigenomics. Nat Rev Genet 9(6):465–476, Epub 2008/05/09
12. Branco MR, Ficz G, Reik W (2012) Uncovering the role of 5-hydroxymethylcytosine in the epigenome. Nat Rev Genet 13(1):7–13, Epub 2011/11/16
13. Bannister AJ, Kouzarides T (2011) Regulation of chromatin by histone modifications. Cell Res 21(3):381–395, Epub 2011/02/16
14. Rakyan VK, Down TA, Balding DJ, Beck S (2011) Epigenome-wide association studies for common human diseases. Nat Rev Genet 12(8):529–541, Epub 2011/07/13
15. Visscher PM, Brown MA, McCarthy MI, Yang J (2012) Five years of GWAS discovery. Am J Hum Genet 90(1):7–24, Epub 2012/01/17
16. Beck S (2010) Taking the measure of the methylome. Nat Biotechnol 28(10):1026–1028, Epub 2010/10/15
17. Laird PW (2010) Principles and challenges of genomewide DNA methylation analysis. Nat Rev Genet 11(3):191–203, Epub 2010/02/04
18. Rakyan VK, Beyan H, Down TA, Hawa MI, Maslau S, Aden D et al (2011) Identification of type 1 diabetes-associated DNA methylation variable positions that precede disease diagnosis. PLoS Genet 7(9):e1002300, Epub 2011/10/08
19. Bell JT, Spector TD (2011) A twin approach to unraveling epigenetics. Trends Genet 27(3):116–125, Epub 2011/01/25
20. Schreiber S, Rosenstiel P, Albrecht M, Hampe J, Krawczak M (2005) Genetics of Crohn disease, an archetypal inflammatory barrier disease. Nat Rev Genet 6(5):376–388, Epub 2005/04/30

21. Felsen J, Wolarsky W (1955) Familial incidence of ulcerative colitis and ileitis. Gastroenterology 28(3):412–417, Epub 1955/03/01
22. Rosenstiel P, Sina C, Franke A, Schreiber S (2009) Towards a molecular risk map—recent advances on the etiology of inflammatory bowel disease. Semin Immunol 21(6):334–345, Epub 2009/11/21
23. Hampe J, Cuthbert A, Croucher PJ, Mirza MM, Mascheretti S, Fisher S et al (2001) Association between insertion mutation in NOD2 gene and Crohn's disease in German and British populations. Lancet 357(9272):1925–1928, Epub 2001/06/27
24. Hugot JP, Chamaillard M, Zouali H, Lesage S, Cezard JP, Belaiche J et al (2001) Association of NOD2 leucine-rich repeat variants with susceptibility to Crohn's disease. Nature 411(6837):599–603, Epub 2001/06/01
25. Duerr RH, Taylor KD, Brant SR, Rioux JD, Silverberg MS, Daly MJ et al (2006) A genome-wide association study identifies IL23R as an inflammatory bowel disease gene. Science 314(5804):1461–1463, Epub 2006/10/28
26. Franke A, Balschun T, Karlsen TH, Hedderich J, May S, Lu T et al (2008) Replication of signals from recent studies of Crohn's disease identifies previously unknown disease loci for ulcerative colitis. Nat Genet 40(6):713–715, Epub 2008/04/29
27. Jostins L, Ripke S, Weersma RK, Duerr RH, McGovern DP, Hui KY et al (2012) Host-microbe interactions have shaped the genetic architecture of inflammatory bowel disease. Nature 491(7422):119–124, Epub 2012/11/07
28. Orholm M, Binder V, Sorensen TI, Rasmussen LP, Kyvik KO (2000) Concordance of inflammatory bowel disease among Danish twins. Results of a nationwide study. Scand J Gastroenterol 35(10):1075–1081, Epub 2000/12/01
29. Maher B (2008) Personal genomes: the case of the missing heritability. Nature 456(7218): 18–21, Epub 2008/11/07
30. Renz H, von Mutius E, Brandtzaeg P, Cookson WO, Autenrieth IB, Haller D (2011) Gene-environment interactions in chronic inflammatory disease. Nat Immunol 12(4):273–277, Epub 2011/03/23
31. Egger G, Liang G, Aparicio A, Jones PA (2004) Epigenetics in human disease and prospects for epigenetic therapy. Nature 429(6990):457–463, Epub 2004/05/28
32. Goelz SE, Vogelstein B, Hamilton SR, Feinberg AP (1985) Hypomethylation of DNA from benign and malignant human colon neoplasms. Science 228(4696):187–190, Epub 1985/04/12
33. Esteller M (2007) Cancer epigenomics: DNA methylomes and histone-modification maps. Nat Rev Genet 8(4):286–298, Epub 2007/03/07
34. Melo SA, Ropero S, Moutinho C, Aaltonen LA, Yamamoto H, Calin GA et al (2009) A TARBP2 mutation in human cancer impairs microRNA processing and DICER1 function. Nat Genet 41(3):365–370, Epub 2009/02/17
35. Saito Y, Liang G, Egger G, Friedman JM, Chuang JC, Coetzee GA et al (2006) Specific activation of microRNA-127 with downregulation of the proto-oncogene BCL6 by chromatin-modifying drugs in human cancer cells. Cancer Cell 9(6):435–443, Epub 2006/06/13
36. Heijmans BT, Tobi EW, Stein AD, Putter H, Blauw GJ, Susser ES et al (2008) Persistent epigenetic differences associated with prenatal exposure to famine in humans. Proc Natl Acad Sci U S A 105(44):17046–17049, Epub 2008/10/29
37. Mann MR, Lee SS, Doherty AS, Verona RI, Nolen LD, Schultz RM et al (2004) Selective loss of imprinting in the placenta following preimplantation development in culture. Development 131(15):3727–3735, Epub 2004/07/09
38. Rakyan VK, Down TA, Thorne NP, Flicek P, Kulesha E, Graf S et al (2008) An integrated resource for genome-wide identification and analysis of human tissue-specific differentially methylated regions (tDMRs). Genome Res 18(9):1518–1529, Epub 2008/06/26
39. van Overveld FJ, Demkow UA, Gorecka D, Zielinski J, De Backer WA (2003) Inhibitory capacity of different steroids on neutrophil migration across a bilayer of endothelial and bronchial epithelial cells. Eur J Pharmacol 477(3):261–267, Epub 2003/10/03

40. Backdahl L, Bushell A, Beck S (2009) Inflammatory signalling as mediator of epigenetic modulation in tissue-specific chronic inflammation. Int J Biochem Cell Biol 41(1):176–184, Epub 2008/09/17
41. Shuto T, Furuta T, Oba M, Xu H, Li JD, Cheung J et al (2006) Promoter hypomethylation of Toll-like receptor-2 gene is associated with increased proinflammatory response toward bacterial peptidoglycan in cystic fibrosis bronchial epithelial cells. FASEB J 20(6):782–784, Epub 2006/02/16
42. Barros SP, Offenbacher S (2009) Epigenetics: connecting environment and genotype to phenotype and disease. J Dent Res 88(5):400–408, Epub 2009/06/06
43. Gervin K, Vigeland MD, Mattingsdal M, Hammero M, Nygard H, Olsen AO et al (2012) DNA methylation and gene expression changes in monozygotic twins discordant for psoriasis: identification of epigenetically dysregulated genes. PLoS Genet 8(1):e1002454, Epub 2012/02/01
44. Lewis BP, Burge CB, Bartel DP (2005) Conserved seed pairing, often flanked by adenosines, indicates that thousands of human genes are microRNA targets. Cell 120(1):15–20, Epub 2005/01/18
45. Nahid MA, Pauley KM, Satoh M, Chan EK (2009) miR-146a is critical for endotoxin-induced tolerance: IMPLICATION IN INNATE IMMUNITY. J Biol Chem 284(50): 34590–34599, Epub 2009/10/21
46. Taganov KD, Boldin MP, Chang KJ, Baltimore D (2006) NF-kappaB-dependent induction of microRNA miR-146, an inhibitor targeted to signaling proteins of innate immune responses. Proc Natl Acad Sci U S A 103(33):12481–12486, Epub 2006/08/04
47. Gloria L, Cravo M, Pinto A, de Sousa LS, Chaves P, Leitao CN et al (1996) DNA hypomethylation and proliferative activity are increased in the rectal mucosa of patients with long-standing ulcerative colitis. Cancer 78(11):2300–2306, Epub 1996/12/01
48. Kellermayer R, Balasa A, Zhang W, Lee S, Mirza S, Chakravarty A et al (2010) Epigenetic maturation in colonic mucosa continues beyond infancy in mice. Hum Mol Genet 19(11):2168–2176, Epub 2010/03/04
49. Gonsky R, Deem RL, Targan SR (2009) Distinct methylation of IFNG in the gut. J Interferon Cytokine Res 29(7):407–414, Epub 2009/05/20
50. Gonsky R, Deem RL, Landers CJ, Derkowski CA, Berel D, McGovern DP et al (2011) Distinct IFNG methylation in a subset of ulcerative colitis patients based on reactivity to microbial antigens. Inflamm Bowel Dis 17(1):171–178, Epub 2010/09/18
51. Dideberg V, Kristjansdottir G, Milani L, Libioulle C, Sigurdsson S, Louis E et al (2007) An insertion-deletion polymorphism in the interferon regulatory factor 5 (IRF5) gene confers risk of inflammatory bowel diseases. Hum Mol Genet 16(24):3008–3016, Epub 2007/09/21
52. Balasa A, Gathungu G, Kisfali P, Smith EO, Cho JH, Melegh B et al (2010) Assessment of DNA methylation at the interferon regulatory factor 5 (IRF5) promoter region in inflammatory bowel diseases. Int J Colorectal Dis 25(5):553–556, Epub 2010/02/04
53. Lin Z, Hegarty JP, Cappel JA, Yu W, Chen X, Faber P et al (2011) Identification of disease-associated DNA methylation in intestinal tissues from patients with inflammatory bowel disease. Clin Genet 80(1):59–67, Epub 2010/10/19
54. Nimmo ER, Prendergast JG, Aldhous MC, Kennedy NA, Henderson P, Drummond HE et al (2012) Genome-wide methylation profiling in Crohn's disease identifies altered epigenetic regulation of key host defense mechanisms including the Th17 pathway. Inflamm Bowel Dis 18(5):889–899, Epub 2011/10/25
55. Gore SD, Jones C, Kirkpatrick P (2006) Decitabine. Nat Rev Drug Discov 5(11):891–892, Epub 2006/11/23
56. Issa JP, Kantarjian HM, Kirkpatrick P (2005) Azacitidine. Nat Rev Drug Discov 4(4): 275–276, Epub 2005/05/03
57. Lee RC, Feinbaum RL, Ambros V (1993) The C. elegans heterochronic gene lin-4 encodes small RNAs with antisense complementarity to lin-14. Cell 75(5):843–854, Epub 1993/12/03
58. John B, Enright AJ, Aravin A, Tuschl T, Sander C, Marks DS (2004) Human microRNA targets. PLoS Biol 2(11):e363, Epub 2004/10/27

59. Sonkoly E, Wei T, Janson PC, Saaf A, Lundeberg L, Tengvall-Linder M et al (2007) MicroRNAs: novel regulators involved in the pathogenesis of psoriasis? PLoS One 2(7):e610, Epub 2007/07/12
60. Suarez Y, Wang C, Manes TD, Pober JS (2010) Cutting edge: TNF-induced microRNAs regulate TNF-induced expression of E-selectin and intercellular adhesion molecule-1 on human endothelial cells: feedback control of inflammation. J Immunol 184(1):21–25, Epub 2009/12/02
61. Worm J, Stenvang J, Petri A, Frederiksen KS, Obad S, Elmen J et al (2009) Silencing of microRNA-155 in mice during acute inflammatory response leads to derepression of c/ebp beta and down-regulation of G-CSF. Nucleic Acids Res 37(17):5784–5792, Epub 2009/07/15
62. Häsler R, Jacobs G, Till A, Grabe N, Cordes C, Nikolaus S, et al (2012) Microbial pattern recognition causes distinct functional microRNA signatures in human primary monocytes PLoS One.7(2):e31151. doi: 10.1371/journal.pone.0031151
63. Wu F, Zhang S, Dassopoulos T, Harris ML, Bayless TM, Meltzer SJ et al (2010) Identification of microRNAs associated with ileal and colonic Crohn's disease. Inflamm Bowel Dis 16(10):1729–1738, Epub 2010/09/18
64. Okubo M, Tahara T, Shibata T, Yamashita H, Nakamura M, Yoshioka D et al (2011) Association study of common genetic variants in pre-microRNAs in patients with ulcerative colitis. J Clin Immunol 31(1):69–73, Epub 2010/09/18
65. Zwiers A, Kraal L, van de Pouw Kraan TC, Wurdinger T, Bouma G, Kraal G (2012) Cutting edge: a variant of the IL-23R gene associated with inflammatory bowel disease induces loss of microRNA regulation and enhanced protein production. J Immunol 188(4):1573–1577, Epub 2012/01/21
66. Duttagupta R, DiRienzo S, Jiang R, Bowers J, Gollub J, Kao J et al (2012) Genome-wide maps of circulating miRNA biomarkers for ulcerative colitis. PLoS One 7(2):e31241, Epub 2012/02/24
67. Fasseu M, Treton X, Guichard C, Pedruzzi E, Cazals-Hatem D, Richard C et al (2010) Identification of restricted subsets of mature microRNA abnormally expressed in inactive colonic mucosa of patients with inflammatory bowel disease. PLoS One 5(10):e13160, Epub 2010/10/20
68. Gent AE, Hellier MD, Grace RH, Swarbrick ET, Coggon D (1994) Inflammatory bowel disease and domestic hygiene in infancy. Lancet 343(8900):766–767, Epub 1994/03/26
69. Hampe J, Heymann K, Krawczak M, Schreiber S (2003) Association of inflammatory bowel disease with indicators for childhood antigen and infection exposure. Int J Colorectal Dis 18(5):413–417, Epub 2003/04/11
70. Dicksved J, Floistrup H, Bergstrom A, Rosenquist M, Pershagen G, Scheynius A et al (2007) Molecular fingerprinting of the fecal microbiota of children raised according to different lifestyles. Appl Environ Microbiol 73(7):2284–2289, Epub 2007/02/13
71. von Mutius E, Fritzsch C, Weiland SK, Roll G, Magnussen H (1992) Prevalence of asthma and allergic disorders among children in united Germany: a descriptive comparison. BMJ 305(6866):1395–1399, Epub 1992/12/05
72. von Mutius E, Martinez FD, Fritzsch C, Nicolai T, Reitmeir P, Thiemann HH (1994) Skin test reactivity and number of siblings. BMJ 308(6930):692–695, Epub 1994/03/12
73. Cho I, Blaser MJ (2012) The human microbiome: at the interface of health and disease. Nat Rev Genet 13(4):260–270, Epub 2012/03/14
74. Turnbaugh PJ, Ley RE, Hamady M, Fraser-Liggett CM, Knight R, Gordon JI (2007) The human microbiome project. Nature 449(7164):804–810, Epub 2007/10/19
75. Arumugam M, Raes J, Pelletier E, Le Paslier D, Yamada T, Mende DR et al (2011) Enterotypes of the human gut microbiome. Nature 473(7346):174–180, Epub 2011/04/22
76. Bamias G, Martin C 3rd, Marini M, Hoang S, Mishina M, Ross WG et al (2003) Expression, localization, and functional activity of TL1A, a novel Th1-polarizing cytokine in inflammatory bowel disease. J Immunol 171(9):4868–4874

77. Turnbaugh PJ, Ley RE, Mahowald MA, Magrini V, Mardis ER, Gordon JI (2006) An obesity-associated gut microbiome with increased capacity for energy harvest. Nature 444(7122):1027–1031, Epub 2006/12/22
78. Hashimoto T, Perlot T, Rehman A, Trichereau J, Ishiguro H, Paolino M et al (2012) ACE2 links amino acid malnutrition to microbial ecology and intestinal inflammation. Nature 487(7408):477–481, Epub 2012/07/28
79. Olszak T, An D, Zeissig S, Vera MP, Richter J, Franke A et al (2012) Microbial exposure during early life has persistent effects on natural killer T cell function. Science 336(6080): 489–493, Epub 2012/03/24
80. Lepage P, Hasler R, Spehlmann ME, Rehman A, Zvirbliene A, Begun A et al (2011) Twin study indicates loss of interaction between microbiota and mucosa of patients with ulcerative colitis. Gastroenterology 141(1):227–236, Epub 2011/05/31
81. Barnett M, Bermingham E, McNabb W, Bassett S, Armstrong K, Rounce J et al (2010) Investigating micronutrients and epigenetic mechanisms in relation to inflammatory bowel disease. Mutat Res 690(1–2):71–80, Epub 2010/03/02
82. Adams D, Altucci L, Antonarakis SE, Ballesteros J, Beck S, Bird A et al (2012) BLUEPRINT to decode the epigenetic signature written in blood. Nat Biotechnol 30(3):224–226, Epub 2012/03/09
83. Rajput S, Wilber A (2010) Roles of inflammation in cancer initiation, progression, and metastasis. Front Biosci (Schol Ed) 2:176–183, Epub 2009/12/29
84. Chan AO, Rashid A (2006) CpG island methylation in precursors of gastrointestinal malignancies. Curr Mol Med 6(4):401–408, Epub 2006/08/12
85. Valinluck V, Sowers LC (2007) Inflammation-mediated cytosine damage: a mechanistic link between inflammation and the epigenetic alterations in human cancers. Cancer Res 67(12):5583–5586, Epub 2007/06/19
86. Azarschab P, Porschen R, Gregor M, Blin N, Holzmann K (2002) Epigenetic control of the E-cadherin gene (CDH1) by CpG methylation in colectomy samples of patients with ulcerative colitis. Genes Chromosomes Cancer 35(2):121–126, Epub 2002/08/31
87. Konishi K, Shen L, Wang S, Meltzer SJ, Harpaz N, Issa JP (2007) Rare CpG island methylator phenotype in ulcerative colitis-associated neoplasias. Gastroenterology 132(4): 1254–1260, Epub 2007/04/06
88. Moriyama T, Matsumoto T, Nakamura S, Jo Y, Mibu R, Yao T et al (2007) Hypermethylation of p14 (ARF) may be predictive of colitic cancer in patients with ulcerative colitis. Dis Colon Rectum 50(9):1384–1392, Epub 2007/08/01
89. Dhir M, Montgomery EA, Glockner SC, Schuebel KE, Hooker CM, Herman JG et al (2008) Epigenetic regulation of WNT signaling pathway genes in inflammatory bowel disease (IBD) associated neoplasia. J Gastrointest Surg 12(10):1745–1753, Epub 2008/08/22
90. Edwards RA, Witherspoon M, Wang K, Afrasiabi K, Pham T, Birnbaumer L et al (2009) Epigenetic repression of DNA mismatch repair by inflammation and hypoxia in inflammatory bowel disease-associated colorectal cancer. Cancer Res 69(16):6423–6429, Epub 2009/07/30
91. Arasaradnam RP, Khoo K, Bradburn M, Mathers JC, Kelly SB (2010) DNA methylation of ESR-1 and N-33 in colorectal mucosa of patients with ulcerative colitis (UC). Epigenetics 5(5):422–426, Epub 2010/05/28
92. Kuester D, Guenther T, Biesold S, Hartmann A, Bataille F, Ruemmele P et al (2010) Aberrant methylation of DAPK in long-standing ulcerative colitis and ulcerative colitis-associated carcinoma. Pathol Res Pract 206(9):616–624, Epub 2010/07/16
93. Olaru AV, Selaru FM, Mori Y, Vazquez C, David S, Paun B et al (2011) Dynamic changes in the expression of MicroRNA-31 during inflammatory bowel disease-associated neoplastic transformation. Inflamm Bowel Dis 17(1):221–231, Epub 2010/09/18
94. Bakirtzi K, Hatziapostolou M, Karagiannides I, Polytarchou C, Jaeger S, Iliopoulos D et al (2011) Neurotensin signaling activates microRNAs-21 and -155 and Akt, promotes tumor growth in mice, and is increased in human colon tumors. Gastroenterology 141(5):1749.e1–1761.e1, Epub 2011/08/03

95. Kanaan Z, Rai SN, Eichenberger MR, Barnes C, Dworkin AM, Weller C et al (2012) Differential microRNA expression tracks neoplastic progression in inflammatory bowel disease-associated colorectal cancer. Hum Mutat 33(3):551–560, Epub 2012/01/14
96. Sawyers CL (2008) The cancer biomarker problem. Nature 452(7187):548–552, Epub 2008/04/04
97. Begley CG, Ellis LM (2012) Drug development: raise standards for preclinical cancer research. Nature 483(7391):531–533, Epub 2012/03/31
98. Zahm AM, Thayu M, Hand NJ, Horner A, Leonard MB, Friedman JR (2011) Circulating microRNA is a biomarker of pediatric Crohn disease. J Pediatr Gastroenterol Nutr 53(1): 26–33, Epub 2011/05/07
99. Wu F, Guo NJ, Tian H, Marohn M, Gearhart S, Bayless TM et al (2011) Peripheral blood microRNAs distinguish active ulcerative colitis and Crohn's disease. Inflamm Bowel Dis 17(1):241–250, Epub 2010/09/03
100. Paraskevi A, Theodoropoulos G, Papaconstantinou I, Mantzaris G, Nikiteas N, Gazouli M (2012) Circulating microRNA in inflammatory bowel disease. J Crohns Colitis 6(9): 900–904, Epub 2012/03/06
101. Rakyan VK, Hildmann T, Novik KL, Lewin J, Tost J, Cox AV et al (2004) DNA methylation profiling of the human major histocompatibility complex: a pilot study for the human epigenome project. PLoS Biol 2(12):e405, Epub 2004/11/20

Part III
Pathogenetic Pathways in Inflammatory Bowel Disease

Chapter 10
Nod1 and Nod2 and the Immune Response to Bacteria

Maria Kaparakis-Liaskos and Dana J. Philpott

Abstract The mammalian host has evolved to develop a diverse array of innate immune receptors and strategies to defend itself against infection by microbial pathogens. These germ-line-encoded and conserved microbial receptors, called pattern recognition receptors (PRRs), are associated with the membranes or within the cytosol of host cells. PRRs enable the host to rapidly respond to pathogen-associated molecular patterns (PAMPs), as a first line of defence against microbial intrusion. Signalling via PAMPs enables the host to mount a rapid and non-specific immune response that results in inflammation and ultimately the activation of the adaptive immune system.

The host has a variety of PRRs, including the membrane-bound toll-like receptors (TLRs) and the cytoplasmic nucleotide-binding oligomerisation domain (Nod)-like receptor (NLR) protein family. In this chapter, we will focus predominantly on Nod1 and Nod2, which are members of the NLR family of proteins, and the role they have in the initiation and development of an immune response to bacteria. We will discuss the various methods whereby bacteria are detected and can induce signalling via Nod receptors and the role of Nod proteins in human disease, especially Nod2's role in Crohn's disease.

M. Kaparakis-Liaskos, Ph.D.
Centre for Innate Immunity and Infectious Diseases, Monash Institute of Medical Research, Monash University, Clayton, VIC 3168, Australia
e-mail: maria.liaskos@monash.edu

D.J. Philpott, Ph.D. (✉)
Department of Immunology, University of Toronto, MSB Room 7318, 1 King's College Circle, Toronto, ON, Canada, M5S 1A8
e-mail: dana.philpott@utoronto.ca

M. D'Amato and J.D. Rioux (eds.), *Molecular Genetics of Inflammatory Bowel Disease*, DOI 10.1007/978-1-4614-8256-7_10, © Springer Science+Business Media New York 2013

Nod-Like Receptors

Mammalian cells express one or more types of cytosolic PRRs that play important roles in host defence against microbial pathogens, in addition to their recognition of "danger signals" from within eukaryotic cells. One family of mammalian PRRs is the nucleotide oligomerisation domain (Nod) family, which have a central role in host defence against microbial pathogens [1]. Since the initial discovery of the first Nod-like receptors (NLRs), being Nod1 and Nod2, this family of PRRs expanded to currently consist of 23 genes in humans and 34 genes in mice [2, 3]. Hence, due to the rapid expansion of this group of cytoplasmic PRRs, it has been renamed the nucleotide-binding domain (NBD) and leucine-rich repeat (LRR) containing, or NLR family [4].

The NLR family is comprised of cytoplasmic proteins that are suggested to be sentinel receptors at front-line mucosal surfaces as well as in immune cells [5]. NLR proteins share some common features, being a C-terminal LRR-containing domain and central NACHT NBD(s) [1, 6]. The NLR family is now divided into four subfamilies based on the composition of the n-terminal effector domain of these receptors [4]. The Nod proteins, Nod1 and Nod2, contain a caspase-activated recruitment domain (CARD) at their n-terminus and are classified within the NLRC subfamily. This chapter will predominantly focus on the expression, detection and immunoregulation initiated by Nod1 and Nod2 in response to bacterial pathogens and their contribution to the regulation of gastrointestinal homeostasis, inflammation and immunity.

Nod1 and Nod2

The mammalian Nod proteins, Nod1 and Nod2, are critical in the regulation of inflammation and host defence against bacterial infections [2]. Nod1 and Nod2 proteins are located within the cytosolic compartment of host cells, and it has been suggested that their role in controlling the intestinal microbiota may have been a major selective pressure throughout evolution [3]. Nods are comprised of three domains: the first being a central nucleotide-binding oligomerisation domain (Nod) that is required for self-oligomerisation of the receptor. Secondly, Nods have a C-terminal LRR domain that contains multiple LRRs whose function is to sense the bacterial pathogen-associated molecular pattern (PAMP), being peptidoglycan. Finally, all Nods have a CARD. The CARD is essential for the homodimerisation of the receptor, and the recruitment of downstream adaptor proteins through homophilic and heterophilic protein interactions that are required to facilitate the pro-inflammatory signalling cascade in response to bacterial PAMP recognition [7]. Nod1 contains only one CARD domain, whereas Nod2 contains two CARD domains.

Initially, Nod1 and Nod2 were proposed to be intracellular sensors of bacterial lipopolysaccharide (LPS) [8, 9] and were suggested to have a role in regulating apoptosis and the NF-kappaB pathway [10]. However, the initial finding that Nods detected LPS was incorrect due to contaminants contained within the LPS preparations used in these studies. Further refinement and purification of bacterial cell preparations resulted in the identification that Nods sense bacterial peptidoglycan fragments contained within the cell wall of bacteria; however, the motifs recognised by Nod1 and Nod2 differ [11, 12].

In the host, Nod1 is expressed ubiquitously by most cell types [13–15]. Nod1 detects a specific and conserved structure of peptidoglycan that is commonly found in almost all Gram-negative bacterial peptidoglycan, as well as some Gram-positive bacteria such as *Bacillus subtilis* and *Listeria monocytogenes* [16, 17]. The muropeptide structure detected by Nod1 is composed of a disaccharide moiety, *N*-acetylglucosamine–*N*-acetylmuramic acid (GlcNAc–MurNAc), linked to a tripeptide of which the terminal amino acid is *meso*-diaminopimelate (mDAP), also known as GM-TriDAP [16, 17]. Most Gram-negative organisms contain mDAP within their cell wall [18]. Furthermore, most Gram-positive bacteria contain a lysine residue at the terminal position of their peptidoglycan, rendering their peptidoglycan incapable of signalling via Nod1. Interestingly, recognition of peptidoglycan by Nod1 is host specific, as human Nod1 specifically detects the GM-TriDAP structure of peptidoglycan [17]. However, murine Nod1 is most responsive to a tetrapeptide muropeptide containing L-alanine–D-glutamate–*meso*DAP–D-alanine (GM-TetraDAP) [19].

Nod2 was first identified approximately 10 years ago [20], and similar to Nod1, it also is composed of a NBD and multiple C-terminal LRRs; however, it has two N-terminal CARDs. Nod2 expression is mainly restricted to leukocytes consisting of T cells [21], neutrophils [22], macrophages [20] and dendritic cells. In addition, Nod2 is expressed at low levels by intestinal epithelial cell lines and primary intestinal epithelial cells [10, 23, 24]. The expression of Nod2 is basal within these cells; however, its expression can be induced by a variety of inflammatory signals such as LPS, tumour necrosis factor (TNF) and interferon gamma (IFN-gamma) [23–25]. In contrast to Nod1, Nod2 is considered to be a sensor of both Gram-negative and Gram-positive bacteria due to its ability to detect muramyl dipeptide (MDP), a component common to the peptidoglycan of both classes of bacteria [11, 12]. Therefore, Nod2 regulates the production of inflammatory mediators in response to all types of bacterial pathogens in order to maintain gut homeostasis [6].

More recently, the importance and impact of Nods on immune responses and in pathogenesis now extends beyond detecting bacteria. Nods have been implicated in the progression and development of gut homeostasis, chronic asthma, arthritis, dermatitis, IBD, multiple sclerosis [26], obesity [27], Chagas disease [28] and malaria [29, 30]. Conversely, over-activation of Nods can result in the development of autoinflammatory diseases such as Blau syndrome and sarcoidosis [31]. Collectively, these studies highlight that Nods may contribute to the progression and development of disease of various aetiologies; however, the mechanisms whereby they contribute to these diseases are not well understood and remain to be elucidated.

Nods Initiate a Pro-inflammatory Signalling Cascade

The detection of bacterial peptidoglycan by Nods initiates a signalling cascade that ultimately results in the production of pro-inflammatory cytokines and the development of an inflammatory innate immune response (see Fig. 10.1). The classical pathway of Nod signalling is as follows. Upon Nod recognition of peptidoglycan via its LRRs, it is speculated that Nod receptors self-oligomerise via their Nod domains. This activation and homodimerisation of Nods enables them to mediate the recruitment and oligomerisation of the RIP-like interacting CLARP kinase (RICK) [9, 32, 33], also known as receptor-interacting protein-2 (RIP-2), a member of the receptor-interacting protein kinase family [34]. RIP-2 subsequently interacts via an electrostatic interaction with the CARD domain(s) contained within Nod receptors [33, 35]. The interaction between the CARD of RIP-2 and Nods is specific and essential to the signalling process, as Nod1 signalling can be abolished when a truncated form of RIP-2 lacking the CARD is transfected into cells [20]. The homophilic CARD–CARD

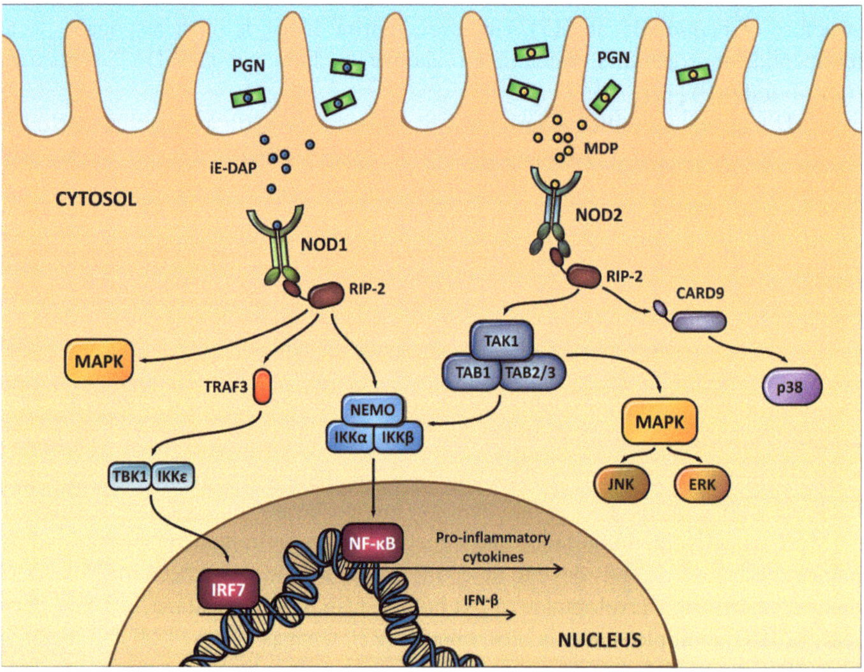

Fig. 10.1 Pro-inflammatory signalling mediated by Nod1 and Nod2. Nod1 and Nod2 detect their peptidoglycan ligands, *N*-acetylglucosamine–*N*-acetylmuramic acid-L-Ala–D-Glu–meso-diaminopimelic acid (GM-triDAP) and muramyl dipeptide (MDP), respectively, to trigger inflammation. Through their recruit of Rip2, Nod1 and Nod2 trigger NFkappaB and MAPK pathways to drive inflammatory cytokine production. Nod1 has also been shown to active IRF7 downstream of TBK1 and IKK epsilon leading to type I interferon production (IFB-beta) (Courtesy of artist: Priya Alwis)

interaction between Nods and RIP-2 results in RIP-2 being subsequently K63-polyubiquitinated within its kinase domain by the E3 ubiquitin ligases cIAP1, cIAP2 and xIAP [36, 37]. This in turn initiates the K63-linked polyubiquitination of NEMO, a scaffolding protein and regulator of the IkappaB kinase (IKK) complex [38, 39]. Therefore, the polyubiquitination of RIP-2 is essential for the activation of IKK, which in the case of Nod2 signalling, subsequently mediates via polyubiquitinated NEMO the recruitment of the transforming growth factor-beta (TGF-beta)-associated kinase (TAK1) [32], in addition to the TAB1 and TAB2 complex. The interaction between Nods, IKK and TAK1 results in the phosphorylation and degradation of the IKK complex, and the degradation of IkappaB proteins by the proteasome [10, 20, 32, 33]. This in turn ultimately facilitates the dissociation of the NF-kappaB p50 and p65 complex and the phosphorylation and translocation of the p65 subunit into the nucleus, enabling it to bind to consensus binding sites within pro-inflammatory genes. This allows the transcription of pro-inflammatory molecules including CXCL5, CXCL8 and its murine homologue CXCL2 (or macrophage inflammatory protein-2, Mip-2) [17, 24, 40].

Although significant advances have been made to broaden our understanding of the mechanisms and pathway(s) of Nod activation and Nod-dependent pro-inflammatory responses, certain key components of this pathway are yet to be elucidated. For example, the mechanism and location(s) where Nods directly interact with peptidoglycan and RIP-2 remain unknown. It has been proposed that when Nods are in an inactive state, their LRR remain folded over the Nod region and upon sensing their ligand undergo conformational changes that allow the homodimerisation of the receptors [1, 41]. Furthermore, our knowledge of Nod binding partners is limited. Two different protein interaction screens identified the LRR- and PDZ domain-containing family member Erbin as a binding partner of Nod2; however, the cellular impact of this interaction and its role in Nod signalling remains unclear [42, 43]. Additional screens have also found other Nod2 interactors, including Grim19 [30] and, more recently, carbamoyl phosphate synthetase/aspartate transcarbamylase/dihydroorotase (CAD [44]), but their overall contribution to bacterial detection and Crohn's disease pathogenesis is still unclear. Moreover, researchers have postulated that additional cytoplasmic host structures may be required for Nod signalling, similar to the inflammasome, and hence, the existence of a "Nodosome" or "Nod signalosome" has been proposed [45, 46]. Further studies are required to address these key steps in the process of Nod activation and ligand recognition.

Outcome of Nod Signalling

The Production of Antimicrobial Peptides

Signalling via the Nod1 and Nod2 receptor pathways results in the production of antimicrobial peptides, known as defensins, by epithelial cells. The production of defensins in response to Nod2 signalling is essential for the regulation of commensal organisms and maintaining gut homeostasis, as Nod2-defficient mice have an impaired regulation of bacterial load in their terminal ileum [47]. This study was performed by Petnicki-Ocwieja and colleagues, who isolated the intestinal crypts of wild-type C57BL/6 and Nod2 knockout mice and subsequently cultured them with bacteria. The supernatant obtained from wild-type crypts cultured with bacteria displayed potent antimicrobial activity against *Escherichia coli*, *Salmonella* and *Listeria monocytogenes* in a dose-dependent manner. Whereas supernatants from crypts isolated from Nod2 or RIP-2 knockout mice and stimulated with bacteria were hindered in their antimicrobial function [47]. The antimicrobial activity was attributed to Nod2 signalling inducing the production of alpha-defensins, which are small, cationic antimicrobial peptides produced by Paneth cells of the intestine [47]. In addition, a second study identified that both Nod2 and RIP-2 expressed by epithelial cells located within the intestinal ileal crypts facilitated the protection of mice against intestinal *Helicobacter hepaticus*-induced inflammation, due to the production of alpha-defensins which function in controlling the pathogen [48]. Furthermore, Nod2-dependent production of alpha-defensins protects mice against the intracellular pathogen *Listeria*, as Nod2 knockout mice display an inability to produce intestinal antimicrobial alpha-defensin peptides, known as cryptdins [49]. These findings have been corroborated in vivo by Wehkamp and colleagues, who identified that the production of alpha-defensins by Paneth cells in the intestine of Crohn's disease patients was reduced, and this decrease was most pronounced in patients with mutations in *Nod2*, suggesting that Nod2 and alpha-defensins may have a role in regulating the integrity and homeostasis of the gastrointestinal tract [50]. It should be noted that a Nod2-independent, MyD88-dependent mechanism for the production of antimicrobial peptides by Paneth cells has been reported [51], and this system may potentially function in unison with Nod2 to regulate the level of microbial flora within the gut, which will be discussed in detail below.

Indeed, a similar role for Nod1 in maintaining the intestinal microbiota homeostasis has been reported. Bouskra et al., identified that Nod1 knockout mice have a greater total number of bacteria in their gut, possibly due to the lack of beta-defensin antimicrobials produced at the intestinal epithelial surface in the absence of Nod1 signalling [52]. The antimicrobial peptides human beta-defensins (HBDs) are small cationic, low molecular weight peptides with immunomodulatory properties required for host defence from bacterial pathogens [53]. HBDs are endogenously produced by epithelial cells and their expression can be upregulated during infection in a Nod1-dependent manner. We, and others, identified that the gastric pathogen *Helicobacter pylori* induced the production of HBD2 by human epithelial cells in an NF-kappaB and Nod1-dependent manner [54, 55]. Furthermore, we demonstrated that culture supernatants obtained from *H. pylori*-stimulated epithelial cells contained HBD2 which exerted potent antimicrobial activity against *H. pylori*, and that Nod1 was essential for the production of this functional antimicrobial [55]. Similarly, the ability of Nod2 to induce the expression of HBDs has been demonstrated using MDP [56]. Collectively, these studies identify the ability of Nods to induce the production of alpha- and beta-defensins that function to regulate the overall number of the intestinal microbiota and reduce the ability of pathogenic bacteria to colonise the gastrointestinal tract.

The Production of Type I Interferons (IFNs)

Work by Watanabe and colleagues identified a novel pathway of Nod1 signalling, resulting in the induction of type I interferons (IFNs), an immune response typically associated with a viral infection [57]. The authors demonstrated that *H. pylori* stimulation of gastrointestinal epithelial cells initiated Nod1 signalling, the activation of RIP-2 and its interaction with the TNF receptor-associated factor 3 (TRAF3). This resulted in the sequential activation of TANK-binding kinase 1 (TBK1), IκB kinase epsilon (IKK epsilon) and the IFN regulatory factor 7 (IRF7). Subsequently, IRF7 can activate the transcription factor complex IFN-stimulated gene factor 3 (ISGF3), composed of Stat1, Stat2 and IRF9, enabling it to bind to an IFN-stimulated response element (ISRE), resulting in the production of the pro-inflammatory cytokines CXCL10, also known as IP-10 and IFN-beta [57] (see Fig. 10.1). The authors propose that this will in turn result in the generation of a pro-inflammatory T helper1 (Th1) response as a result of *H. pylori* infection.

The Production of Inflammatory Cytokines and the Recruitment of Innate Immune Cells

One of the key outcomes of Nod signalling is the production of cytokines, resulting in the recruitment and activation of pro-inflammatory innate immune cells. Studies using knockout animals have clarified the contribution of Nods to pathogen-initiated inflammation. Work by Masumoto and colleagues identified that administration of a Nod1 ligand intraperitoneally to wild-type mice induced neutrophil recruitment and the production of CCL2, also known as monocyte chemotactic protein-1 (MCP-1), and CXCL2 in the serum of these animals. However, Nod1 knockout mice displayed an inability to produce CCL2 in their serum in response to Nod1 ligand administration. Their findings clearly identified a role for Nod1 in the production of pro-inflammatory cytokines that functions to recruit monocytes and dendritic cells to the site of infection, further enhancing the development of a cellular innate and adaptive immune response [58]. In addition, Nod1-mediated neutrophil recruitment is an important immune response against the enteric Gram-positive pathogen *Clostridium difficile* [59]. *C. difficile* is normally located within the intestinal tract of healthy individuals, where its levels are maintained by the intestinal microbiota; however, in antibiotic-treated individuals, it is the causative agent of pseudomembranous colitis [60]. Work by Hasegawa and colleagues identified that Nod1 knockout mice infected with *C. difficile* in their intestinal tract were more prone to lethality due to an impaired clearance of the pathogen, compared to wild-type controls [59]. The impaired clearance of *C. difficile* in Nod1 knockout animals was dependent on a defect in the ability of these animals to produce CXCL1 and induce the recruitment of neutrophils to the infected site.

Similarly, Nod2 signalling by pathogens results in the secretion of pro-inflammatory cytokines and the recruitment of inflammatory cells, and some examples are listed below. Clearance of the enteric pathogen *Citrobacter rodentium* is regulated by Nod2-induced production of CCL2, which enables the recruitment of inflammatory monocytes into the colon and the induction of an adaptive immune response [61]. *Streptococcus pneumoniae* infection of phagocytes results in the production of CCL2, which propagates the inflammatory response by inducing the recruitment of macrophages to the site of infection to assist in the clearance of the pathogen [62]. Furthermore, it was shown in vivo using Nod2 knockout mice that Nod2 was also required for the generation of an antibody response specific for *S. pneumoniae* [62]. Also, infection with the Nod2-signalling bacterium *Mycobacterium*, influences the production of TNFα and IL1-β by macrophages in addition to regulating the ability of macrophages to control the intracellular growth of this pathogen [63]. This finding was validated in vitro using siRNA to knockdown of Nod2 and observing that the lack of Nod2 enhanced growth of *Mycobacterium* in macrophages [63]. A genetic association of Nod2 in regulating *M. leprae* infection was identified by performing a genome analysis of patients with leprosy, revealing that these infected individuals had a single-nucleotide polymorphism (SNP) in Nod2 that may be attributed to the disease outcome [64]. Moreover, using RIP-2 knockout animals, and peripheral blood mononuclear cells from individuals homozygous for a Nod2 polymorphism, it was determined that Nod2 plays a key role in the production of the pro-inflammatory cytokines IL-10, IL-6 and IL-1β in response to *Borrelia*, the causative agent of Lyme disease [65]. Finally, *Legionella pneumophila* has been reported to activate both Nod1 and Nod2, resulting in the induction of NF-kappaB and IFN-beta [66]. These researchers also showed that Nod1 is essential for the clearance of *L. pneumophila* in vivo [66].

The Production of Reactive Oxygen Species

In addition to driving innate immune cells to produce pro-inflammatory cytokines and facilitate the further recruitment of innate immune cells, Nod signalling can enhance the production of reactive oxygen species by innate immune cells. Moreover, Nod expression can be upregulated in the presence of pro-inflammatory cytokines, further enhancing the innate immune response initiated via Nod signalling. A study by Totemeyer and colleagues identified that IFN-gamma increased the expression of Nod2 within macrophages, heightening the production of the antimicrobial nitric oxide (NO) [67]. Furthermore, Nod1 stimulation with Gram-negative peptidoglycan or bacteria resulted in the expression of inducible nitric oxide synthase and NO production in combination with IFN-gamma in a diverse range of host cell types including bone marrow-derived dendritic cells [68] macrophages [19, 69], hepatocytes [70], mesothelial cells and smooth muscle cells [71]. An in vivo example of the requirement of Nods in the clearance of pathogens via NO production has been reported using *Chlamydophila pneumoniae* [72]. Clearance of *C. pneumoniae* in Nod1, Nod2 and RIP-2 knockout animals was impaired, due to an inability to induce iNOS expression and NO production, which subsequently resulted in delayed neutrophil recruitment to the lungs [72].

Enhanced Phagocytosis by Innate Immune Cells

Nod1 stimulation can promote and enhance the ability of innate immune cells such as macrophages and neutrophils to phagocytose pathogenic organisms. Indeed, peptidoglycan fragments from the Gram-negative organism *Haemophilus influenzae* were capable of inducing neutrophils to phagocytose opsonised Gram-positive *S pneumoniae* [73]. This finding was further corroborated using a murine infection model, whereby neutrophils from mice treated with purified peptidoglycan ligands or Nod1-signalling *H. influenzae* displayed increased killing of *S. pneumonia* in a Nod1-dependent manner. Moreover, the requirement for Nod1 to facilitate phagocytosis of bacterial pathogens by neutrophils was further validated in vivo, as Nod1 knockout mice administered with *H. influenzae* prior to infection with *S. pneumoniae* had an impaired ability to clear the pathogen [73]. In addition, a second study reported that peptidoglycan originating from the intestinal microbiota may facilitate in priming bone marrow-derived neutrophils to display enhanced killing of *S. pneumoniae* and *Staphylococcus aureus* [22]. The authors demonstrated by colonising germ-free mice with *Escherichia coli* containing radiolabelled peptidoglycan that during colonisation, peptidoglycan from intestinal organisms can translocate across the intestinal mucosa, entering the circulation where it can facilitate in the development of neutrophil function [22].

Autophagy

Autophagy is a cellular cytoplasmic process that targets intracellular components for degradation and occurs downstream of the early endosome pathway [74, 75]. The process of autophagy is essential for the clearance of cytosolic cargo, being either damaged host organelles or proteins, or as a defence mechanism for the degradation of internalised bacterial or viral pathogens [74, 75].

Three studies have recently reported the ability of Nods to regulate the intracellular degradation process of autophagy in response to bacterial pathogens. We identified that both Nod1 and Nod2 are required for autophagy in response to bacteria, using the invasive pathogen *Shigella flexneri* [76]. Furthermore, we identified that Nods interacted with ATG16L1, a component of the autophagosome, enabling its recruitment to the cellular site of bacterial entry into host cells to establish autophagy. Indeed, mutations in ATG16L1 are linked to susceptibility of Crohn's disease, providing a possible physiological relevance for the requirement of Nods in bacterial-induced autophagy, discussed in further detail below. Other research groups have also established a requirement for Nods in the development of bacteria-induced autophagy and the regulation of an inflammatory response in Nod-stimulated human dendritic and colonic epithelial cells [77, 78]. However, there are some key differences between the findings reported by all three groups. Cooney et al. identified that Nod2 was required for the induction of autophagy in dendritic cells and promoting the generation of an adaptive immune response as a result of autophagy-induced increased antigen presentation within MHC II complexes [77]. This study also identified that Nod2-dependent autophagy required the autophagy-related proteins ATG5, ATG7 and ATG16L and was dependent on RIP-2 [77]. Similarly, Homer and colleagues also identified that Nod-2 dependent autophagy required RIP-2 [78]. Whereas, contrary to these findings, work by Travassos and colleagues showed that Nod1-induced autophagy of intracellular bacteria was RIP-2 independent, as ATG16 could co-localise with Nod1-signalling *Shigella* in RIP-2-deficient mouse embryonic fibroblasts (MEFs). This clear discrepancy in findings for the requirement of RIP-2 between groups may potentially be a difference in the cell type examined, or the pathogen model used, as Travassos focused on epithelial cells and murine macrophages using a *Shigella* pathogen model [76]. Whereas work by Cooney and Homer focused specifically on human dendritic cells stimulated with Nod-ligands [77] and *Salmonella*-infected colonic cells, respectively [78]. Further work is required to elucidate the exact role of RIP-2 in Nod-induced autophagy.

Development of the Gut Microbiota

The importance of Nods in the development of the intestinal microbiota is in part due to their location at the mucosal surfaces, in addition to their rapid ability to sense the presence of bacteria and produce antimicrobial peptides that function to control the bacterial burden at these sites. Using Nod2 knockout animals, it has been identified that this pattern recognition molecule (PRM) plays a key role in the composition of the intestinal microbiota during development [79]. Rehman and colleagues examined the faecal and ileal microbiota compositions in wild-type C57BL/6 and Nod2 knockout animals by generating a 16s ribosomal RNA clone library. This study identified that there was a shift in the composition of the ileal and faecal microbiota composition in Nod2 knockout animals when compared to C57BL/6 control mice [79]. Indeed, they identified that in the absence of Nod2, elevated total bacterial numbers were present within the faeces and terminal ileum of mice compared to their wild-type controls, and that Nod2 knockout mice displayed increased numbers of *Bacteroidetes* and *Firmicutes* compared to control mice [79]. In addition, a second study reported of a similar increase in the numbers of *Bacteroides*, *Firmicutes* and *Bacillus* spp. present in the terminal ilea of Nod2-deficient or RIP-2-deficient animals [47]. Interestingly, the regulation of Nod2 in the development of the gastrointestinal microbiota occurred early in the developmental stage of these animals, as an altered microbial composition was evident upon weaning of these mice [79].

The importance of the increased number of *Firmicutes* and *Bacteroides spp.* in these knockout animals is apparent when comparisons are made to the microbiota of Crohn's disease patients. Individuals who were homozygous for the Nod2 SNP13, commonly associated with Crohn's disease, displayed elevated numbers of *Firmicutes* and *Bacteroides* in their ileum compared to their healthy counterparts, suggesting that Nod2 regulation of the intestinal microbiota is associated with a genetic predisposition to Crohn's disease [79]. This genetic alteration may account for the dysregulation in the intestinal microbiota of Crohn's disease patients, as Nod2 may be required to suppress the levels of opportunistic pathogens in these individuals [79].

Role in Intestinal Development

In addition to controlling the bacterial composition of the intestinal tract, Nod signalling induced by the microbial flora also contributes to the development of lymphoid follicles within the intestinal tract. Bouskra et al. identified that Nod1 signalling by the microbiota present within intestinal crypts resulted in the production of defensins and CCR6 signalling, ultimately facilitating the development and formation of lymphoid follicles [52]. Using bone marrow chimaeras, it was determined that Gram-negative bacterial commensals signalling via Nod1 present within intestinal epithelial cells, and not haematopoietic cells, were responsible for the development of intestinal lymphoid follicles within animals [52]. The result of Nod1 signalling in epithelial cells by the intestinal microbiota subsequently enabled the host to generate polymeric IgA antibodies, that are immunoreactive against the intestinal flora, and progress the development of Peyer's patches and mesenteric lymph nodes that drain the intestinal tissue [52]. Ultimately, these studies identified that the impaired development of intestinal lymphoid follicles in Nod1 knockout animals resulted in an altered microbial flora, in addition to identifying a direct function of microbial Nod1 signalling in the development of the lymphoid compartment and the generation of secondary lymphoid tissues [52]. These findings have been implicated as having a role in shaping the development of the mucosal lymphoid compartment of Crohn's disease individuals.

The Development of Adaptive Immune Responses

Innate immune responses initiated by PRRs such as Nods are broad, have been conserved throughout evolution and ultimately result in the recruitment of proinflammatory cells such as dendritic cells, macrophages or neutrophils to the site of infection. These activated innate immune cells produce cytokines, as described in the aforementioned sections, and facilitate further the recruitment of adaptive inflammatory immune cells. Activation of the adaptive immune system results in the generation of a pathogen-specific response and involves T cells that are responsible for the progression of inflammation, or B cells that are required for the generation of a humoral, antibody-mediated response.

To date, most of the studies examining Nod-dependent adaptive immune responses have used Nod ligands as an adjuvant in conjunction with model antigens, such as the chicken egg ovalbumin protein, OVA. A study by Fritz and colleagues was one of the first to report the requirement for Nod1 in priming antigen-specific T cell development and humoral antibody responses in vivo [80]. Indeed, using Nod1-deficient mice, and Nod1-deficient animals reconstituted with bone marrow from Nod1-competent animals, this study reported that Nod1 stimulation in non-haematopoietic cells was responsible for priming antigen-specific Th2 immunity in response to injection with the Nod1 ligand FK156 and ovalbumin as an antigen [80]. The Th2 immune response observed was characterised by the presence of IL-4- and IL-5-producing $CD4^+$ T cells and antibodies of the IgG1 subtype [80]. This study has subsequently been validated and expanded upon by Magalhães and colleagues, who showed that both Nod1 and Nod2 activation results in the development of a Th2-dependent immune response in vivo and that RIP-2 is essential for the establishment of Nod-dependent adaptive immune responses [81]. Again, this study relied on the injection of Nod agonists into wild-type and Nod knockout animals, and adaptive immune responses to OVA were determined. The authors identified that wild-type mice administered with both FK156 and OVA displayed elevated IL-4- and IL-5-producing cells in their spleens in addition to IgG1 antibody responses, which were abrogated in RIP-2-deficient animals [81]. Indeed, using bone marrow chimeric mice, a second study by the same authors identified that these responses are dependent on Nod1 and Nod2 expression by cells within the stromal compartment, and not by dendritic cells, considered as the most potent antigen-presenting cell in the host [82]. Therefore, the initial Nod1-signalling response triggered in epithelial cells can direct the hosts dendritic cells to initiate the development of a T helper 1 (Th1) inflammatory response and T helper 2 (Th2) humoral response, resulting in antibody production. Furthermore, Nod1 can function in combination with toll-like receptor (TLR) stimulation to enhance the adaptive immune response and initiate the development of a Th1, Th2 and Th17 responses [80].

In addition to Nod1, Nod2 signalling is also capable of facilitating the development of the adaptive arm of the immune response. Nod2-deficient T cells have an impaired ability to induce the production of IL-2 and IFN-gamma and function similarly to T cells that lack CD28, a costimulatory molecule required to enable the clonal expansion of T cells and the generation of Th1 immunity [83]. However, it should be noted that this finding was observed using a model of the intracellular parasite, *Toxoplasma gondii* [83]. Furthermore, Nod2 expressed by human dendritic cells seems to be required for bacterial processing and handling, via autophagy, and the generation of $CD4^+$ T cell responses, possibly due to presentation of antigen via MHC class II to T cells [77]. This finding was validated using human dendritic cells with impaired Nod2 function displaying an inferior ability to induce antigen-specific T cell responses [77].

Th17 $CD4^+$ T cells are one of the most recently discovered adaptive immune cells and are characterised by their secretion of IL-17 which functions in recruiting neutrophils to the site of inflammation [84]. In addition, Th17 cells also produce IL-22, which facilitates the production of antimicrobial peptides and tissue repair factors by epithelial cells [85]. Recently, the role of Nods in the generation of innate Th17 immune cells during microbial infection was determined [86]. Using the animal models of *Citrobacter rodentium*- and *S. typhimurium*-induced colitis, it was identified that both Nod1 and Nod2 were essential for the induction of mucosal Th17 immune responses early during infection, being 4 days and 24 h postinfection with the respective pathogens. The Th17 immune response initiated by innate Th17 cells (iTh17) was dependent on the expression of IL-6, and these cells were essential for the development of mucosal immunity against both bacterial pathogens [86]. Using Nod1 and Nod2 knockout animals, it was determined that Nods were essential for the generation of Th17 responses to mucosal pathogens, as these knockout animals displayed an increase in the burden of infection and reduced pathology, when compared to wild-type control mice [86]. This is the first identification of the requirement for Nods in the development of the third arm of the adaptive immune response, being the rapid development of iTh17 immunity, which may function to fill the immune gap until the mature adaptive immune response develops [86].

Finally, Nods have been identified as having a role in regulating the development of an adaptive immune response, via the generation and regulation of Foxp3 expressing suppressive T regulatory cells (Tregs) [21]. Tregs function to maintain self-tolerance in the host and suppress autoreactive T cells in the periphery via their production of the immunosuppressive cytokine IL-10 [87]. Peripheral blood mononuclear cells from Crohn's disease patients expressing a mutation in their Nod2 gene had a low production of the immunosuppressive cytokine IL-10, indicating a potential role for Nod2 in facilitating the suppression of immunoreactive T cells located within the periphery [88]. Furthermore, Rahman and colleagues reported the ability of the Nod2 agonist MDP to activate NF-kappaB in Tregs, hence protecting them from Fas-mediated apoptosis or programmed cell death [21]. Moreover, this study reported of a deficiency in Tregs located within the lamina propria of patients who were homozygous for *nod2* variants, providing a possible explanation for the chronic inflammatory response observed in these Crohn's disease patients [21].

How Bacteria Signal via Nod1 and Nod2

Bacteria utilise numerous mechanisms to release their peptidoglycan and facilitate its entry into host cells. Similarly, the host uses a variety of mechanisms to enable the uptake of peptidoglycan into the cytoplasm to initiate Nod signalling. These mechanisms are discussed in detail below and are summarised in Fig. 10.2.

Invasion

A number of bacteria have been reported to be capable of inducing Nod signalling, with a dependence on the bacteria being viable or actively invading host cells (see Fig. 10.2). One of the earliest examples identifying that invasive pathogens could initiate NF-kappaB-dependent IL-8 production potentially via intracellular Nod1

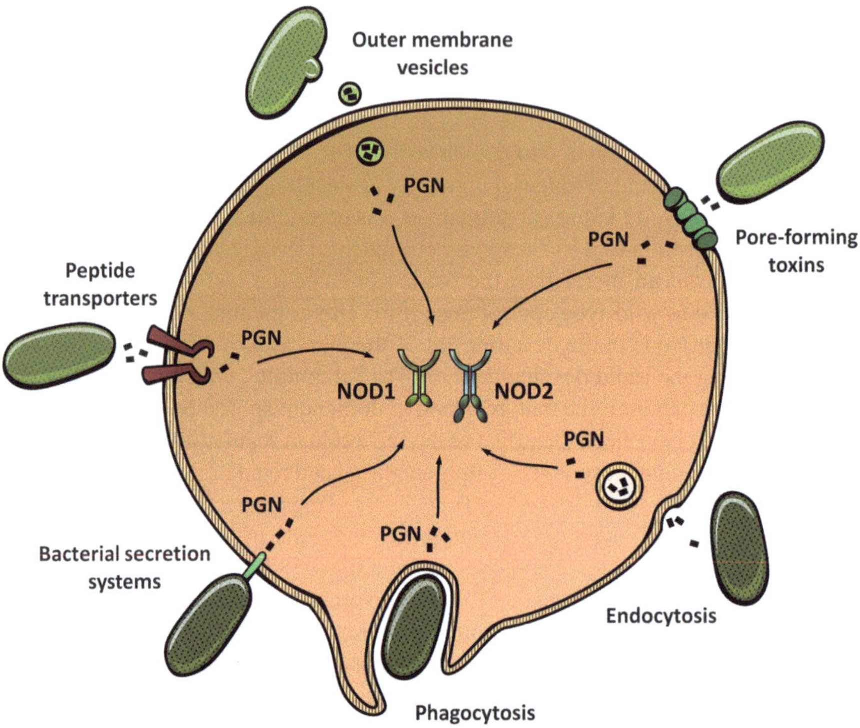

Fig. 10.2 Mechanisms that facilitate entry of Nod ligands into the cytoplasm of target cells. There are a number of ways that peptidoglycan has been shown to enter into the cytoplasm to interact with either Nod1 or Nod2. These include (1) phagocytosis, (2) endocytosis, (3) entry through pores made by pore-forming toxins, (4) delivery through outer membrane vesicles (OMVs), (5) peptide transporters and (6) bacterial secretion systems (Courtesy of artist: Priya Alwis)

was demonstrated using invasive *Shigella flexneri* [9]. This study also reported that NF-kappaB-dependent IL-8 production was not induced by noninvasive nor heat-killed *Shigella* [9]. This finding was subsequently confirmed, as viable and invasive *Chlamydophila pneumoniae* could induce Nod1 signalling and the secretion of IL-8 by human endothelial cells, whereas this was not evident when epithelial cells were stimulated with heat-inactivated organisms [89]. Furthermore, a third study demonstrated that transfection of heat-killed bacteria into epithelial cells enabled Nod1 signalling to occur [89]. This same group also subsequently identified that invasive *Listeria monocytogenes* was capable of inducing IL-8 production by human endothelial cells, in a Nod1 and p38 MARK-dependent manner [90]. Since then, numerous studies have reported that Nod1 signalling, resulting in NF-kappaB-dependent IL-8 production, can be induced in host epithelial and haematopoietic cells by the invasive enteric pathogens *Escherichia coli* [91], *Listeria* [92], *Salmonella enterica* serovar Typhimurium [68] and *Mycobacterium avium* ssp. *paratuberculosis*, which is associated with Crohn's disease [93]. In addition, *Moraxella catarrhalis*, a lung pathogen, is capable of invading bronchial epithelial and primary small airway epithelial cells, resulting in the generation and secretion of IL-8 in a Nod1-dependent manner [94].

Similarly, invasive pathogens can signal via Nod2, as is the case for *Mycobacterium tuberculosis* [95, 96] and *Mycobacterium bovis* infected macrophages [96]. The requirement for Nod2 in the recognition of *M. tuberculosis* and cytokine production was demonstrated using mononuclear cells isolated from individuals homozygous for the 3020insC Nod2 mutation, revealing a defective cytokine production in response to infection [95]. Intracellular *Salmonella* can also activate Nod2, resulting in killing of the pathogen, and mutations in Nod2 enable intracellular *Salmonella* to survive within host cells [23]. In addition, *S. pneumoniae* can transiently invade epithelial and endothelial cells, resulting in further upregulation of Nod2 expression in vivo and in vitro [97]. The authors reported that RIP2 and the signal-transducing molecules IRAK, IRAK2, TRAF6, NIK, TAB2, and TAK1 are involved in this process [97].

Uptake of Bacteria by Phagocytosis and the Degradation of Their Peptidoglycan by Lysosomes

The cellular process of phagocytosis enables the uptake of large particles by host phagocytic cells, predominantly by macrophages. Phagocytosis of Gram-negative and Gram-positive pathogens by host phagocytes enables intracellular Nods capable of detecting bacterial peptidoglycan and initiating a pro-inflammatory response, via the degradation of the internalised pathogen by phagolysosomal fusion. The host produces enzymes that can degrade peptidoglycan to subunits that are capable of being recognised by Nod1 and Nod2. The most prevalent host enzyme responsible for peptidoglycan degradation is lysozyme, which is found in host mucosal secretions and in granules contained within phagocytes. An analysis of lysosomal extracts obtained from human innate immune cells revealed that lysosome-degraded peptidoglycan components were capable of activating the Nod-signalling pathway in host cells [98]. Therefore, the authors of this study proposed that intracellular peptidoglycan traffics to the lysosome, where it is degraded into smaller soluble subunits, enabling more efficient recognition by the intracellular Nod receptors [98].

Release of Peptidoglycan by Hydrolases or Remodelling That Can Be Internalised by Host Cells

Bacteria are required to constantly remodel their peptidoglycan layer during the process of bacterial growth and division. During the process of cellular remodelling and biosynthesis, bacteria shed their peptidoglycan into their extracellular environment, and this shedding is particularly high during the stage of exponential growth

[99]. Furthermore, bacterial MDP can be released after degradation of ingested bacteria by host lysozyme [100].

Some bacteria are very efficient at recycling their peptidoglycan and minimising the amount that is released into the surrounding environment during this remodelling process. For example, *E. coli* is very efficient at recycling its peptidoglycan as it only releases approximately 6 % of its total peptidoglycan content [101], whereas *Bacilli* can release between 30 and 50 % of their peptidoglycan during remodelling, which may have a function in its pathogenesis [102]. Indeed, it has been suggested that some pathogens such as *Neisseria gonorrhoeae* and *Bordetella pertussis* release peptidoglycan to facilitate invasive disease due to destruction of the epithelial cell barrier [2]. Released peptidoglycan can ultimately be taken up by host cells via endocytosis or peptide transporters, as discussed below.

Bacterial Secretion Systems

Some of the more virulent *H. pylori* strains harbour a cag pathogenicity island (cagPAI), which encodes for a type 4 secretion system (T4SS). It was well established that *H. pylori cag*PAI-positive bacteria harbouring a T4SS are able to induce IκB degradation and the nuclear translocation of NF-kappaB [103, 104]. Work by Viala and colleagues identified that *H. pylori cag*PAI-positive bacteria are able to translocate their peptidoglycan into host epithelial cells via the T4SS [40] (see Fig. 10.2). We have furthered these findings and identified that *H. pylori cag*PAI-positive bacteria can also activate p38 and extracellular signal-related kinase (ERK) MAPKs and AP-1 in a Nod1-dependent manner, identifying a novel pathway of Nod signalling in response to Gram-negative bacterial infection [105]. Moreover, we reported that the T4SS of *H. pylori* interacts with integrins located within lipid rafts on the host cell membrane to facilitate Nod1 signalling [106]. A second pathogen, *C. rodentium*, which expresses a functional type III secretion system, has similarly been reported to initiate the production of cytokines via both Nod1 and Nod2 pathways [107].

Endocytosis and Peptide Transporters

Non-phagocytic epithelial cells are capable of endocytosing peptidoglycan in a clathrin-mediated and dynamin-dependent process. Lee and colleagues used HEK293T cells to show that the internalisation of Nod1 ligands was pH dependent and was optimal at pH 5.5–6, suggesting that the intracellular location of peptidoglycan was within early endosomes. Similarly, Nod2-stimulating peptidoglycan was capable of entering human epithelial cell lines in an identical manner to Nod1-signalling peptidoglycan [108]. In addition, a putative transporter for Nod1 ligands was suggested to exist within early endosomes, being SLC15A4, which was confirmed by knocking down expression of this transporter using siRNA. Indeed, the expression of this putative transporter, whose optimal function is at pH 5.5–6, was highly expressed in tissues obtained from IBD patients, suggesting a potential role for Nod detection of peptidoglycan in IBD-affected individuals [108]. Furthermore, a second study also reported the mechanism of Nod2-signalling peptidoglycan entry into host cells, being macrophages [109]. The authors expanded on the current knowledge that MDP is internalised into acidified vesicles in macrophages and identified that MDP enters macrophages in a clathrin- and dynamin-mediated manner [109]. It was previously identified that MDP could cross the plasma membrane of host cells and enter the cytoplasm via a plasma membrane transporter hPepT1 [110, 111], initially identified as a transporter of oligopeptides [112]. Two studies reported that the hPepT1 transporter was responsible for transporting Nod2-stimulating MDP, but not Nod1-inducing peptidoglycan, into the cytoplasm of intestinal epithelial cells [110, 111]. However, a second group reported that MDP uptake and subsequent Nod2-dependent signalling in macrophages did not require the peptide transporter PepT1 [109]. The contribution of the hPepT1 transporter in facilitating MDP translocation into the cytoplasm may be a cell-specific phenomenon, and further research is required to clarify the requirement of this transporter within different cells. Collectively, these studies indicate that the endocytic pathway enables the uptake of Nod-stimulating peptidoglycan ligands into host cells in a pH-dependent manner.

Outer Membrane Vesicles

Almost all Gram-negative bacteria secrete outer membrane vesicles (OMVs) or "blebs" as part of their normal growth. OMVs are spherical, bilayered membrane nano-structures ranging from 20 to 300 nm in size, which are released naturally both in vitro and in vivo [113]. We have recently identified that OMVs from *H. pylori*, *Pseudomonas aeruginosa* and *N. gonorrhoea* contain peptidoglycan [114]. We have shown that Gram-negative bacterial OMVs enter non-phagocytic host epithelial cells via lipid rafts, rendering their peptidoglycan-containing cargo accessible to Nod1. Indeed, depletion of lipid rafts on the surface of epithelial cells impaired the ability of OMVs to enter and signal via Nod1, and this was restored once lipid rafts were replenished on the host cellular membrane [114]. Similarly, a second study also has shown the ability of OMVs from *E. coli* to enter host cells in a lipid raft-dependent manner and release toxin within the host cell [115]. In addition, work by Bielig and colleagues confirmed the ability of Gram-negative bacterial OMVs to induce Nod-dependent responses using the pathogen *Vibrio cholerae* [116].

Furthermore, oral administration of *H. pylori* OMVs to wild-type and Nod1 knockout animals revealed that OMVs could initiate rapid innate immune responses within the gastric tissue of immunocompetent animals, whereas no inflammatory responses were observed in Nod1 knockout mice [114]. Moreover, oral OMV administration to wild-type animal resulted in the development of an OMV-specific antibody response that was absent in Nod1 knockout animals. Collectively, these findings identified a mechanism whereby Gram-negative mucosal pathogens can initiate Nod1-dependent innate and adaptive immune responses in the absence of cellular invasion or bacterial secretion systems.

Direct Binding of Ligands to Nod Receptors

Since the discovery of Nod1 and Nod2 and the identification of peptidoglycan as the bacterial product triggering the activation of these receptors, scientists have debated whether or not these molecules directly bound their respective ligands. Early work had suggested that this was likely the case. Indeed, mouse Nod1 prefers the tetrapeptide, L-Ala–D-Glu–mesoDAP–D-Ala, whereas human Nod1 is better triggered by L-Ala–D-Glu–mesoDAP; therefore, by swapping their LRR domains, the sensing specificities of these two molecules could be switched [19].

Recently, three papers demonstrated direct in vitro binding of their ligands. The first paper showed using surface plasmon resonance and atomic force microscopy that Nod1 can directly bind to L-Ala-γ–D-Glu–*meso*-diaminopimelic acid but not MDP, the Nod2 ligand. Following this, two papers showed the direct interaction of Nod2 with MDP, one using chips coated with MDP self-assembled monolayers and surface plasmon resonance to measure Nod2–MDP interactions [117] and the other, binding of recombinant Nod2 to biotinylated MDP [118]. The significance of these studies is that now these kinds of assays can be used to develop screens for identifying novel inhibitors or activators of Nod proteins, which might reveal interesting small molecules for the treatment of Crohn's disease.

Role of Nods in Crohn's Disease

Nod2 was the first susceptibility gene linked to Crohn's disease. In 2001, two groups reported that mutations in *Nod2*, and in particular, a frame-shift mutation caused by a cytosine insertion at nucleotide position 3020, were linked to Crohn's disease susceptibility [8, 119]. For Nod1, its link to inflammatory bowel disease has been less clear. Although some studies have shown associations of *Nod1* polymorphisms with Crohn's disease [120], this has not been supported in other populations [121].

Nod2 Genetics and Crohn's Disease

The original paper by Hugot et al., which first described *Nod2* as susceptibility loci for Crohn's disease, identified two variants of *Nod2*, termed SNP8 and SNP12, in addition to the frame-shift mutation, termed SNP13. While SNP13 lies within the

LRR region of Nod2 and results in impaired sensing of MDP [11, 12], SNP12 is proximal to SNP13 within the LRR and SNP8 is within the NOD domain. Interestingly, these two mutations have a variable affect on MDP sensing and at high MDP doses, approach wild-type levels of cell activation [122]. On the other hand, individuals homozygous for the frame-shift mutation are completely unable to detect MDP at any dose [122]. With this in mind, these findings call into question whether a lack of ability to detect MDP really underlies Crohn's disease pathogenesis. Alternatively, these findings may suggest that treating patients with MDP, especially *Nod2* heterozygous and compound heterozygous individuals, might ameliorate disease.

More recently, deep sequencing of GWAS-identified loci found five new variants of *Nod2* associated with Crohn's disease [123]. Functional analyses of two of these *Nod2* variants showed diminished responsiveness to MDP compared to wild-type Nod2 transfected cells, yet not as profound as cells transfected with the Nod2 frame-shift mutant [123]. Further studies into how these *Nod2* variants are impaired in their activity will certainly shed light onto our understanding of Crohn's disease pathogenesis.

Lessons Learned from Colitis Models

No animal model can mimic all aspects of human disease. Nevertheless, much insight into how Nod2 potentially regulates intestinal homeostasis has been gained from models of colitis in Nod2-deficient mice. The most commonly used chemical models are the dextran sulphate sodium (DSS) and trinitrobenzene sulfonic acid (TNBS)-induced models, where the chemical insult damages the epithelial layer leading to intestinal inflammation that is driven by exposure of damaged mucosal tissue to the commensal microbiota. Bacteria-induced models of colitis are milder and perhaps more physiologically relevant models of induced colitis, which have been extensively used in the past few years. Again, while no model is perfect, different aspects governing the regulation of intestinal inflammation can be revealed by these different models.

In the DSS and TNBS models, Nod2 triggering has been shown to be important to protect mice from severe disease. Treatment of wild-type mice with the Nod2 ligand, MDP, ameliorates colitis induced by DSS, and this effect was gone in DSS-treated Nod2-deficient animals [36, 124]. Moreover, in a TNBS colitis model, a *Lactobacillus* strain producing a highly active Nod2 ligand within its peptidoglycan was also able to ameliorate colitis in a Nod2-dependent fashion [125]. In the TNBS model, Nod2-deficient mice have been reported to develop more severe colitis [126, 127]. Interestingly, one group showed that the protective Nod2-dependent signals emanate from the bone marrow since chimeric mice with a Nod2-deficient hematopoietic compartment were as susceptible as Nod2-deficient mice [127]. However, it is not yet known how these bone marrow-derived Nod2 signals are protective in colitis and, indeed, what these factors might be.

The role of Nod2 in host defence against microbial infection and subsequent colitis has been examined in different bacterial infection models. Nod2-deficient mice are more susceptible to *Helicobacter hepaticus* infection, demonstrating increased intestinal inflammation and increased frequency of IFN-gamma secreting Th1 cells in Peyer's patches. Interestingly, over-expression of alpha-defensins in Nod2-deficient Paneth cells was able to dampen intestinal inflammation [48]. Nod2-deficient mice were also shown to display a delayed intestinal clearance of *Citrobacter rodentium* due to reduced CCL2 expression in the colon and resulting in impaired recruitment of inflammatory monocytes [61]. Our group, on the other hand, found only in the background of Nod1 deficiency that Nod2-deficient mice were more susceptible to *C. rodentium* infection [86] and, indeed, *Salmonella enterica* serovar Typhimurium infection [128]. The reason for this discrepancy is unclear but likely points to the ever-growing understanding that differences in commensal microbiota between animal facilities impact disease susceptibility. Finally, a knock-in mouse strain carrying the frame-shift mutation in *Nod2*, which is associated with human Crohn's disease, was generated recently and, similar to Nod2-deficient mice, this mutant Nod2 strain exhibits severely impaired sensing of MDP and increased susceptibility to the enteric organism *Enterococcus faecalis* [129]. It will be interesting in the future to explore the mechanisms of intestinal homeostasis dysregulation in this mouse model.

Conclusion and Perspectives

Nod1 and Nod2 were the first characterised NLR family of cytosolic PRRs. As described in this chapter, much research has focused on how these receptors detect their ligand, peptidoglycan, and how this triggers an inflammatory response. More recent findings have highlighted that Nod signalling at the intestinal mucosa, especially Nod2, is critical for the maintenance of the integrity of the gut barrier and the regulation of inflammatory pathways that control both homeostasis and protection against intrusion by microbial pathogens. The challenge for the future will be to understand how dysregulation of Nod2 signalling leads to inflammation in Crohn's disease patients. What is still unclear is what cell type is critical for Nod2-dependent homeostatic regulation, be it hematopoietic cells or epithelial cells of the intestine, including Paneth cells. Moreover, a key question for the future will be to understand whether Nod2's ability to detect MDP is linked to intestinal homeostasis. As pointed out above, while cells derived from patients with the homozygous frame-shift mutation in *Nod2* cannot detect MDP, cells with other mutations still retain some ability to respond to this bacterial product. In the future, understanding the pathogenic implications of these Crohn's disease-associated mutations in *Nod2*, beyond the frame-shift mutation, will help to unravel the mysteries behind this disease as well as point to new avenues for treatment.

References

1. Strober W, Murray PJ, Kitani A, Watanabe T (2006) Signalling pathways and molecular interactions of NOD1 and NOD2. Nat Rev Immunol 6(1):9–20, Epub 2006/02/24
2. Clarke TB, Weiser JN (2011) Intracellular sensors of extracellular bacteria. Immunol Rev 243(1):9–25, Epub 2011/09/03
3. Ausubel FM (2005) Are innate immune signaling pathways in plants and animals conserved? Nat Immunol 6(10):973–979
4. Ting JP, Lovering RC, Alnemri ES, Bertin J, Boss JM, Davis BK et al (2008) The NLR gene family: a standard nomenclature. Immunity 28(3):285–287, Epub 2008/03/18
5. Philpott DJ, Girardin SE (2010) Nod-like receptors: sentinels at host membranes. Curr Opin Immunol 22(4):428–434
6. Werts C, Rubino S, Ling A, Girardin SE, Philpott DJ (2011) Nod-like receptors in intestinal homeostasis, inflammation, and cancer. J Leukoc Biol 90(3):471–482, Epub 2011/06/10
7. Inohara N, Nunez G (2003) NODs: intracellular proteins involved in inflammation and apoptosis. Nat Rev Immunol 3(5):371–382
8. Ogura Y, Bonen DK, Inohara N, Nicolae DL, Chen FF, Ramos R et al (2001) A frameshift mutation in NOD2 associated with susceptibility to Crohn's disease. Nature 411(6837): 603–606
9. Girardin SE, Tournebize R, Mavris M, Page AL, Li X, Stark GR et al (2001) CARD4/Nod1 mediates NF-kappaB and JNK activation by invasive Shigella flexneri. EMBO Rep 2(8): 736–742
10. Inohara N, Koseki T, del Peso L, Hu Y, Yee C, Chen S et al (1999) Nod1, an Apaf-1-like activator of caspase-9 and nuclear factor-kappaB. J Biol Chem 274(21):14560–14567
11. Girardin SE, Boneca IG, Viala J, Chamaillard M, Labigne A, Thomas G et al (2003) Nod2 is a general sensor of peptidoglycan through muramyl dipeptide (MDP) detection. J Biol Chem 278(11):8869–8872, Epub 2003/01/16
12. Inohara N, Ogura Y, Fontalba A, Gutierrez O, Pons F, Crespo J et al (2003) Host recognition of bacterial muramyl dipeptide mediated through NOD2. Implications for Crohn's disease. J Biol Chem 278(8):5509–5512
13. Fritz JH, Girardin SE, Fitting C, Werts C, Mengin-Lecreulx D, Caroff M et al (2005) Synergistic stimulation of human monocytes and dendritic cells by toll-like receptor 4 and NOD1- and NOD2-activating agonists. Eur J Immunol 35(8):2459–2470
14. Enoksson M, Ejendal KF, McAlpine S, Nilsson G, Lunderius-Andersson C (2011) Human cord blood-derived mast cells are activated by the Nod1 agonist M-TriDAP to release pro-inflammatory cytokines and chemokines. J Innate Immun 3(2):142–149, Epub 2010/11/26
15. Qiu F, Maniar A, Diaz MQ, Chapoval AI, Medvedev AE (2011) Activation of cytokine-producing and antitumor activities of natural killer cells and macrophages by engagement of Toll-like and NOD-like receptors. Innate Immun 17(4):375–387, Epub 2010/08/05
16. Chamaillard M, Hashimoto M, Horie Y, Masumoto J, Qiu S, Saab L et al (2003) An essential role for NOD1 in host recognition of bacterial peptidoglycan containing diaminopimelic acid. Nat Immunol 4(7):702–707, Epub 2003/06/11
17. Girardin SE, Boneca IG, Carneiro LA, Antignac A, Jehanno M, Viala J et al (2003) Nod1 detects a unique muropeptide from gram-negative bacterial peptidoglycan. Science 300(5625):1584–1587, Epub 2003/06/07
18. Chamaillard M, Girardin SE, Viala J, Philpott DJ (2003) Nods, Nalps and Naip: intracellular regulators of bacterial-induced inflammation. Cell Microbiol 5(9):581–592
19. Magalhaes JG, Philpott DJ, Nahori MA, Jehanno M, Fritz J, Bourhis LL et al (2005) Murine Nod1 but not its human orthologue mediates innate immune detection of tracheal cytotoxin. EMBO Rep 6(12):1201–1207
20. Ogura Y, Inohara N, Benito A, Chen FF, Yamaoka S, Nunez G (2001) Nod2, a Nod1/Apaf-1 family member that is restricted to monocytes and activates NF-kappaB. J Biol Chem 276(7):4812–4818

21. Rahman MK, Midtling EH, Svingen PA, Xiong Y, Bell MP, Tung J et al (2010) The pathogen recognition receptor NOD2 regulates human FOXP3+ T cell survival. J Immunol 184(12):7247–7256
22. Clarke TB, Davis KM, Lysenko ES, Zhou AY, Yu Y, Weiser JN (2010) Recognition of peptidoglycan from the microbiota by Nod1 enhances systemic innate immunity. Nat Med 16(2): 228–231
23. Hisamatsu T, Suzuki M, Reinecker HC, Nadeau WJ, McCormick BA, Podolsky DK (2003) CARD15/NOD2 functions as an antibacterial factor in human intestinal epithelial cells. Gastroenterology 124(4):993–1000
24. Gutierrez O, Pipaon C, Inohara N, Fontalba A, Ogura Y, Prosper F et al (2002) Induction of Nod2 in myelomonocytic and intestinal epithelial cells via nuclear factor-kappa B activation. J Biol Chem 277(44):41701–41705
25. Rosenstiel P, Fantini M, Brautigam K, Kuhbacher T, Waetzig GH, Seegert D et al (2003) TNF-alpha and IFN-gamma regulate the expression of the NOD2 (CARD15) gene in human intestinal epithelial cells. Gastroenterology 124(4):1001–1009, Epub 2003/04/03
26. Shaw PJ, Barr MJ, Lukens JR, McGargill MA, Chi H, Mak TW et al (2011) Signaling via the RIP2 adaptor protein in central nervous system-infiltrating dendritic cells promotes inflammation and autoimmunity. Immunity 34(1):75–84
27. Zhao L, Hu P, Zhou Y, Purohit J, Hwang D (2011) NOD1 activation induces proinflammatory gene expression and insulin resistance in 3T3-L1 adipocytes. Am J Physiol Endocrinol Metab 301(4):E587–E598, Epub 2011/06/23
28. Silva GK, Gutierrez FR, Guedes PM, Horta CV, Cunha LD, Mineo TW et al (2010) Cutting edge: nucleotide-binding oligomerization domain 1-dependent responses account for murine resistance against Trypanosoma cruzi infection. J Immunol 184(3):1148–1152, Epub 2010/01/01
29. Finney CA, Lu Z, LeBourhis L, Philpott DJ, Kain KC (2009) Disruption of Nod-like receptors alters inflammatory response to infection but does not confer protection in experimental cerebral malaria. Am J Trop Med Hyg 80(5):718–722
30. Sorbara MT, Philpott DJ (2011) Peptidoglycan: a critical activator of the mammalian immune system during infection and homeostasis. Immunol Rev 243(1):40–60, Epub 2011/09/03
31. Fritz JH, Ferrero RL, Philpott DJ, Girardin SE (2006) Nod-like proteins in immunity, inflammation and disease. Nat Immunol 7(12):1250–1257
32. Hasegawa M, Fujimoto Y, Lucas PC, Nakano H, Fukase K, Nunez G et al (2008) A critical role of RICK/RIP2 polyubiquitination in Nod-induced NF-kappaB activation. EMBO J 27(2):373–383, Epub 2007/12/15
33. Inohara N, Koseki T, Lin J, del Peso L, Lucas PC, Chen FF et al (2000) An induced proximity model for NF-kappa B activation in the Nod1/RICK and RIP signaling pathways. J Biol Chem 275(36):27823–27831
34. Meylan E, Tschopp J (2005) The RIP kinases: crucial integrators of cellular stress. Trends Biochem Sci 30(3):151–159
35. Kobayashi K, Inohara N, Hernandez LD, Galan JE, Nunez G, Janeway CA et al (2002) RICK/Rip2/CARDIAK mediates signalling for receptors of the innate and adaptive immune systems. Nature 416(6877):194–199
36. Bertrand MJ, Doiron K, Labbe K, Korneluk RG, Barker PA, Saleh M (2009) Cellular inhibitors of apoptosis cIAP1 and cIAP2 are required for innate immunity signaling by the pattern recognition receptors NOD1 and NOD2. Immunity 30(6):789–801, Epub 2009/05/26
37. Krieg A, Correa RG, Garrison JB, Le Negrate G, Welsh K, Huang Z et al (2009) XIAP mediates NOD signaling via interaction with RIP2. Proc Natl Acad Sci U S A 106(34): 14524–14529, Epub 2009/08/12
38. Yang Y, Yin C, Pandey A, Abbott D, Sassetti C, Kelliher MA (2007) NOD2 pathway activation by MDP or Mycobacterium tuberculosis infection involves the stable polyubiquitination of Rip2. J Biol Chem 282(50):36223–36229, Epub 2007/10/20
39. Abbott DW, Wilkins A, Asara JM, Cantley LC (2004) The Crohn's disease protein, NOD2, requires RIP2 in order to induce ubiquitinylation of a novel site on NEMO. Curr Biol 14(24):2217–2227, Epub 2004/12/29

40. Viala J, Chaput C, Boneca IG, Cardona A, Girardin SE, Moran AP et al (2004) Nod1 responds to peptidoglycan delivered by the Helicobacter pylori cag pathogenicity island. Nat Immunol 5(11):1166–1174, Epub 2004/10/19
41. Tanabe T, Chamaillard M, Ogura Y, Zhu L, Qiu S, Masumoto J et al (2004) Regulatory regions and critical residues of NOD2 involved in muramyl dipeptide recognition. EMBO J 23(7):1587–1597, Epub 2004/03/27
42. Kufer TA, Kremmer E, Banks DJ, Philpott DJ (2006) Role for Erbin in bacterial activation of Nod2. Infect Immun 74(6):3115–3124
43. McDonald C, Chen FF, Ollendorff V, Ogura Y, Marchetto S, Lecine P et al (2005) A role for Erbin in the regulation of Nod2-dependent NF-kappaB signaling. J Biol Chem 280(48): 40301–40309, Epub 2005/10/06
44. Richmond AL, Kabi A, Homer CR, Marina-Garcia N, Nickerson KP, Nesvizhskii AI et al (2012) The nucleotide synthesis enzyme CAD inhibits NOD2 antibacterial function in human intestinal epithelial cells. Gastroenterology 142(7):1483–1492.e6. Epub 2012/03/06
45. Magalhaes JG, Sorbara MT, Girardin SE, Philpott DJ (2010) What is new with Nods? Curr Opin Immunol 23(1):29–34
46. Tattoli I, Travassos LH, Carneiro LA, Magalhaes JG, Girardin SE (2007) The Nodosome: Nod1 and Nod2 control bacterial infections and inflammation. Semin Immunopathol 29(3):289–301
47. Petnicki-Ocwieja T, Hrncir T, Liu YJ, Biswas A, Hudcovic T, Tlaskalova-Hogenova H et al (2009) Nod2 is required for the regulation of commensal microbiota in the intestine. Proc Natl Acad Sci U S A 106(37):15813–15818, Epub 2009/10/07
48. Biswas A, Liu YJ, Hao L, Mizoguchi A, Salzman NH, Bevins CL et al (2010) Induction and rescue of Nod2-dependent Th1-driven granulomatous inflammation of the ileum. Proc Natl Acad Sci U S A 107(33):14739–14744, Epub 2010/08/04
49. Kobayashi KS, Chamaillard M, Ogura Y, Henegariu O, Inohara N, Nunez G et al (2005) Nod2-dependent regulation of innate and adaptive immunity in the intestinal tract. Science 307(5710):731–734, Epub 2005/02/05
50. Wehkamp J, Harder J, Weichenthal M, Schwab M, Schaffeler E, Schlee M et al (2004) NOD2 (CARD15) mutations in Crohn's disease are associated with diminished mucosal alpha-defensin expression. Gut 53(11):1658–1664
51. Vaishnava S, Behrendt CL, Ismail AS, Eckmann L, Hooper LV (2008) Paneth cells directly sense gut commensals and maintain homeostasis at the intestinal host-microbial interface. Proc Natl Acad Sci U S A 105(52):20858–20863, Epub 2008/12/17
52. Bouskra D, Brezillon C, Berard M, Werts C, Varona R, Boneca IG et al (2008) Lymphoid tissue genesis induced by commensals through NOD1 regulates intestinal homeostasis. Nature 456(7221):507–510, Epub 2008/11/07
53. Ganz T (2003) Defensins: antimicrobial peptides of innate immunity. Nat Rev Immunol 3(9):710–720, Epub 2003/09/02
54. Boughan PK, Argent RH, Body-Malapel M, Park JH, Ewings KE, Bowie AG et al (2006) Nucleotide-binding oligomerization domain-1 and epidermal growth factor receptor: critical regulators of beta-defensins during Helicobacter pylori infection. J Biol Chem 281(17): 11637–11648, Epub 2006/03/04
55. Grubman A, Kaparakis M, Viala J, Allison C, Badea L, Karrar A et al (2010) The innate immune molecule, NOD1, regulates direct killing of Helicobacter pylori by antimicrobial peptides. Cell Microbiol 12(5):626–639, Epub 2009/12/31
56. Voss E, Wehkamp J, Wehkamp K, Stange EF, Schroder JM, Harder J (2006) NOD2/CARD15 mediates induction of the antimicrobial peptide human beta-defensin-2. J Biol Chem 281(4):2005–2011
57. Watanabe T, Asano N, Fichtner-Feigl S, Gorelick PL, Tsuji Y, Matsumoto Y et al (2010) NOD1 contributes to mouse host defense against Helicobacter pylori via induction of type I IFN and activation of the ISGF3 signaling pathway. J Clin Invest 120(5):1645–1662, Epub 2010/04/15

58. Masumoto J, Yang K, Varambally S, Hasegawa M, Tomlins SA, Qiu S et al (2006) Nod1 acts as an intracellular receptor to stimulate chemokine production and neutrophil recruitment in vivo. J Exp Med 203(1):203–213, Epub 2006/01/19
59. Hasegawa M, Yamazaki T, Kamada N, Tawaratsumida K, Kim YG, Nunez G et al (2011) Nucleotide-binding oligomerization domain 1 mediates recognition of Clostridium difficile and induces neutrophil recruitment and protection against the pathogen. J Immunol 186(8):4872–4880
60. Rupnik M, Wilcox MH, Gerding DN (2009) Clostridium difficile infection: new developments in epidemiology and pathogenesis. Nat Rev Microbiol 7(7):526–536, Epub 2009/06/17
61. Kim YG, Kamada N, Shaw MH, Warner N, Chen GY, Franchi L et al (2011) The Nod2 sensor promotes intestinal pathogen eradication via the chemokine CCL2-dependent recruitment of inflammatory monocytes. Immunity 34(5):769–780, Epub 2011/05/14
62. Davis KM, Nakamura S, Weiser JN (2011) Nod2 sensing of lysozyme-digested peptidoglycan promotes macrophage recruitment and clearance of S. pneumoniae colonization in mice. J Clin Invest 121(9):3666–3676
63. Brooks MN, Rajaram MVS, Azad AK, Amer AO, Valdivia-Arenas MA, Park J-H et al (2010) NOD2 controls the nature of the inflammatory response and subsequent fate of Mycobacterium tuberculosis and M. bovis BCG in human macrophages. Cell Microbiol 13(3):402–418
64. Imielinski M, Baldassano RN, Griffiths A, Russell RK, Annese V, Dubinsky M et al (2009) Common variants at five new loci associated with early-onset inflammatory bowel disease. Nat Genet 41(12):1335–1340, Epub 2009/11/17
65. Oosting M, Berende A, Sturm P, ter Hofstede HJM, de Jong DJ, Kanneganti T-D et al (2010) Recognition of Borrelia burgdorferi by NOD2 is central for the induction of an inflammatory reaction. J Infect Dis 201(12):1849–1858
66. Berrington WR, Iyer R, Wells RD, Smith KD, Skerrett SJ, Hawn TR (2010) NOD1 and NOD2 regulation of pulmonary innate immunity to Legionella pneumophila. Eur J Immunol 40(12):3519–3527
67. Totemeyer S, Sheppard M, Lloyd A, Roper D, Dowson C, Underhill D et al (2006) IFN-gamma enhances production of nitric oxide from macrophages via a mechanism that depends on nucleotide oligomerization domain-2. J Immunol 176(8):4804–4810, Epub 2006/04/06
68. Le Bourhis L, Magalhaes JG, Selvanantham T, Travassos LH, Geddes K, Fritz JH et al (2009) Role of Nod1 in mucosal dendritic cells during Salmonella pathogenicity island 1-independent Salmonella enterica serovar Typhimurium infection. Infect Immun 77(10):4480–4486
69. Mosa A, Trumstedt C, Eriksson E, Soehnlein O, Heuts F, Janik K et al (2009) Nonhematopoietic cells control the outcome of infection with Listeria monocytogenes in a nucleotide oligomerization domain 1-dependent manner. Infect Immun 77(7):2908–2918
70. Scott MJ, Chen C, Sun Q, Billiar TR (2010) Hepatocytes express functional NOD1 and NOD2 receptors: A role for NOD1 in hepatocyte CC and CXC chemokine production. J Hepatol 53(4):693–701
71. Moreno L, McMaster SK, Gatheral T, Bailey LK, Harrington LS, Cartwright N et al (2010) NOD1 is a dominant pathway for NOS2 induction in vascular smooth muscle cells: comparison with TLR4 responses in macrophages. Br J Pharmacol 160:1997–2007
72. Shimada K, Chen S, Dempsey PW, Sorrentino R, Alsabeh R, Slepenkin AV et al (2009) The NOD/RIP2 pathway is essential for host defenses against Chlamydophila pneumoniae lung infection. PLoS Pathog 5(4):e1000379, Epub 2009/04/11
73. Lysenko ES, Clarke TB, Shchepetov M, Ratner AJ, Roper DI, Dowson CG et al (2007) Nod1 signaling overcomes resistance of S. pneumoniae to opsonophagocytic killing. PLoS Pathog 3(8):e118
74. Simonsen A, Tooze SA (2009) Coordination of membrane events during autophagy by multiple class III PI3-kinase complexes. J Cell Biol 186(6):773–782, Epub 2009/10/03
75. Deretic V, Levine B (2009) Autophagy, immunity, and microbial adaptations. Cell Host Microbe 5(6):527–549

76. Travassos LH, Carneiro LA, Ramjeet M, Hussey S, Kim YG, Magalhaes JG et al (2009) Nod1 and Nod2 direct autophagy by recruiting ATG16L1 to the plasma membrane at the site of bacterial entry. Nat Immunol 11(1):55–62
77. Cooney R, Baker J, Brain O, Danis B, Pichulik T, Allan P et al (2010) NOD2 stimulation induces autophagy in dendritic cells influencing bacterial handling and antigen presentation. Nat Med 16(1):90–97, Epub 2009/12/08
78. Homer CR, Richmond AL, Rebert NA, Achkar JP, McDonald C (2010) ATG16L1 and NOD2 interact in an autophagy-dependent antibacterial pathway implicated in Crohn's disease pathogenesis. Gastroenterology 139(5):1630–1641.e2
79. Rehman A, Sina C, Gavrilova O, Hasler R, Ott S, Baines JF et al (2011) Nod2 is essential for temporal development of intestinal microbial communities. Gut 60(10):1354–1362
80. Fritz JH, Le Bourhis L, Sellge G, Magalhaes JG, Fsihi H, Kufer TA et al (2007) Nod1-mediated innate immune recognition of peptidoglycan contributes to the onset of adaptive immunity. Immunity 26(4):445–459
81. Magalhães JG, Lee J, Geddes K, Rubino S, Philpott DJ, Girardin SE (2011) Essential role of Rip2 in the modulation of innate and adaptive immunity triggered by Nod1 and Nod2 ligands. Eur J Immunol 41(5):1445–1455
82. Magalhaes JG, Rubino SJ, Travassos LH, Le Bourhis L, Duan W, Sellge G et al (2011) Nucleotide oligomerization domain-containing proteins instruct T cell helper type 2 immunity through stromal activation. Proc Natl Acad Sci USA 108(36):14896–14901, Epub 2011/08/23
83. Shaw MH, Reimer T, Sanchez-Valdepenas C, Warner N, Kim YG, Fresno M et al (2009) T cell-intrinsic role of Nod2 in promoting type 1 immunity to Toxoplasma gondii. Nat Immunol 10(12):1267–1274, Epub 2009/11/03
84. Ouyang W, Kolls JK, Zheng Y (2008) The biological functions of T helper 17 cell effector cytokines in inflammation. Immunity 28(4):454–467, Epub 2008/04/11
85. Colonna M (2009) Interleukin-22-producing natural killer cells and lymphoid tissue inducer-like cells in mucosal immunity. Immunity 31(1):15–23, Epub 2009/07/17
86. Geddes K, Rubino SJ, Magalhães JG, Streutker C, Le Bourhis L, Cho JH et al (2011) Identification of an innate T helper type 17 response to intestinal bacterial pathogens. Nat Med 17(7):837–844
87. Wing K, Sakaguchi S (2010) Regulatory T cells exert checks and balances on self tolerance and autoimmunity. Nat Immunol 11(1):7–13, Epub 2009/12/18
88. Noguchi E, Homma Y, Kang X, Netea MG, Ma X (2009) A Crohn's disease-associated NOD2 mutation suppresses transcription of human IL10 by inhibiting activity of the nuclear ribonucleoprotein hnRNP-A1. Nat Immunol 10(5):471–479, Epub 2009/04/08
89. Opitz B, Forster S, Hocke AC, Maass M, Schmeck B, Hippenstiel S et al (2005) Nod1-mediated endothelial cell activation by Chlamydophila pneumoniae. Circ Res 96(3):319–326
90. Opitz B, Puschel A, Beermann W, Hocke AC, Forster S, Schmeck B et al (2006) Listeria monocytogenes activated p38 MAPK and induced IL-8 secretion in a nucleotide-binding oligomerization domain 1-dependent manner in endothelial cells. J Immunol 176(1):484–490
91. Kim JG, Lee SJ, Kagnoff MF (2004) Nod1 is an essential signal transducer in intestinal epithelial cells infected with bacteria that avoid recognition by toll-like receptors. Infect Immun 72(3):1487–1495
92. Boneca IG, Dussurget O, Cabanes D, Nahori MA, Sousa S, Lecuit M et al (2007) A critical role for peptidoglycan N-deacetylation in Listeria evasion from the host innate immune system. Proc Natl Acad Sci U S A 104(3):997–1002
93. Pott J, Basler T, Duerr CU, Rohde M, Goethe R, Hornef MW (2009) Internalization-dependent recognition of Mycobacterium avium ssp. paratuberculosis by intestinal epithelial cells. Cellular microbiology 11(12):1802–1815
94. Slevogt H, Seybold J, Tiwari KN, Hocke AC, Jonatat C, Dietel S et al (2007) Moraxella catarrhalis is internalized in respiratory epithelial cells by a trigger-like mechanism and initiates a TLR2- and partly NOD1-dependent inflammatory immune response. Cell Microbiol 9(3):694–707

95. Ferwerda G, Girardin SE, Kullberg BJ, Le Bourhis L, de Jong DJ, Langenberg DM et al (2005) NOD2 and toll-like receptors are nonredundant recognition systems of Mycobacterium tuberculosis. PLoS Pathog 1(3):279–285
96. Brooks MN, Rajaram MV, Azad AK, Amer AO, Valdivia-Arenas MA, Park JH et al (2011) NOD2 controls the nature of the inflammatory response and subsequent fate of Mycobacterium tuberculosis and M. bovis BCG in human macrophages. Cell Microbiol 13(3):402–418
97. Opitz B, Puschel A, Schmeck B, Hocke AC, Rosseau S, Hammerschmidt S et al (2004) Nucleotide-binding oligomerization domain proteins are innate immune receptors for internalized Streptococcus pneumoniae. J Biol Chem 279(35):36426–36432
98. Iyer JK, Coggeshall KM (2011) Cutting edge: primary innate immune cells respond efficiently to polymeric peptidoglycan, but not to peptidoglycan monomers. J Immunol 186(7):3841–3845
99. Park JT, Uehara T (2008) How bacteria consume their own exoskeletons (turnover and recycling of cell wall peptidoglycan). Microbiol Mol Biol Rev 72(2):211–227. table of contents, Epub 2008/06/07
100. Meylan E, Tschopp J, Karin M (2006) Intracellular pattern recognition receptors in the host response. Nature 442(7098):39–44
101. Goodell EW, Schwarz U (1985) Release of cell wall peptides into culture medium by exponentially growing Escherichia coli. J Bacteriol 162(1):391–397, Epub 1985/04/01
102. Mauck J, Chan L, Glaser L (1971) Turnover of the cell wall of Gram-positive bacteria. J Biol Chem 246(6):1820–1827, Epub 1971/03/25
103. Hirata Y, Maeda S, Ohmae T, Shibata W, Yanai A, Ogura K et al (2006) Helicobacter pylori induces IkappaB kinase alpha nuclear translocation and chemokine production in gastric epithelial cells. Infect Immun 74(3):1452–1461
104. Maeda S, Yoshida H, Ogura K, Mitsuno Y, Hirata Y, Yamaji Y et al (2000) H. pylori activates NF-kappaB through a signaling pathway involving IkappaB kinases, NF-kappaB-inducing kinase, TRAF2, and TRAF6 in gastric cancer cells. Gastroenterology 119(1):97–108
105. Allison CC, Kufer TA, Kremmer E, Kaparakis M, Ferrero RL (2009) Helicobacter pylori induces MAPK phosphorylation and AP-1 activation via a NOD1-dependent mechanism. J Immunol 183(12):8099–8109, Epub 2009/12/17
106. Hutton ML, Kaparakis-Liaskos M, Turner L, Cardona A, Kwok T, Ferrero RL (2010) Helicobacter pylori exploits cholesterol-rich microdomains for induction of NF-kappaB-dependent responses and peptidoglycan delivery in epithelial cells. Infect Immun 78(11): 4523–4531, Epub 2010/08/18
107. LeBlanc PM, Yeretssian G, Rutherford N, Doiron K, Nadiri A, Zhu L et al (2008) Caspase-12 modulates NOD signaling and regulates antimicrobial peptide production and mucosal immunity. Cell Host Microbe 3(3):146–157, Epub 2008/03/11
108. Lee J, Tattoli I, Wojtal KA, Vavricka SR, Philpott DJ, Girardin SE (2009) pH-dependent internalization of muramyl peptides from early endosomes enables Nod1 and Nod2 signaling. J Biol Chem 284(35):23818–23829
109. Marina-Garcia N, Franchi L, Kim YG, Hu Y, Smith DE, Boons GJ et al (2009) Clathrin- and dynamin-dependent endocytic pathway regulates muramyl dipeptide internalization and NOD2 activation. J Immunol 182(7):4321–4327, Epub 2009/03/21
110. Vavricka SR, Musch MW, Chang JE, Nakagawa Y, Phanvijhitsiri K, Waypa TS et al (2004) hPepT1 transports muramyl dipeptide, activating NF-kappaB and stimulating IL-8 secretion in human colonic Caco2/bbe cells. Gastroenterology 127(5):1401–1409, Epub 2004/11/03
111. Ismair MG, Vavricka SR, Kullak-Ublick GA, Fried M, Mengin-Lecreulx D, Girardin SE (2006) hPepT1 selectively transports muramyl dipeptide but not Nod1-activating muramyl peptides. Can J Physiol Pharmacol 84(12):1313–1319, Epub 2007/05/10
112. Fei YJ, Kanai Y, Nussberger S, Ganapathy V, Leibach FH, Romero MF et al (1994) Expression cloning of a mammalian proton-coupled oligopeptide transporter. Nature 368(6471):563–566, Epub 1994/04/07
113. Kuehn MJ, Kesty NC (2005) Bacterial outer membrane vesicles and the host-pathogen interaction. Genes Dev 19(22):2645–2655

114. Kaparakis M, Turnbull L, Carneiro L, Firth S, Coleman HA, Parkington HC et al (2010) Bacterial membrane vesicles deliver peptidoglycan to NOD1 in epithelial cells. Cell Microbiol 12(3):372–385, Epub 2009/11/06
115. Kesty NC, Mason KM, Reedy M, Miller SE, Kuehn MJ (2004) Enterotoxigenic Escherichia coli vesicles target toxin delivery into mammalian cells. EMBO J 23(23):4538–4549
116. Bielig H, Rompikuntal PK, Dongre M, Zurek B, Lindmark B, Ramstedt M et al (2011) NOD-like receptor activation by outer membrane vesicles from Vibrio cholerae non-O1 non-O139 strains is modulated by the quorum-sensing regulator HapR. Infect Immun 79(4):1418–1427, Epub 2011/01/26
117. Grimes CL, Ariyananda Lde Z, Melnyk JE, O'Shea EK (2012) The innate immune protein Nod2 binds directly to MDP, a bacterial cell wall fragment. J Am Chem Soc 134(33): 13535–13537, Epub 2012/08/04
118. Mo J, Boyle JP, Howard CB, Monie TP, Davis BK, Duncan JA (2012) Pathogen sensing by nucleotide-binding oligomerization domain-containing protein 2 (NOD2) is mediated by direct binding to muramyl dipeptide and ATP. J Biol Chem 287(27):23057–23067, Epub 2012/05/03
119. Hugot JP, Chamaillard M, Zouali H, Lesage S, Cezard JP, Belaiche J et al (2001) Association of NOD2 leucine-rich repeat variants with susceptibility to Crohn's disease. Nature 411(6837):599–603, Epub 2001/06/01
120. McGovern DP, Hysi P, Ahmad T, van Heel DA, Moffatt MF, Carey A et al (2005) Association between a complex insertion/deletion polymorphism in NOD1 (CARD4) and susceptibility to inflammatory bowel disease. Hum Mol Genet 14(10):1245–1250, Epub 2005/03/26
121. Van Limbergen J, Nimmo ER, Russell RK, Drummond HE, Smith L, Anderson NH et al (2007) Investigation of NOD1/CARD4 variation in inflammatory bowel disease using a haplotype-tagging strategy. Hum Mol Genet 16(18):2175–2186, Epub 2007/07/07
122. van Heel DA, Ghosh S, Butler M, Hunt KA, Lundberg AM, Ahmad T et al (2005) Muramyl dipeptide and toll-like receptor sensitivity in NOD2-associated Crohn's disease. Lancet 365(9473):1794–1796, Epub 2005/05/25
123. Rivas MA, Beaudoin M, Gardet A, Stevens C, Sharma Y, Zhang CK et al (2011) Deep resequencing of GWAS loci identifies independent rare variants associated with inflammatory bowel disease. Nat Genet 43(11):1066–1073, Epub 2011/10/11
124. Watanabe T, Asano N, Murray PJ, Ozato K, Tailor P, Fuss IJ et al (2008) Muramyl dipeptide activation of nucleotide-binding oligomerization domain 2 protects mice from experimental colitis. J Clin Invest 118(2):545–559, Epub 2008/01/12
125. Macho Fernandez E, Valenti V, Rockel C, Hermann C, Pot B, Boneca IG et al (2011) Anti-inflammatory capacity of selected lactobacilli in experimental colitis is driven by NOD2-mediated recognition of a specific peptidoglycan-derived muropeptide. Gut 60(8):1050–1059, Epub 2011/04/08
126. Barreau F, Meinzer U, Chareyre F, Berrebi D, Niwa-Kawakita M, Dussaillant M et al (2007) CARD15/NOD2 is required for Peyer's patches homeostasis in mice. PLoS One 2(6):e523, Epub 2007/06/15
127. Penack O, Smith OM, Cunningham-Bussel A, Liu X, Rao U, Yim N et al (2009) NOD2 regulates hematopoietic cell function during graft-versus-host disease. J Exp Med 206(10):2101–2110, Epub 2009/09/10
128. Geddes K, Rubino S, Streutker C, Cho JH, Magalhaes JG, Le Bourhis L et al (2010) Nod1 and Nod2 regulation of inflammation in the Salmonella colitis model. Infect Immun 78(12):5107–5115, Epub 2010/10/06
129. Kim YG, Shaw MH, Warner N, Park JH, Chen F, Ogura Y et al (2011) Cutting edge: Crohn's disease-associated Nod2 mutation limits production of proinflammatory cytokines to protect the host from Enterococcus faecalis-induced lethality. J Immunol 187(6):2849–2852, Epub 2011/08/19

Chapter 11
The IL23-Th17 Axis in Intestinal Inflammation

Kevin J. Maloy

Abstract Over the past 5 years, a wealth of data has emerged from experimental models linking the IL23/Th17 axis with chronic intestinal inflammation. Clinical studies have also reported elevated levels of IL23 and Th17 cytokines in the inflamed intestine of IBD patients, and recent GWAS have associated polymorphisms in *IL23R*, and in other Th17-related genes, with susceptibility to IBD. However, the precise mechanisms through which the IL23-Th17 axis contributes to intestinal homeostasis are not fully understood. Recent studies have revealed that IL23 drives intestinal inflammation by stimulating conserved effector responses, characterised by the production of IL-17A, IFN-γ and IL-22, from several populations of innate and adaptive intestinal leukocytes. However, the effects of individual Th17 cytokines are complex, ranging from disease protective to highly pathogenic, and are governed by the context in which they are expressed and by the presence of additional factors in the intestine. More recently, it has been shown that distinct members of the intestinal microbiota can modulate the Th17 axis and can regulate the balance between Th17 and Treg cells in the intestine. In addition, several approaches using pharmacological or biological inhibitors of the IL23/Th17 axis have been demonstrated to alleviate autoimmune pathology. These findings suggest that strategies targeting the IL23/Th17 axis could constitute novel therapies for IBD. However, a better understanding of how the IL23/Th17 axis interacts with host genetic factors and the intestinal microbiota in the normal and diseased intestine is necessary to ensure that these novel therapies are applied to appropriate patient cohorts.

K.J. Maloy, Ph.D. (✉)
Sir William Dunn School of Pathology, University of Oxford, South Parks Road, Oxford OX1 3RE, UK
e-mail: kevin.maloy@path.ox.ac.uk

M. D'Amato and J.D. Rioux (eds.), *Molecular Genetics of Inflammatory Bowel Disease*, DOI 10.1007/978-1-4614-8256-7_11, © Springer Science+Business Media New York 2013

Introduction

Chronic inflammatory disorders, including inflammatory bowel diseases (IBD) that affect the gastrointestinal tract, are an important cause of morbidity in the developed world [1]. In the last decade, technical innovations such as genome-wide association studies (GWAS) and deep sequencing of the intestinal microbiota, together with continued progress in understanding the regulation of immune and inflammatory networks, have driven tremendous advances in identification of genetic, environmental and immune factors that together determine IBD susceptibility [2]. Despite these advances, our understanding of IBD pathogenesis is far from complete, and there is a growing acceptance that the heterogeneous clinical features of IBD may reflect the expression of distinct combinations of risk factors in individual patients [3]. A corollary of this concept is the realisation that a global 'cure' for IBD may be an unrealistic aspiration and instead, the focus should be on developing new therapies that can be employed in appropriate patient subgroups exhibiting shared pathogenic pathways. The successful employment of biological therapies targeting TNF-α, which have been shown to have potent disease-ameliorating effects in a significant proportion of IBD patients, illustrates the utility of this approach. However, additional therapies need to be developed to target alternative inflammatory pathways that may predominate in other patient subgroups that respond poorly to conventional therapies [4].

In parallel to more detailed characterization of IBD patients, complementary studies on intestinal immunity have revealed a plethora of new cellular and molecular factors that can contribute to protective and pathogenic immune responses in the gut [3]. These findings have refined our view of intestinal immune homeostasis and have led to a re-evaluation of IBD pathogenesis by identifying new players in intestinal inflammatory responses. One key finding to emerge from these studies was an association of the pro-inflammatory cytokine IL23 with chronic intestinal inflammation, both in preclinical mouse models of the disease and in IBD patient cohorts GWAS [2, 3]. Furthermore, IL23 has been strongly associated with Th17 $CD4^+$ T cell responses, which have been implicated in immune pathology in several inflammatory diseases. As such, there is currently enormous interest in the development of therapies targeting the IL23/Th17 axis for the treatment of autoimmune and inflammatory disease. In this review, I will summarise the data linking the IL23/Th17 axis with intestinal inflammation as well as the mechanisms responsible, and I will consider how manipulation of this axis may be achieved and whether this strategy may prove useful in the treatment of IBD.

Linking the IL23-Th17 Axis to Intestinal Pathology

Early molecular profiling of cytokines and T cells present in intestinal lesions in IBD gave rise to the paradigm that Crohn's disease (CD) was associated with IFN-γ and IL-12 and was therefore a Th1-mediated disorder, whereas ulcerative colitis

(UC) was predominantly associated with Th2 responses, characterised by IL-5 and IL-13 [5]. This paradigm was reinforced by studies in experimental IBD models where depletion of IL-12, through genetic ablation of the IL-12p40 subunit, or using anti-p40 monoclonal antibodies (mAb), led to amelioration of colitis [5]. However, the discovery that the IL-12p40 subunit could also pair with a p19 subunit to form the heterodimeric cytokine IL23 [6] led to a reappraisal of the relative contributions of IL-12 and IL23 to inflammatory pathology.

Seminal studies in mouse models of autoimmune disease, such as experimental autoimmune encephalitis (EAE) and collagen-induced arthritis (CIA), demonstrated that IL-12 was not required for disease, whereas IL23 was indispensable for tissue immune pathology [7]. In addition, IL23-driven autoimmune pathology was associated with the accumulation of a new inflammatory $CD4^+$ T cell subset, termed Th17 cells on account of their secretion of the cytokines IL-17A and IL-17F, but which have subsequently been shown to be highly heterogeneous and to produce additional cytokines, such as IL-22 and IL-21 [7–9]. A series of parallel studies in experimental models of IBD reported similar findings, identifying IL23 as the central driver of chronic intestinal pathology. Selective depletion of IL23, either through genetic ablation or using a p19-specific mAb, was found to prevent spontaneous colitis in *IL-10*$^{-/-}$ mice [10], to attenuate intestinal inflammation in *Rag*$^{-/-}$ recipients of naïve $CD45RB^{hi}CD4^+$ T cells [11] and to inhibit the typhlocolitis that was triggered in wild-type mice by infection with the intestinal bacterium *Helicobacter hepaticus* and concomitant blockade of IL-10R signals [12]. In addition to these models of T cell-mediated IBD, IL23 was also shown to drive innate immune-mediated intestinal inflammation in *Rag*$^{-/-}$ mice induced either by infection with *Helicobacter hepaticus* [11] or by injection of agonistic anti-CD40 mAb [14]. Similar to what had been described in autoimmune tissue pathology, IL23-mediated intestinal pathology was found to be associated with high levels of Th17 cytokines in the inflamed intestine [11, 12]. Interestingly, although intestinal pathology in these diverse IBD models was accompanied by a range of systemic inflammatory responses such as splenomegaly and liver inflammation, elevated IL23 and Th17 responses were largely restricted to the inflamed gut, with the systemic immune pathology often driven by IL-12 [11, 13], suggesting that IL23 and IL-12 differentially regulate inflammatory responses at different sites.

The data derived from the various mouse IBD models were consistent with reports of elevated levels of Th17 cytokines in human IBD tissue samples [14–16]. However, the key finding that cemented the link between IL23 and human IBD resulted from GWAS studies of large cohorts of IBD patients, which identified single nucleotide polymorphisms (SNPs) in the *IL23R* gene locus that were associated with either susceptibility or resistance to CD [17, 18]. These findings were replicated by subsequent studies, which also extended the associations to UC, indicating that *IL23R* SNPs may contribute to susceptibility to both major forms of IBD [19–21]. The mechanisms through which *IL23R* polymorphisms control susceptibility to IBD are not yet clear and are complicated by the findings that both protective and susceptibility alleles have been identified and that some of the SNPs are in non-coding regions [17–21]. However, a major protective *IL23R* allele contains a non-synonymous SNP (R381Q) in the cytoplasmic domain of the IL23R, and T cells expressing this *IL23R* variant exhibited reduced STAT3 phosphorylation and decreased IL-17A production in response to IL23 stimulation [22, 23], suggesting that impaired IL23R signalling confers a significant degree of protection from IBD. Alternatively, protective SNPs in non-coding regions may cause alternative splicing leading to the production of soluble truncated forms of the IL23R that can inhibit signalling by the full-length receptor [24]. These findings are consistent with a pathogenic role of IL23R signalling in IBD and with clinical evidence of elevated production of IL23 in the inflamed lesions of IBD patients [25, 26]. It remains to be established whether risk-associated *IL23R* SNPs result in constitutively active IL23R signalling or in attenuated production of decoy receptors. It is important to consider that IL23R expression is not restricted to effector T cells, but has also been reported for NK cells and various mononuclear phagocyte cells, including macrophages and dendritic cells [27]. Thus, further studies on both innate and adaptive leukocytes expressing risk-associated or protective *IL23R* alleles should illuminate additional pathways through which IL23R signals control inflammatory responses in the intestine.

IL23 Stimulates Conserved Effector Responses from Innate and Adaptive Intestinal Leukocytes

IL23 is secreted by activated macrophages and DC in response to microbial stimulation [28] and is constitutively expressed by a population of lamina propria DC in the ileum, a site that is often affected in CD patients [29]. IL23 expression was increased in the inflamed intestine in mouse IBD models, particularly within colonic $CD11c^+$ DC, although $CD11b^+CD11c^-$ monocytic populations also expressed high levels of IL23 [11, 14]. A population of myeloid DC present in the mesenteric lymph node of CD patients was shown to secrete elevated levels of IL23 upon bacterial stimulation, when compared to the same DC population isolated from control subjects or UC patients [30]. However, CD patients have also been reported to harbour a subset of $CD14^+$ macrophages in the intestinal lamina propria that secreted high levels of IL23 and could drive the differentiation of Th17 and Th1 cells [26, 31]. Moreover, a recent study in mice reported that a $CD103^+CD11b^+$ lamina propria DC subset rapidly responded to bacterial flagellin stimulation by producing high amounts of IL23 [32]. Thus, it appears that both mucosal DC and macrophages may contribute to IL23 production in the intestine, especially during infection or chronic inflammation. However, as pattern recognition receptor (PRR) stimulation by the same microbial agonists also drives IL-12 secretion, it is not yet clear what determines whether particular populations of intestinal macrophages and DC predominantly secrete IL23 or IL-12.

Early studies on IL23 in autoimmune disease models established a strong association between IL23-driven pathology and the development of Th17 $CD4^+$ T cell responses [7–9]. Although Th17 cells were first described as a population of $IL23R^+$ memory $CD4^+$ T cells, Th17 cell differentiation from naïve T cells is driven by TCR activation in the presence of STAT-3-activating cytokines, such as IL-6 and IL-21, together with TGF-β and IL-1β, and is dependent on the transcription factor RORγt (called RORC in humans) [7–9]. These factors promote IL23R expression on activated T cells, and IL23R signals enhance Th17 differentiation and survival [7–9]. Th17 cells are highly heterogeneous and individual Th17 cells may secrete various combinations of IL-17A, IL-17F, IL-22, IL-21 as well as other cytokines [7–9]. In addition, Th17 cells exhibit considerable plasticity, meaning that they can further differentiate and acquire functions attributed to other effector Th cell subsets, such as the production of IFN-γ or IL-10 [9, 33]. The instability of the Th17 lineage is thought to be due to epigenetic modifications that permit the reactivation of key transcription factors that can drive reprogramming of Th17 cells [34, 35]. Th17 cells preferentially accumulate at mucosal surfaces, particularly the intestine, but they share this niche with other leukocyte populations, including immune suppressive $FoxP3^+$ regulatory cells that exhibit a reciprocal relationship with Th17 cells [36]. Maintenance of an appropriate balance between Th17 and Treg cells in the intestine plays a key role in regulating intestinal immune homeostasis [36].

Mouse IBD models have identified several mechanisms through which IL23 promotes pathogenic T cell response in the intestine. In the T cell transfer model of IBD, where naïve $CD45RB^{hi}$ $CD4^+$ T cells induce severe chronic colitis and wasting disease following adoptive transfer into lymphopenic (SCID or *Rag*$^{-/-}$) recipients [37], *IL23*$^{-/-}$*Rag*$^{-/-}$mice did not develop colitis following reconstitution with WT naïve $CD45RB^{hi}$ $CD4^+$ T cells [11]. The absence of intestinal pathology in *IL23*$^{-/-}$*Rag*$^{-/-}$recipients correlated with decreased IL-17A and IFN-γ levels in the intestine and with reduced frequencies of both Th1 and Th17 cells in the MLN [11]. Similar findings were reported in models of *Helicobacter hepaticus*-triggered T cell-mediated typhlocolitis, where a lack of intestinal pathology in *IL23*$^{-/-}$ mice again correlated with a reduction in both Th1 and Th17 responses [12]. In humans, a population of gut-homing $CD161^+CD4^+$ Th17 cells was identified and found to be enriched within intestinal lesions of CD patients [38]. Furthermore, in contrast to the $CD161^+CD4^+$ T cells present in healthy controls, those from CD patients secreted high levels of IL-17A and IFN-γ in response to IL23, suggesting that IL23 drives similar T cell effector responses in the inflamed intestine in humans [38]. Direct IL23 signalling into $CD4^+$ T cells is required for the development of T cell-mediated intestinal inflammation because *IL23R*$^{-/-}$ naïve $CD4^+$ T cells were unable to induce colitis in *Rag*$^{-/-}$ recipients [39]. In addition, T cell intrinsic IL23R signals promoted the accumulation of a population of IL-$17A^+$IFN-γ^+ double producer (DP) T cells within the inflamed colon, suggesting that IL23 may potentiate the pathogenic potential of Th17 cells [39]. Although the precise role of these IL-$17A^+$IFN-γ^+ DP cells in intestinal pathology remains to be determined, similar T cells have been isolated from inflammatory lesions of CD patients [40]. IL23 also potentiates inflammatory T cell responses in the gut by limiting the development of Treg responses. For example, in the naïve $CD4^+$ T cell transfer colitis model, a higher frequency of $FoxP3^+$Treg cells developed when the *Rag*$^{-/-}$ recipients could not produce IL23 [41]. Similarly, *IL23R*$^{-/-}$naïve $CD4^+$ T cells showed an enhanced ability to differentiate into $FoxP3^+$Treg cells and to secrete IL-10, compared to WT naïve $CD4^+$ T cells following transfer into *Rag*$^{-/-}$ mice [39]. Thus, IL23 drives chronic intestinal inflammation by promoting a range of pathogenic effector Th cell responses in the inflamed intestine, encompassing Th1, Th17 and IL-$17A^+$IFN-γ^+ DP cells, and by inhibiting the development of $FoxP3^+$Treg cells.

IL23 was also shown to drive colitis in T cell-deficient mice, and this innate immune intestinal pathology was also associated with elevated production of IL-17A, IFN-γ and IL-22 [11, 13, 42]. In colitic *Rag*$^{-/-}$ mice, a population of Thy^{Hi}Sca-1^{+}RORγt^{+}IL23R^{+} cells was identified as the key source of IL-17A, IFN-γ and IL-22 in the inflamed intestinal lamina propria, and depletion of these cells using anti-Thy1 mAb ameliorated intestinal pathology [42]. Phenotypic analysis revealed that these cells were negative for myeloid and lymphocyte lineage markers (Lin^{-}) but that they shared some markers with NK cells and LTi cells and were therefore termed innate lymphoid cells (ILC) [42]. A similar population of ILC that produced Th17 cytokines in response to IL23 was increased in the inflamed intestine of CD patients [43]. Furthermore, a series of parallel studies have described a variety of additional RORγt^{+}IL23R^{+} innate leukocyte populations found in the intestine and skin of mice and humans that secrete Th17 cytokines in response to IL23, including γδ T cells, NK T cells, NKp46^{+} cells and LTi cells [44]. These innate populations rapidly respond to signals associated with stress, infection or injury by secreting IL-17A and IL-22 and other mediators that are important for early tissue-protective and antimicrobial responses at mucosal surfaces [44, 45]. It is not yet clear whether these other innate populations can contribute to IBD pathogenesis, although NKp46^{+} cell populations secreting IFN-γ were increased in the intestinal mucosa in CD patients [46].

It has become apparent that the colitis-promoting effects of IL23 can be exacerbated by IL-1β, another pro-inflammatory cytokine that has been linked to Th17 responses. A recent study found that IL-1β synergised with IL23 to promote innate and adaptive pathogenic Th17 responses in the gut [47]. In innate immune colitis, IL-1β enhanced IL23R expression by ILC and promoted the accumulation of IL-17A-producing ILC in the colon [47]. Conversely, IL23R signals enhanced IL-1R1 expression on CD4^{+} T cells which enhanced Th17 cell accumulation and survival in the gut [47]. Similarly, IL-1β also synergized with IL23 to induce maximal production of IL-17 and IFN-γ from CD161^{+} CD4^{+} T cells isolated from CD patients [38]. In summary, it has become clear that IL23 mediates chronic intestinal inflammation by driving the expression of conserved inflammatory cytokine responses from both innate leukocytes and T cells in the intestine and that this activity can be amplified by additional inflammatory mediators.

Protective and Pathogenic Functions of Th17 Cytokines in the Intestine

IBD arises as a consequence of aberrant or chronic over-expression of protective immune mechanisms, and Th17 cells have been shown to be important for protection against pathogenic infections of mucosal surfaces, particularly by extracellular bacterial and fungal pathogens [48]. The heterogeneous nature of Th17 cells, together with the increased focus on biological therapies for the treatment of chronic inflammatory disorders, has prompted efforts to dissect the roles of distinct Th17 cytokines in intestinal homeostasis. These studies have revealed that the activities of individual cytokines are complex and may encompass both pathogenic and protective effects in the intestine. In one sense this is not surprising, since inflammation and tissue repair are tightly linked processes and the intestine is often considered a site of controlled inflammation. Tissue-protective effects may reflect direct fortification or repair of the epithelial barrier or may be related to host-protective immune responses that limit pathogen colonisation or invasion in the gut [49, 50]. The latter activity is exemplified by studies on *Citrobacter rodentium* infection in mice, which serves as a mouse surrogate of enteropathogenic/enterohaemorrhagic *E. coli* (EPEC/EHEC) infection of humans [51]. *C. rodentium* is an attaching-effacing (A/E) bacterial pathogen, forming pedestals on infected intestinal epithelial cells (IEC), and in WT mice disease manifests as transient weight loss, diarrhoea and mild intestinal inflammation, which abate with clearance of the pathogen after around 3 weeks [51]. IL23 is essential for protection from *C. rodentium* as *IL23*$^{-/-}$ mice succumbed to this normally self-limiting infection [52]. Several Th17 cytokines have been shown to play a role in protection, as *IL-17A*$^{-/-}$ mice, *IL-17F*$^{-/-}$mice and *IL-22*$^{-/-}$ mice all exhibited increased *C. rodentium* colonisation that was frequently associated with marked systemic bacterial penetration [53, 54]. Impaired antibacterial immunity correlated with decreased production of β-defensins in *IL-17A*$^{-/-}$and *IL-17F*$^{-/-}$ mice [53] and with a reduction in RegIIIβ and RegIIIγ expression in *IL-22*$^{-/-}$ mice [54], confirming that these Th17 cytokines were potent inducers of antimicrobial peptides [55]. Importantly, this host-protective type 17 response is not an exclusive function of adaptive CD4^{+} Th17 cells, because production of IL-22 by a CD90^{+}LTi-like ILC population has been shown to provide protection against *C. rodentium* during the very early stages of infection [56]. These findings are consistent with the hypothesis that IL23 co-ordinates both early and long-term protection to extracellular mucosal pathogens by triggering immediate type 17 responses from tissue-resident ILC and by promoting the development of adaptive Th17 responses which endow memory and specificity to the response [42, 44].

Studies in the acute colitis model induced by administration of dextran sulphate sodium (DSS) to mice have identified divergent protective and pathogenic activities of distinct Th17 cytokines during acute disruption of the intestinal epithelial barrier. DSS, administered in the drinking water for 5–7 days, is toxic to IEC, resulting in barrier disruption and translocation of intestinal bacteria into the lamina propria. This triggers an acute inflammatory response and transient weight loss, which are followed by restitution and repair of the epithelial barrier. Blockade of IL-17A, using neutralising mAbs or by genetic ablation, resulted in exacerbated intestinal inflammation in DSS-treated mice, indicating a protective role for IL-17A [57, 58]. Similarly, IL-22 also has a protective role during DSS colitis as *IL-22*$^{-/-}$mice exhibited increased morbidity and mortality following DSS administration [59]. Although the mechanisms responsible for the protective effects of IL-17A and IL-22 in DSS colitis are incompletely understood, these cytokines have been shown to fortify intestinal barrier integrity in several ways: by stimulating the production of the tight junction protein claudin [60], by stimulating IEC proliferation and mucosal wound healing [16, 61], and by promoting goblet cell restitution and mucin production [62]. In addition, as noted above, IL-17A and IL-22 have been reported to induce AMP production [53, 54], which may also limit bacterial colonisation and translocation across the damaged epithelium and thus temper local and systemic inflammatory responses.

In contrast to the protective activity of IL-17A and IL-22, other Th17 cytokines have been found to have a pathogenic role in DSS colitis. For example, *IL-17F*$^{-/-}$mice exhibited attenuated intestinal pathology after DSS administration when compared with wild-type mice [58], suggesting that IL-17F promotes acute intestinal inflammation in this model. This finding was somewhat surprising, since IL-17F is highly homologous to IL-17A and has overlapping functions, particularly in terms of induction of cytokines and chemokines that promote recruitment of neutrophils to inflammatory sites [63]. In addition, both IL-17A and IL-17F signal through the same IL-17RA/RC receptor complex, although the downstream signal transduction pathways are still poorly defined [64]. Moreover, the demonstration that Th17 cells can also produce a functional IL-17A/F heterodimer [65, 66], further complicates dissection of their respective roles. Interestingly, *IL-17RA*$^{-/-}$ mice phenocopy *IL-17F*$^{-/-}$ mice in showing attenuated colitis after DSS administration, suggesting that the pathogenic effects of IL-17F may dominate the protective effects of IL-17A during acute colitis [67]. Further studies on the precise interactions of IL-17A and IL-17F with their functional receptors, the cellular distributions of IL-17RA and IL-17RC, and the downstream signalling cascades triggered by receptor engagement may illuminate the mechanisms through which IL-17A and IL-17F have differential effects [64]. IL-21 is another Th17 cytokine that appears to play a pathogenic role in acute colitis, as *IL-21*$^{-/-}$ mice exhibited reduced intestinal inflammation following administration of DSS or TNBS and treatment of wild type with an IL-21R-Fc fusion protein led to an attenuation of DSS colitis [68]. Based on its ascribed functions, there are several potential ways in which IL-21 might contribute to inflammatory responses in the intestine, including promoting the secretion of chemokines such as MIP3α from IEC [69], inducing tissue-degrading enzymes such as matrix metalloproteinases (MMP) from intestinal fibroblasts [70], and amplifying effector Th17 and Th1 responses while inhibiting Treg responses [71–75]. Taken together, these results indicate different Th17 cytokines play distinct roles following acute disruption of the intestinal epithelial barrier, with IL-21 and IL-17F exacerbating acute inflammatory responses, whereas IL-17A and IL-22 co-ordinate tissue-protective antimicrobial responses and wound healing.

Additional insights into the dichotomous functions of Th17 cytokines in the inflamed intestine have been derived from studies with mouse models of chronic intestinal inflammation. In the T cell transfer model of colitis, *IL-17A*$^{-/-}$ naïve CD45RB^{hi} CD4^{+} T cells induced severe colitis that was indistinguishable from that observed in *Rag*$^{-/-}$ recipients of wild-type naïve CD45RB^{hi} CD4^{+} T cells, showing that T cell-derived IL-17A was not required for chronic intestinal pathology [41, 76]. These studies were confirmed and extended by Leppkes et al., who reported that naïve CD45RB^{hi} CD4^{+} T cells isolated from *IL-17F*$^{-/-}$ or *IL-22*$^{-/-}$ mice had equivalent colitogenic potential to their WT counterparts following transfer into *Rag*$^{-/-}$ recipients, demonstrating that T cell-derived IL-17F or IL-22 was also dispensable for chronic intestinal pathology [77]. Although these results argued against an essential role for Th17 cytokines in T cell-mediated colitis, the same study reported that naïve CD45RB^{hi}CD4^{+} T cells lacking the RORγt transcription factor, which drives Th17 differentiation, were unable to induce colitis in *Rag*$^{-/-}$ recipients, suggesting that Th17 responses were required for T cell-mediated colitis [77]. They provided a potential explanation for these puzzling findings by showing that treatment with an anti-IL-17A depleting mAb completely ameliorated colitis in *Rag*$^{-/-}$ mice that were reconstituted with *IL-17F*$^{-/-}$ naïve CD45RB^{hi} CD4^{+} T cells, suggesting that IL-17A or IL-17F could play a redundant pathogenic role in T cell transfer colitis [77]. A number of further studies have demonstrated a pathogenic role for Th17 cells and IL-17A in chronic intestinal inflammation. For example, when polarised Th1 or Th17 cell lines reactive with enteric bacterial antigen were adoptively transferred into SCID mice, the Th17 cell line induced much more severe disease than the Th1 cell line [78]. Furthermore, colonisation of multiple intestinal neoplasia (Min) mice with an enterotoxigenic strain of the human commensal *Bacteroides fragilis* rapidly induced colitis and colon cancer development that was dominated by Th17 cells and could be attenuated by anti-IL-17A mAb [79]. Neutralisation of IL-17A also attenuated innate immune-mediated intestinal pathology in *Helicobacter hepaticus*-infected *Rag*$^{-/-}$ mice [42]. Moreover, transgenic mice with selective ablation of STAT3 or IL-10R in FoxP3^{+} Treg cells developed spontaneous colitis due to dysregulated Th17 cell responses, with IL-17A again shown to play a key pathogenic role [80, 81]. In contrast to IL-17A, most studies thus far suggest that IL-22 plays a protective role in chronic intestinal inflammation. Local gene delivery of IL-22 ameliorated Th2-mediated UC-like colitis in *TCRα*$^{-/-}$ mice, by restoring mucus secretion from goblet cells [62]. Furthermore, in the naïve CD4^{+} T cell transfer colitis model, accelerated onset of pathology was observed in *IL-22*$^{-/-}$*Rag*$^{-/-}$ recipients, suggesting that innate sources of IL-22 conferred some protection against intestinal inflammation [59]. However, a recent study by the same group found that IL-22 played a pathogenic role in colitis mediated by adoptive transfer of CD45RB^{lo} effector CD4^{+} T cells into *Rag*$^{-/-}$recipients [82]. The reasons for this difference are unclear and further studies are required to more clearly define the activities of IL-22 in the chronically inflamed intestine.

It is not clear why in some cases Th17 cells and cytokines appear to play a predominantly pathogenic role, while in others they may contribute to protection from intestinal inflammation. It is likely to depend on several factors, including the combination of cytokines that these cells produce, the nature of the inflammatory stimulus in the gut and the composition of the local cytokine microenvironment. These factors shape the functional plasticity and heterogeneity observed in Th17 cells, leading to a spectrum of potential Th17 effector responses [33, 34]. For example, studies in an experimental model of airway inflammation demonstrated that, in the absence of IL-17A, IL-22 mediated a tissue-protective response in the lung, whereas when IL-17A was present, IL-22 contributed to tissue pathology [83]. As noted earlier, the inflamed colon of both mice and CD patients contains a small population of IL-17A^{+}IFN-γ^{+} DP T cells that have been identified [39, 40], but whether they represent a more pathogenic version of Th17 cells remains to be determined [50]. Elegant fate-mapping studies using IL-17A-eYFP reporter mice subjected to EAE induction demonstrated that these IL-17A^{+}IFN-γ^{+} DP T cells arose from IL-17A^{+}IFNγ^{-} classical Th17 cells and that their acquisition of IFN-γ expression was dependent on IL23 [84]. Similarly, the failure of *IL23R*$^{-/-}$ CD4^{+} T cells to induce T cell transfer colitis was associated with reduced development of IL-17A^{+}IFN-γ^{+} DP T cells in the intestine, consistent with the hypothesis that they represent a pathogenic Th17 phenotype [39]. It is also possible that while Th17 cytokines may identify pathogenic leukocyte populations in the inflamed intestine, the pathogenic activities of these cells may be mediated by additional factors. For example, in the case of EAE, recent reports indicated that GM-CSF was an essential factor required for the induction of autoimmune pathology in the brain by Th17 cells [85, 86]. However, in the T cell transfer model of IBD, it was found that *Rag*$^{-/-}$ recipients of *GM-CSF*$^{-/-}$ CD4^{+} T cells developed equivalent colitis to recipients of WT CD4^{+} T cells, indicating that T cell-derived GM-SCF is not essential for T cell-mediated IBD [47]. Thus, factors present within the intestinal tissue microenvironment may modulate the expression profile and pathogenic potential of Th17 cells, with IL23R and IL-1R stimulation associated with a more pro-inflammatory phenotype, whereas TGF-β may promote a more regulated IL-10-secreting Th17 population [33, 34]. The divergent tissue-protective and pathological activities of Th17 cytokines in the gut are summarised in Fig. 11.1.

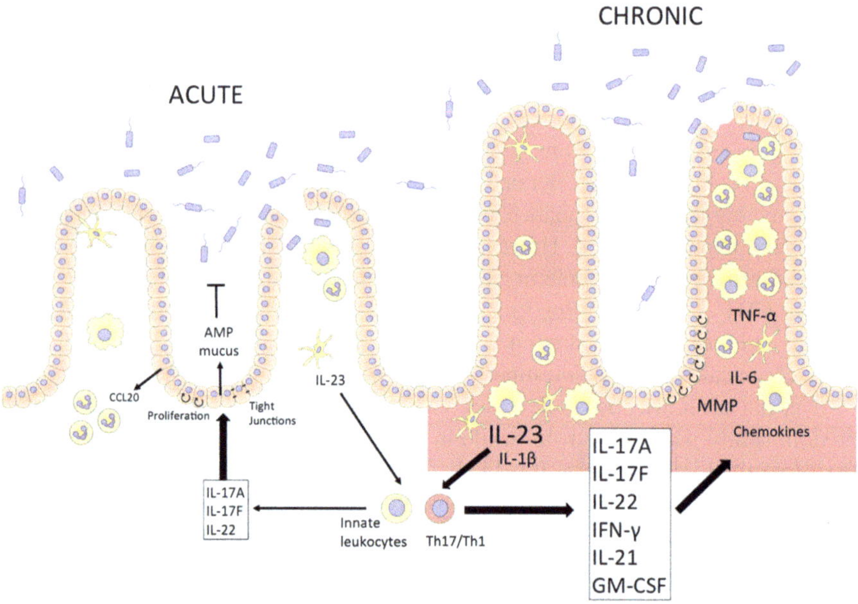

Fig. 11.1 Protective and pathogenic activities of the IL23/Th17 axis in the intestine. Acute damage or infection of the intestinal epithelium stimulates intestinal DC to secrete IL23, which induces rapid responses from resident RORγt^{+}IL23R^{+} innate leukocyte populations, such as ILC, LTi and NKp46^{+} cells. These innate populations secrete the Th17 cytokines IL-17A, IL-17F and IL-22, which co-ordinate tissue-protective responses from the epithelium. This includes fortification of the epithelial barrier by increasing AMP and mucus secretion and expression of tight junction proteins and stimulating IEC proliferation. In addition, CCL20 secretion recruits neutrophils to combat any translocated intestinal bacteria. During chronic intestinal inflammation, increased IL23 secretion by intestinal DC and macrophages, in concert with IL-1β, induces differentiation and activation of Th17, Th1 and IL-17A^{+}IFN-γ^{+} DP T cells, resulting in sustained secretion of high levels of a potent cocktail of pro-inflammatory cytokines. These promote recruitment and overactivation of myeloid cells leading to the production of further inflammatory cytokines that mediate chronic intestinal inflammation. In addition, the excessive induction of MMP secretion and IEC proliferation is indicative of a hyperactivated wound-healing response leading to dysregulated epithelial hyperplasia and tissue remodelling

Modulation of the Th17 Axis by Intestinal Microbiota

The relative enrichment of Th17 responses in the intestine suggests that local environmental factors can drive their differentiation or accumulation in the gut. Indeed, in the last few years, emerging evidence has implicated a key role for the intestinal microbiota in the stimulation of Th17 responses. The paucity of Th17 cells present in germ-free (GF) mice indicated a requirement for the intestinal microflora, although the bacteria involved were not identified [87–90]. A key breakthrough was reported by two parallel studies which demonstrated that colonisation of mice with segmented filamentous bacteria (SFB) led to the differentiation of Th17 cells in the intestine [91, 92]. In one study, a disparity in intestinal Th17 levels between genetically identical C57BL/6 strains from different commercial vendors was shown to correlate with the presence of SFB [91]. Subsequently, mono-colonisation of GF mice with SFB was shown to drive the differentiation of intestinal Th17 cells [91]. The other study reported that the intestinal immune maturation induced in GF mice following colonisation with intestinal microbiota was largely dependent on the presence of nonculturable, *Clostridia*-related, sporulated bacteria [92]. This led them to SFB as a potential candidate, and they demonstrated that mono-colonisation of GF mice with SFB could recapitulate the induction of intestinal T cell responses observed in mice with a conventional diverse microbiota [92]. Exactly how SFB drive Th17 differentiation is incompletely understood, although it was associated with the induction of serum amyloid A which conditioned DC to promote Th17 cell differentiation from naïve $CD4^{+}$T cells [91]. The microbial factors that are responsible for SFB-induced Th17 differentiation have not been characterised, although the recent complete genome sequencing of SFB may offer a starting point from which to identify such factors [93, 94]. In contrast to the free-living microbiota that dominate the intestinal lumen, SFB are intimately embedded within the epithelial cell layer in the ileum, raising the question as to whether they are viewed as innocuous commensals or as potentially pathogenic agents [95]. Indeed, although mono-colonisation of GF mice with SFB was associated with improved resistance to the mucosal pathogen *Citrobacter rodentium* [91], it was recently shown to promote systemic Th17 responses that accelerated autoimmune arthritis and to exacerbate the development of EAE [96, 97]. In addition, it should be noted that SFB colonisation of GF mice did not only induce the differentiation of Th17 cells but also simultaneously promoted the development of intestinal Th1 cells, Th2 cells and $FoxP3^{+}$Treg cells and induced potent IgA responses [92]. Thus, rather than exclusively promoting intestinal Th17 cell development, SFB stimulates a diverse range of adaptive immune responses in and beyond the gut.

Although the studies on SFB colonisation have provided an excellent illustrative example of how an individual microbe can promote Th17 cell differentiation, a human equivalent of SFB has yet to be identified, suggesting that distinct microbes drive intestinal Th17 responses in humans. Indeed, mounting evidence indicates that additional intestinal microbes and environmental factors may also regulate intestinal Th17 development. This includes ATP derived from the intestinal microflora, which has been shown to indirectly promote Th17 differentiation in the intestine by stimulating a population of CD70⁺CD11c⁺ lamina propria DC to produce Th17 polarising factors, such as IL-6, IL23 and TGF-β [87]. In addition, sensing of microflora-derived DNA through TLR9 was reported to promote Th1 and Th17 accumulation and to limit Treg cell numbers in the small intestine [88]. The key role of Th17 cells in anti-fungal responses, together with evidence that stimulation of dectin receptors and the associated signalling adapter CARD9 can promoteTh17 differentiation, indicates that fungi can also potently induce Th17 responses [98].

Other components of the intestinal microbiota may negatively regulate Th17 responses through the induction of Treg cells. For example, *Bacteroides fragilis*, a prominent human commensal, expresses a unique surface polysaccharide (PSA) that has been shown to bind to TLR2 on $CD4^+$ T cells to promote murine Treg maturation, to promote IL-10 production and to suppress Th17 responses [99]. Furthermore, colonisation of GF mice with a cocktail of 46 indigenous commensal *Clostridium* strains belonging to clusters XIVa and IV led to a dramatic induction of $FoxP3^+$ Treg in the colon [100]. In addition, mice with increased levels of *Clostridia* clusters XIVa and IV exhibited reduced susceptibility to DSS colitis [100], and it is interesting that both *Clostridium* cluster XIVa (Lachnospiraceae) and *Clostridium* cluster IV (*Faecalibacterium prausnitzii*) have been reported to be significantly reduced in the intestines of IBD patients [101–103]. Another recent study described how colonisation of GF mice with the limited altered Schaedler flora (ASF), a benign microbiota comprising eight of the most common intestinal bacterial commensal species but lacking SFB, induced $Foxp3^+$ Treg cell development in the colon [104]. However, when Treg induction or activation was blocked during the colonisation of GF mice with ASF, potent Th17 cell responses were induced in the colonic lamina propria [104]. Finally, TCR sequencing of colonic Treg cells indicates that many of them are specific for antigens derived from the intestinal microbiota [105]. Thus, Treg cell induction by the commensal microbiota plays a key role in inhibiting Th17 cell differentiation in the intestine, and additional studies are required in order to define the microbiota-derived molecules that drive the differentiation of distinct $CD4^+$ T cell subsets in the gut.

Therapeutic Strategies for Targeting the IL23/Th17 Axis

The strong association of the IL23/Th17 axis with a range of autoimmune diseases has stimulated the development of biological agents targeting this pathway for the treatment of chronic immune-mediated disorders. As illustrated with TNF-α blockers, it is likely that some biologicals developed to target the IL23/Th17 axis in systemic autoimmune disease will prove also to be useful for the treatment of IBD [4]. Consistent with this concept of shared pathogenic pathways in different types of chronic inflammatory diseases, more than 50 % of IBD susceptibility loci have also been associated with other inflammatory diseases, including the *IL-23R* SNPs which have been linked with susceptibility to psoriasis and ankylosing spondylitis [2,

106]. The early studies in mouse IBD models showing that targeted ablation of the shared IL-12/23p40 subunit ameliorate intestinal pathology have provided a rationale for IL-12/23p40 blockade in human IBD. A humanised mAb [107] specific for the shared IL-12/23p40 subunit, which had been previously shown to be effective in psoriasis [108, 109], was found to have beneficial effects in a phase II trial in patients with moderate-to-severe active CD [107]. In mouse chronic IBD models, treatment with a specific anti-IL23p19 mAb has been shown to prevent the development of IBD and also to reverse active disease when given therapeutically [11, 78]. Further clinical trials with antibodies or drugs that specifically target IL23 are required in order to test the effectiveness of IL23 blockade in IBD patients.

With respect to targeting Th17 responses, numerous clinical trials are under way to evaluate blockade of IL-17A in chronic inflammatory disorders. Early results in autoimmune diseases have been very promising, with anti-IL-17A mAb treatment reported to have beneficial effects in psoriasis, rheumatoid arthritis and autoimmune uveitis [110, 111]. However, the heterogeneous nature of pathogenic T cell effector responses observed in IBD may necessitate blockade of multiple pathways in order to significantly attenuate disease. This concept is supported by studies in mouse IBD models. For example, combined treatment with neutralising anti-IL-17A plus anti-IL-6 mAb was shown to more effectively attenuate T cell transfer colitis than blocking either cytokine alone [10], and treatment with anti-IL-6R mAb significantly reduced intestinal pathology in $Rag^{-/-}$ mice that had been reconstituted with $IL\text{-}17A^{-/-}$ CD45RB^{hi} CD4^{+} T cells [76]. Interestingly, a humanised anti-IL-6R mAb (tocilizumab) with proven efficacy in rheumatoid arthritis was also reported to have beneficial effects in CD patients [1]. The synergistic effects of IL-1β and IL23 on intestinal Th17 responses and the attenuating effects of IL-1β ablation in mouse IBD models, together with the high levels of IL-1β present in IBD lesions, suggest that IL-1β depletion may have beneficial effects on chronic intestinal inflammation [47]. However, although biological targeting of IL-1β using the soluble IL-1R antagonist (anakinra) has been shown to be very effective in controlling IL-1β-mediated systemic auto-inflammatory disorders [112], it has not yet been evaluated in IBD. Despite high levels of local and systemic IL-22 correlating with the severity of intestinal inflammation in CD [113], the documented epithelial protective effects of this cytokine caution against targeting IL-22 in IBD. However, chronic overexpression of IL-22 in the inflamed gut may stimulate an excessive epithelial wound-healing response that could predispose to the development of colon cancer, and IBD is a known risk factor for colon cancer development [1, 3].

Alternative strategies to block Th17 responses could involve targeting the transcription factors that drive Th17 differentiation, such as RORγt and STAT3. For example, pioglitazone, an agonist of the nuclear receptor PPARγ, selectively inhibited Th17 differentiation by suppressing the induction of RORγt by TGF-β/IL-6 and blocked the development of EAE autoimmune pathology [114]. Similarly, the cardiac glycoside digoxin was also shown to specifically inhibit Th17 differentiation by antagonising RORγt/RORC activity and to suppress autoimmune disease in the mouse CNS [115]. Importantly, although high concentrations of digoxin are toxic for human cells, non-toxic synthetic derivatives of digoxin were shown to similarly suppress Th17 differentiation [115]. Another recent study described a synthetic RORγt ligand that repressed RORγt activity, which inhibited Th17 cell development and was also able to attenuate autoimmune pathology in mice [116]. Furthermore, zinc, a trace element that is required by many enzymes and transcription factors, was shown to inhibit Th17 responses by suppressing STAT3 activation, and zinc treatment could suppress the development of EAE and CIA in mice [117]. Thus, a variety of pharmacological agents have been identified that can suppress deleterious Th17 responses in mice, by antagonising key transcription factors involved in Th17 differentiation. However, it remains to be determined whether these small molecule compounds are able to ameliorate chronic intestinal pathology. Alternatively, other 'natural' pharmacological agents may also be employed. For example, low levels of vitamin D have been associated with an increased incidence of multiple sclerosis (MS) and IBD [118], and polymorphisms in the vitamin D receptor (VDR) gene have been associated with susceptibility to IBD [119]. Furthermore, VDR-deficient mice exhibited increased susceptibility in both acute and chronic (*IL-10*$^{-/-}$ mice) IBD models, and treatment with active vitamin D resulted in attenuation of intestinal inflammation [118]. The suppressive effects of vitamin D treatment on EAE pathology were associated with VDR-mediated suppression of IL-17A expression in T cells, by blocking the Th17-promoting transcription factors Runx1 and NFAT [120]. Thus, stimulation of the VDR pathway could represent a potential therapeutic approach to target pathogenic Th17 responses in the gut [118].

Therapies that preferentially induce regulatory T cell populations in the intestine may represent an alternative way to control pathogenic Th17 responses. The critical role of FoxP3^{+} Treg cells in suppressing chronic intestinal inflammation is well documented and has been shown to be dependent on their production of immune suppressive factors, especially IL-10 [121, 122]. Using a modified version of the T cell transfer colitis model in which pathogenic Th17 cells were injected into *Rag*$^{-/-}$ mice, it was recently demonstrated that Th17-mediated colitis was suppressed by cotransfer of Foxp3^{+} or Tr1 Treg cells in an IL-10-dependent fashion [123]. IL-10 signals through STAT3 and STAT3 signalling in Treg cells was essential for their control of colitogenic Th17 responses, because mice with selective ablation of STAT3 in FoxP3^{+} Treg cells developed lethal Th17-mediated colitis [80]. These observations were complemented by another recent study which found that *IL-10R*$^{-/-}$ Foxp3^{+} Treg cells were unable to suppress Th17-mediated intestinal inflammation [81]. Thus, IL-10 appears to play a dual role in suppressing colitogenic Th17 responses, by directly inhibiting proliferation of Th17 cells and by promoting the regulatory functions of Foxp3^{+} Treg [80, 81, 123]. The critical role of IL-10 signals in regulating intestinal homeostasis in humans is illustrated by clinical reports showing that mutations in the *IL-10RA*, *IL-10RB* or *IL-10* genes can result in severe, early-onset IBD [124, 125]. It is worth noting that Th17 cells themselves can produce IL-10 and that this was enhanced by TGF-β and was associated with abrogated pathogenicity in EAE, suggesting that IL-10 can also act in an autocrine-negative feedback loop to inhibit Th17 cell-mediated inflammatory responses [126]. Taken together, these findings suggest that boosting IL-10 regulatory circuits may constitute a powerful means of suppressing pathogenic Th17 responses in the intestine.

As reported with biological therapies targeting TNF-α, one important caveat of blocking host cytokines may be a loss of efficient protective immunity against certain pathogens [4]. Indeed, a number of recent human genetic studies have demonstrated that deficiencies in Th17 responses are associated with impaired protective immunity to extracellular fungal and bacterial pathogens. Thus, mutations in *STAT3*, *DECTIN1* or *CARD9* have all been reported to lead to reduced numbers of Th17 cells and to increased susceptibility to chronic mucocutaneous candidiasis (CMC) [127–129]. Similarly, patients suffering from autoimmune polyendocrine syndrome-1 (APS-1) caused by mutations in the autoimmune regulator gene *AIRE* also develop neutralising autoantibodies to Th17 cytokines and frequently suffer from CMC [130, 131]. Finally, patients harbouring rare loss-of-function mutations in *IL-17RA* or dominant negative mutations in *IL-17F* are also highly susceptible to CMC and staphylococcal infections [132]. Taken together, these studies highlight the key protective role played by Th17 responses against the opportunistic pathogens *Candida albicans* and *Staphylococcus aureus*, but the limited range of infections to which patients with Th17 deficiencies are predisposed suggests that manageable therapeutic manipulation of this pathway may be possible.

Conclusions

There is now solid evidence that the IL23-Th17 axis plays a key role in chronic intestinal inflammation, but the cellular and molecular pathways involved remain incompletely characterised. Recent advances have enhanced our knowledge of Th17 differentiation, have highlighted the extreme heterogeneity and plasticity of this $CD4^+$ T cell subset and have indicated that many different types of innate and adaptive leukocytes have the potential to express Th17 cytokines in the gut. The plasticity of Th17 responses and their modulation by host and environmental factors indicate that a broad range of approaches could potentially be employed to suppress or modulate Th17 responses in the gut. In addition, data from preclinical IBD models, as well as from clinical trials in patients with autoimmune diseases, have demonstrated the potential benefits of targeting the IL23-Th17 axis to prevent immune pathology. However, the effects of individual Th17 cytokines are complex, ranging from disease protective to highly pathogenic, and are governed by the context in which they are expressed and by the presence of additional factors in the intestine. Thus, a more complete characterisation of the pathological lesions in individual IBD patients, coupled with an improved understanding of how genetic factors control susceptibility to IBD, is necessary in order develop tailored therapies targeted to defined IBD patient subgroups who will be most likely to benefit.

Acknowledgments The author would like to acknowledge The Wellcome Trust and CRUK for research grants support and George Song-Zhao for help with preparing the figure.

References

1. Kaser A, Zeissig S, Blumberg RS (2010) Inflammatory bowel disease. Annu Rev Immunol 28:573–621, Epub 2010/03/03
2. Khor B, Gardet A, Xavier RJ (2011) Genetics and pathogenesis of inflammatory bowel disease. Nature 474(7351):307–317, Epub 2011/06/17
3. Maloy KJ, Powrie F (2011) Intestinal homeostasis and its breakdown in inflammatory bowel disease. Nature 474(7351):298–306, Epub 2011/06/17
4. Melmed GY, Targan SR (2010) Future biologic targets for IBD: potentials and pitfalls. Nat Rev Gastroenterol Hepatol 7(2):110–117, Epub 2010/02/06
5. Bouma G, Strober W (2003) The immunological and genetic basis of inflammatory bowel disease. Nat Rev Immunol 3(7):521–533
6. Oppmann B, Lesley R, Blom B, Timans JC, Xu Y, Hunte B et al (2000) Novel p19 protein engages IL-12p40 to form a cytokine, IL-23, with biological activities similar as well as distinct from IL-12. Immunity 13(5):715–725
7. McGeachy MJ, Cua DJ (2007) The link between IL-23 and Th17 cell-mediated immune pathologies. Semin Immunol 19(6):372–376
8. Ouyang W, Kolls JK, Zheng Y (2008) The biological functions of T helper 17 cell effector cytokines in inflammation. Immunity 28(4):454–467, Epub 2008/04/11
9. Korn T, Bettelli E, Oukka M, Kuchroo VK (2009) IL-17 and Th17 Cells. Annu Rev Immunol 27:485–517
10. Yen D, Cheung J, Scheerens H, Poulet F, McClanahan T, McKenzie B et al (2006) IL-23 is essential for T cell-mediated colitis and promotes inflammation via IL-17 and IL-6. J Clin Invest 116(5):1310–1316
11. Hue S, Ahern P, Buonocore S, Kullberg MC, Cua DJ, McKenzie BS et al (2006) Interleukin-23 drives innate and T cell-mediated intestinal inflammation. J Exp Med 203(11):2473–2483
12. Kullberg MC, Jankovic D, Feng CG, Hue S, Gorelick PL, McKenzie BS et al (2006) IL-23 plays a key role in Helicobacter hepaticus-induced T cell-dependent colitis. J Exp Med 203(11):2485–2494
13. Uhlig HH, McKenzie BS, Hue S, Thompson C, Joyce-Shaikh B, Stepankova R et al (2006) Differential activity of IL-12 and IL-23 in mucosal and systemic innate immune pathology. Immunity 25(2):309–318
14. Fujino S, Andoh A, Bamba S, Ogawa A, Hata K, Araki Y et al (2003) Increased expression of interleukin 17 in inflammatory bowel disease. Gut 52(1):65–70
15. Monteleone G, Monteleone I, Fina D, Vavassori P, Del Vecchio Blanco G, Caruso R et al (2005) Interleukin-21 enhances T-helper cell type I signaling and interferon-gamma production in Crohn's disease. Gastroenterology 128(3):687–694
16. Brand S, Beigel F, Olszak T, Zitzmann K, Eichhorst ST, Otte JM et al (2006) IL-22 is increased in active Crohn's disease and promotes proinflammatory gene expression and intestinal epithelial cell migration. Am J Physiol Gastrointest Liver Physiol 290(4):G827–G838
17. Duerr RH, Taylor KD, Brant SR, Rioux JD, Silverberg MS, Daly MJ et al (2006) A genome-wide association study identifies IL23R as an inflammatory bowel disease gene. Science 314(5804):1461–1463, Epub 2006/10/28
18. Wellcome Trust Case Control Consortium (2007) Genome-wide association study of 14,000 cases of seven common diseases and 3,000 shared controls. Nature 447(7145):661–678, Epub 2007/06/08
19. Tremelling M, Cummings F, Fisher SA, Mansfield J, Gwilliam R, Keniry A et al (2007) IL23R variation determines susceptibility but not disease phenotype in inflammatory bowel disease. Gastroenterology 132(5):1657–1664
20. Franke A, Balschun T, Sina C, Ellinghaus D, Hasler R, Mayr G et al (2010) Genome-wide association study for ulcerative colitis identifies risk loci at 7q22 and 22q13 (IL17REL). Nat Genet 42(4):292–294, Epub 2010/03/17

21. Anderson CA, Boucher G, Lees CW, Franke A, D'Amato M, Taylor KD et al (2011) Meta-analysis identifies 29 additional ulcerative colitis risk loci, increasing the number of confirmed associations to 47. Nat Genet 43(3):246–252, Epub 2011/02/08
22. Pidasheva S, Trifari S, Phillips A, Hackney JA, Ma Y, Smith A et al (2011) Functional studies on the IBD susceptibility gene IL23R implicate reduced receptor function in the protective genetic variant R381Q. PLoS One 6(10):e25038, Epub 2011/10/25
23. Di Meglio P, Di Cesare A, Laggner U, Chu CC, Napolitano L, Villanova F et al (2011) The IL23R R381Q gene variant protects against immune-mediated diseases by impairing IL-23-induced Th17 effector response in humans. PLoS One 6(2):e17160, Epub 2011/03/03
24. Yu RY, Gallagher G (2010) A naturally occurring, soluble antagonist of human IL-23 inhibits the development and in vitro function of human Th17 cells. J Immunol 185(12):7302–7308, Epub 2010/11/16
25. Fuss IJ, Becker C, Yang Z, Groden C, Hornung RL, Heller F et al (2006) Both IL-12p70 and IL-23 are synthesized during active Crohn's disease and are down-regulated by treatment with anti-IL-12 p40 monoclonal antibody. Inflamm Bowel Dis 12(1):9–15
26. Kamada N, Hisamatsu T, Okamoto S, Chinen H, Kobayashi T, Sato T et al (2008) Unique CD14 intestinal macrophages contribute to the pathogenesis of Crohn disease via IL-23/IFN-gamma axis. J Clin Invest 118(6):2269–2280, Epub 2008/05/24
27. Kastelein RA, Hunter CA, Cua DJ (2007) Discovery and biology of IL-23 and IL-27: related but functionally distinct regulators of inflammation. Annu Rev Immunol 25:221–242
28. Langrish CL, McKenzie BS, Wilson NJ, de Waal Malefyt R, Kastelein RA, Cua DJ (2004) IL-12 and IL-23: master regulators of innate and adaptive immunity. Immunol Rev 202:96–105
29. Becker C, Wirtz S, Blessing M, Pirhonen J, Strand D, Bechthold O et al (2003) Constitutive p40 promoter activation and IL-23 production in the terminal ileum mediated by dendritic cells. J Clin Invest 112(5):693–706
30. Sakuraba A, Sato T, Kamada N, Kitazume M, Sugita A, Hibi T (2009) Th1/Th17 immune response is induced by mesenteric lymph node dendritic cells in Crohn's disease. Gastroenterology 137(5):1736–1745, Epub 2009/07/28
31. Kamada N, Hisamatsu T, Honda H, Kobayashi T, Chinen H, Kitazume MT et al (2009) Human CD14+ macrophages in intestinal lamina propria exhibit potent antigen-presenting ability. J Immunol 183(3):1724–1731, Epub 2009/07/14
32. Kinnebrew MA, Buffie CG, Diehl GE, Zenewicz LA, Leiner I, Hohl TM et al (2012) Interleukin 23 production by intestinal CD103(+)CD11b(+) dendritic cells in response to bacterial flagellin enhances mucosal innate immune defense. Immunity 36(2):276–287, Epub 2012/02/07
33. Peters A, Lee Y, Kuchroo VK (2011) The many faces of Th17 cells. Curr Opin Immunol 23(6):702–706, Epub 2011/09/09
34. Wei G, Wei L, Zhu J, Zang C, Hu-Li J, Yao Z et al (2009) Global mapping of H3K4me3 and H3K27me3 reveals specificity and plasticity in lineage fate determination of differentiating CD4+ T cells. Immunity 30(1):155–167, Epub 2009/01/16
35. Mukasa R, Balasubramani A, Lee YK, Whitley SK, Weaver BT, Shibata Y et al (2010) Epigenetic instability of cytokine and transcription factor gene loci underlies plasticity of the T helper 17 cell lineage. Immunity 32(5):616–627, Epub 2010/05/18
36. Littman DR, Rudensky AY (2010) Th17 and regulatory T cells in mediating and restraining inflammation. Cell 140(6):845–858, Epub 2010/03/23
37. Powrie F, Leach MW, Mauze S, Caddle LB, Coffman RL (1993) Phenotypically distinct subsets of CD4+ T cells induce or protect from chronic intestinal inflammation in C. B-17 scid mice. Int Immunol 5(11):1461–1471
38. Kleinschek MA, Boniface K, Sadekova S, Grein J, Murphy EE, Turner SP et al (2009) Circulating and gut-resident human Th17 cells express CD161 and promote intestinal inflammation. J Exp Med 206(3):525–534, Epub 2009/03/11
39. Ahern PP, Schiering C, Buonocore S, McGeachy MJ, Cua DJ, Maloy KJ et al (2010) Interleukin-23 drives intestinal inflammation through direct activity on T cells. Immunity 33(2):279–288, Epub 2010/08/25

40. Annunziato F, Cosmi L, Santarlasci V, Maggi L, Liotta F, Mazzinghi B et al (2007) Phenotypic and functional features of human Th17 cells. J Exp Med 204(8):1849–1861
41. Izcue A, Hue S, Buonocore S, Arancibia-Carcamo CV, Ahern PP, Iwakura Y et al (2008) Interleukin-23 restrains regulatory T cell activity to drive T cell-dependent colitis. Immunity 28(4):559–570
42. Buonocore S, Ahern PP, Uhlig HH, Ivanov II, Littman DR, Maloy KJ et al (2010) Innate lymphoid cells drive interleukin-23-dependent innate intestinal pathology. Nature 464(7293): 1371–1375, Epub 2010/04/16
43. Geremia A, Arancibia-Carcamo CV, Fleming MP, Rust N, Singh B, Mortensen NJ et al (2011) IL-23-responsive innate lymphoid cells are increased in inflammatory bowel disease. J Exp Med 208(6):1127–1133, Epub 2011/05/18
44. Cua DJ, Tato CM (2010) Innate IL-17-producing cells: the sentinels of the immune system. Nat Rev Immunol 10(7):479–489, Epub 2010/06/19
45. Spits H, Di Santo JP (2011) The expanding family of innate lymphoid cells: regulators and effectors of immunity and tissue remodeling. Nat Immunol 12(1):21–27, Epub 2010/11/30
46. Takayama T, Kamada N, Chinen H, Okamoto S, Kitazume MT, Chang J et al (2010) Imbalance of NKp44(+)NKp46(−) and NKp44(−)NKp46(+) natural killer cells in the intestinal mucosa of patients with Crohn's disease. Gastroenterology 139(3):882–892, 892.e1-3. Epub 2010/07/20
47. Coccia M, Harrison OJ, Schiering C, Asquith MJ, Becher B, Powrie F et al (2012) IL-1beta mediates chronic intestinal inflammation by promoting the accumulation of IL-17A secreting innate lymphoid cells and CD4+ Th17 cells. J Exp Med 209(9):1595–1609, Epub 2012/08/15
48. Khader SA, Gaffen SL, Kolls JK (2009) Th17 cells at the crossroads of innate and adaptive immunity against infectious diseases at the mucosa. Mucosal Immunol 2(5):403–411, Epub 2009/07/10
49. Maloy KJ, Kullberg MC (2008) IL-23 and Th17 cytokines in intestinal homeostasis. Mucosal Immunol 1(5):339–349, Epub 2008/12/17
50. Morrison PJ, Ballantyne SJ, Kullberg MC (2011) Interleukin-23 and T helper 17-type responses in intestinal inflammation: from cytokines to T-cell plasticity. Immunology 133(4):397–408, Epub 2011/06/03
51. Mundy R, MacDonald TT, Dougan G, Frankel G, Wiles S (2005) Citrobacter rodentium of mice and man. Cell Microbiol 7(12):1697–1706, Epub 2005/11/29
52. Mangan PR, Harrington LE, O'Quinn DB, Helms WS, Bullard DC, Elson CO et al (2006) Transforming growth factor-beta induces development of the T(H)17 lineage. Nature 441(7090):231–234
53. Ishigame H, Kakuta S, Nagai T, Kadoki M, Nambu A, Komiyama Y et al (2009) Differential roles of interleukin-17A and -17F in host defense against mucoepithelial bacterial infection and allergic responses. Immunity 30(1):108–119, Epub 2009/01/16
54. Zheng Y, Valdez PA, Danilenko DM, Hu Y, Sa SM, Gong Q et al (2008) Interleukin-22 mediates early host defense against attaching and effacing bacterial pathogens. Nat Med 14(3):282–289
55. Liang SC, Tan XY, Luxenberg DP, Karim R, Dunussi-Joannopoulos K, Collins M et al (2006) Interleukin (IL)-22 and IL-17 are coexpressed by Th17 cells and cooperatively enhance expression of antimicrobial peptides. J Exp Med 203(10):2271–2279
56. Sonnenberg GF, Monticelli LA, Elloso MM, Fouser LA, Artis D (2011) CD4(+) lymphoid tissue-inducer cells promote innate immunity in the gut. Immunity 34(1):122–134, Epub 2011/01/05
57. Ogawa A, Andoh A, Araki Y, Bamba T, Fujiyama Y (2004) Neutralization of interleukin-17 aggravates dextran sulfate sodium-induced colitis in mice. Clin Immunol 110(1):55–62
58. Yang XO, Chang SH, Park H, Nurieva R, Shah B, Acero L et al (2008) Regulation of inflammatory responses by IL-17F. J Exp Med 205(5):1063–1075
59. Zenewicz LA, Yancopoulos GD, Valenzuela DM, Murphy AJ, Stevens S, Flavell RA (2008) Innate and adaptive interleukin-22 protects mice from inflammatory bowel disease. Immunity 29(6):947–957, Epub 2008/12/23

60. Kinugasa T, Sakaguchi T, Gu X, Reinecker HC (2000) Claudins regulate the intestinal barrier in response to immune mediators. Gastroenterology 118(6):1001–1011
61. Pickert G, Neufert C, Leppkes M, Zheng Y, Wittkopf N, Warntjen M et al (2009) STAT3 links IL-22 signaling in intestinal epithelial cells to mucosal wound healing. J Exp Med 206(7): 1465–1472
62. Sugimoto K, Ogawa A, Mizoguchi E, Shimomura Y, Andoh A, Bhan AK et al (2008) IL-22 ameliorates intestinal inflammation in a mouse model of ulcerative colitis. J Clin Invest 118(2):534–544
63. Iwakura Y, Ishigame H, Saijo S, Nakae S (2011) Functional specialization of interleukin-17 family members. Immunity 34(2):149–162, Epub 2011/02/26
64. Gaffen SL (2009) Structure and signalling in the IL-17 receptor family. Nat Rev Immunol 9(8):556–567, Epub 2009/07/04
65. Liang SC, Long AJ, Bennett F, Whitters MJ, Karim R, Collins M et al (2007) An IL-17F/A heterodimer protein is produced by mouse Th17 cells and induces airway neutrophil recruitment. J Immunol 179(11):7791–7799
66. Wright JF, Guo Y, Quazi A, Luxenberg DP, Bennett F, Ross JF et al (2007) Identification of an interleukin 17F/17A heterodimer in activated human CD4+ T cells. J Biol Chem 282(18): 13447–13455
67. Zhang Z, Zheng M, Bindas J, Schwarzenberger P, Kolls JK (2006) Critical role of IL-17 receptor signaling in acute TNBS-induced colitis. Inflamm Bowel Dis 12(5):382–388
68. Fina D, Sarra M, Fantini MC, Rizzo A, Caruso R, Caprioli F et al (2008) Regulation of gut inflammation and th17 cell response by interleukin-21. Gastroenterology 134(4):1038–1048
69. Caruso R, Fina D, Peluso I, Stolfi C, Fantini MC, Gioia V et al (2007) A functional role for interleukin-21 in promoting the synthesis of the T-cell chemoattractant, MIP-3alpha, by gut epithelial cells. Gastroenterology 132(1):166–175
70. Monteleone G, Caruso R, Fina D, Peluso I, Gioia V, Stolfi C et al (2006) Control of matrix metalloproteinase production in human intestinal fibroblasts by interleukin 21. Gut 55(12): 1774–1780
71. Strengell M, Matikainen S, Siren J, Lehtonen A, Foster D, Julkunen I et al (2003) IL-21 in synergy with IL-15 or IL-18 enhances IFN-gamma production in human NK and T cells. J Immunol 170(11):5464–5469
72. Korn T, Bettelli E, Gao W, Awasthi A, Jager A, Strom TB et al (2007) IL-21 initiates an alternative pathway to induce proinflammatory T(H)17 cells. Nature 448(7152):484–487
73. Nurieva R, Yang XO, Martinez G, Zhang Y, Panopoulos AD, Ma L et al (2007) Essential autocrine regulation by IL-21 in the generation of inflammatory T cells. Nature 448(7152):480–483
74. Fantini MC, Rizzo A, Fina D, Caruso R, Becker C, Neurath MF et al (2007) IL-21 regulates experimental colitis by modulating the balance between Treg and Th17 cells. Eur J Immunol 37(11):3155–3163
75. Peluso I, Fantini MC, Fina D, Caruso R, Boirivant M, MacDonald TT et al (2007) IL-21 counteracts the regulatory T cell-mediated suppression of human CD4+ T lymphocytes. J Immunol 178(2):732–739
76. Noguchi D, Wakita D, Tajima M, Ashino S, Iwakura Y, Zhang Y et al (2007) Blocking of IL-6 signaling pathway prevents CD4+ T cell-mediated colitis in a T(h)17-independent manner. Int Immunol 19(12):1431–1440
77. Leppkes M, Becker C, Ivanov II, Hirth S, Wirtz S, Neufert C et al (2009) RORgamma-expressing Th17 cells induce murine chronic intestinal inflammation via redundant effects of IL-17A and IL-17F. Gastroenterology 136(1):257–267
78. Elson CO, Cong Y, Weaver CT, Schoeb TR, McClanahan TK, Fick RB et al (2007) Monoclonal anti-interleukin 23 reverses active colitis in a T cell-mediated model in mice. Gastroenterology 132(7):2359–2370
79. Wu S, Rhee KJ, Albesiano E, Rabizadeh S, Wu X, Yen HR et al (2009) A human colonic commensal promotes colon tumorigenesis via activation of T helper type 17 T cell responses. Nat Med 15(9):1016–1022, Epub 2009/08/25

80. Chaudhry A, Rudra D, Treuting P, Samstein RM, Liang Y, Kas A et al (2009) CD4+ regulatory T cells control TH17 responses in a Stat3-dependent manner. Science 326(5955): 986–991, Epub 2009/10/03
81. Chaudhry A, Samstein RM, Treuting P, Liang Y, Pils MC, Heinrich JM et al (2011) Interleukin-10 signaling in regulatory T cells is required for suppression of Th17 cell-mediated inflammation. Immunity 34(4):566–578, Epub 2011/04/23
82. Kamanaka M, Huber S, Zenewicz LA, Gagliani N, Rathinam C, O'Connor W Jr et al (2011) Memory/effector (CD45RB(lo)) CD4 T cells are controlled directly by IL-10 and cause IL-22-dependent intestinal pathology. J Exp Med 208(5):1027–1040, Epub 2011/04/27
83. Sonnenberg GF, Nair MG, Kirn TJ, Zaph C, Fouser LA, Artis D (2010) Pathological versus protective functions of IL-22 in airway inflammation are regulated by IL-17A. J Exp Med 207(6):1293–1305, Epub 2010/05/26
84. Hirota K, Duarte JH, Veldhoen M, Hornsby E, Li Y, Cua DJ et al (2011) Fate mapping of IL-17-producing T cells in inflammatory responses. Nat Immunol 12(3):255–263, Epub 2011/02/01
85. Codarri L, Gyulveszi G, Tosevski V, Hesske L, Fontana A, Magnenat L et al (2011) RORgammat drives production of the cytokine GM-CSF in helper T cells, which is essential for the effector phase of autoimmune neuroinflammation. Nat Immunol 12(6):560–567, Epub 2011/04/26
86. El-Behi M, Ciric B, Dai H, Yan Y, Cullimore M, Safavi F et al (2011) The encephalitogenicity of T(H)17 cells is dependent on IL-1- and IL-23-induced production of the cytokine GM-CSF. Nat Immunol 12(6):568–575, Epub 2011/04/26
87. Atarashi K, Nishimura J, Shima T, Umesaki Y, Yamamoto M, Onoue M et al (2008) ATP drives lamina propria T(H)17 cell differentiation. Nature 455(7214):808–812, Epub 2008/08/22
88. Hall JA, Bouladoux N, Sun CM, Wohlfert EA, Blank RB, Zhu Q et al (2008) Commensal DNA limits regulatory T cell conversion and is a natural adjuvant of intestinal immune responses. Immunity 29(4):637–649, Epub 2008/10/07
89. Ivanov II, Frutos Rde L, Manel N, Yoshinaga K, Rifkin DB, Sartor RB et al (2008) Specific microbiota direct the differentiation of IL-17-producing T-helper cells in the mucosa of the small intestine. Cell Host Microbe 4(4):337–349, Epub 2008/10/16
90. Niess JH, Leithauser F, Adler G, Reimann J (2008) Commensal gut flora drives the expansion of proinflammatory CD4 T cells in the colonic lamina propria under normal and inflammatory conditions. J Immunol 180(1):559–568, Epub 2007/12/22
91. Ivanov II, Atarashi K, Manel N, Brodie EL, Shima T, Karaoz U et al (2009) Induction of intestinal Th17 cells by segmented filamentous bacteria. Cell 139(3):485–498, Epub 2009/10/20
92. Gaboriau-Routhiau V, Rakotobe S, Lecuyer E, Mulder I, Lan A, Bridonneau C et al (2009) The key role of segmented filamentous bacteria in the coordinated maturation of gut helper T cell responses. Immunity 31(4):677–689, Epub 2009/10/17
93. Prakash T, Oshima K, Morita H, Fukuda S, Imaoka A, Kumar N et al (2011) Complete genome sequences of rat and mouse segmented filamentous bacteria, a potent inducer of th17 cell differentiation. Cell Host Microbe 10(3):273–284, Epub 2011/09/20
94. Sczesnak A, Segata N, Qin X, Gevers D, Petrosino JF, Huttenhower C et al (2011) The genome of th17 cell-inducing segmented filamentous bacteria reveals extensive auxotrophy and adaptations to the intestinal environment. Cell Host Microbe 10(3):260–272, Epub 2011/09/20
95. Cerf-Bensussan N, Gaboriau-Routhiau V (2010) The immune system and the gut microbiota: friends or foes? Nat Rev Immunol 10(10):735–744, Epub 2010/09/25
96. Wu HJ, Ivanov II, Darce J, Hattori K, Shima T, Umesaki Y et al (2010) Gut-residing segmented filamentous bacteria drive autoimmune arthritis via T helper 17 cells. Immunity 32(6):815–827, Epub 2010/07/14
97. Lee YK, Menezes JS, Umesaki Y, Mazmanian SK (2011) Proinflammatory T-cell responses to gut microbiota promote experimental autoimmune encephalomyelitis. Proc Natl Acad Sci U S A 108(Suppl 1):4615–4622, Epub 2010/07/28

98. Hernandez-Santos N, Gaffen SL (2012) Th17 cells in immunity to Candida albicans. Cell Host Microbe 11(5):425–435, Epub 2012/05/23
99. Round JL, Mazmanian SK (2010) Inducible Foxp3+ regulatory T-cell development by a commensal bacterium of the intestinal microbiota. Proc Natl Acad Sci U S A 107(27):12204–12209, Epub 2010/06/23
100. Atarashi K, Tanoue T, Shima T, Imaoka A, Kuwahara T, Momose Y et al (2011) Induction of colonic regulatory T cells by indigenous Clostridium species. Science 331(6015):337–341, Epub 2011/01/06
101. Frank DN, St Amand AL, Feldman RA, Boedeker EC, Harpaz N, Pace NR (2007) Molecular-phylogenetic characterization of microbial community imbalances in human inflammatory bowel diseases. Proc Natl Acad Sci U S A 104(34):13780–13785, Epub 2007/08/19
102. Sokol H, Pigneur B, Watterlot L, Lakhdari O, Bermudez-Humaran LG, Gratadoux JJ et al (2008) Faecalibacterium prausnitzii is an anti-inflammatory commensal bacterium identified by gut microbiota analysis of Crohn disease patients. Proc Natl Acad Sci U S A 105(43):16731–16736, Epub 2008/10/22
103. Willing BP, Dicksved J, Halfvarson J, Andersson AF, Lucio M, Zheng Z et al (2010) A pyrosequencing study in twins shows that gastrointestinal microbial profiles vary with inflammatory bowel disease phenotypes. Gastroenterology 139(6):1844–1854.e1. Epub 2010/09/08
104. Geuking MB, Cahenzli J, Lawson MA, Ng DC, Slack E, Hapfelmeier S et al (2011) Intestinal bacterial colonization induces mutualistic regulatory T cell responses. Immunity 34(5): 794–806, Epub 2011/05/21
105. Lathrop SK, Bloom SM, Rao SM, Nutsch K, Lio CW, Santacruz N et al (2011) Peripheral education of the immune system by colonic commensal microbiota. Nature 478(7368): 250–254, Epub 2011/09/23
106. Cho JH, Gregersen PK (2011) Genomics and the multifactorial nature of human autoimmune disease. N Engl J Med 365(17):1612–1623, Epub 2011/10/28
107. Sandborn WJ, Feagan BG, Fedorak RN, Scherl E, Fleisher MR, Katz S et al (2008) A randomized trial of ustekinumab, a human interleukin-12/23 monoclonal antibody, in patients with moderate-to-severe Crohn's disease. Gastroenterology 135(4):1130–1141
108. Leonardi CL, Kimball AB, Papp KA, Yeilding N, Guzzo C, Wang Y et al (2008) Efficacy and safety of ustekinumab, a human interleukin-12/23 monoclonal antibody, in patients with psoriasis: 76-week results from a randomised, double-blind, placebo-controlled trial (PHOENIX 1). Lancet 371(9625):1665–1674, Epub 2008/05/20
109. Papp KA, Langley RG, Lebwohl M, Krueger GG, Szapary P, Yeilding N et al (2008) Efficacy and safety of ustekinumab, a human interleukin-12/23 monoclonal antibody, in patients with psoriasis: 52-week results from a randomised, double-blind, placebo-controlled trial (PHOENIX 2). Lancet 371(9625):1675–1684, Epub 2008/05/20
110. Hueber W, Patel DD, Dryja T, Wright AM, Koroleva I, Bruin G et al (2010) Effects of AIN457, a fully human antibody to interleukin-17A, on psoriasis, rheumatoid arthritis, and uveitis. Sci Transl Med 2(52):52ra72, Epub 2010/10/12
111. Genovese MC, Van den Bosch F, Roberson SA, Bojin S, Biagini IM, Ryan P et al (2010) LY2439821, a humanized anti-interleukin-17 monoclonal antibody, in the treatment of patients with rheumatoid arthritis: a phase I randomized, double-blind, placebo-controlled, proof-of-concept study. Arthritis Rheum 62(4):929–939, Epub 2010/02/05
112. Gabay C, Lamacchia C, Palmer G (2010) IL-1 pathways in inflammation and human diseases. Nat Rev Rheumatol 6(4):232–241, Epub 2010/02/24
113. Schmechel S, Konrad A, Diegelmann J, Glas J, Wetzke M, Paschos E et al (2008) Linking genetic susceptibility to Crohn's disease with Th17 cell function: IL-22 serum levels are increased in Crohn's disease and correlate with disease activity and IL23R genotype status. Inflamm Bowel Dis 14(2):204–212
114. Klotz L, Burgdorf S, Dani I, Saijo K, Flossdorf J, Hucke S et al (2009) The nuclear receptor PPAR gamma selectively inhibits Th17 differentiation in a T cell-intrinsic fashion and suppresses CNS autoimmunity. J Exp Med 206(10):2079–2089, Epub 2009/09/10

115. Huh JR, Leung MW, Huang P, Ryan DA, Krout MR, Malapaka RR et al (2011) Digoxin and its derivatives suppress TH17 cell differentiation by antagonizing RORgammat activity. Nature 472(7344):486–490, Epub 2011/03/29
116. Solt LA, Kumar N, Nuhant P, Wang Y, Lauer JL, Liu J et al (2011) Suppression of TH17 differentiation and autoimmunity by a synthetic ROR ligand. Nature 472(7344):491–494, Epub 2011/04/19
117. Kitabayashi C, Fukada T, Kanamoto M, Ohashi W, Hojyo S, Atsumi T et al (2010) Zinc suppresses Th17 development via inhibition of STAT3 activation. Int Immunol 22(5): 375–386, Epub 2010/03/11
118. Cantorna MT (2012) Vitamin D, multiple sclerosis and inflammatory bowel disease. Arch Biochem Biophys 523(1):103–106, Epub 2011/11/17
119. Simmons JD, Mullighan C, Welsh KI, Jewell DP (2000) Vitamin D receptor gene polymorphism: association with Crohn's disease susceptibility. Gut 47(2):211–214, Epub 2000/07/18
120. Joshi S, Pantalena LC, Liu XK, Gaffen SL, Liu H, Rohowsky-Kochan C et al (2011) 1,25-dihydroxyvitamin D(3) ameliorates Th17 autoimmunity via transcriptional modulation of interleukin-17A. Mol Cell Biol 31(17):3653–3669, Epub 2011/07/13
121. Izcue A, Coombes JL, Powrie F (2009) Regulatory lymphocytes and intestinal inflammation. Annu Rev Immunol 27:313–338, Epub 2009/03/24
122. Rubtsov YP, Rasmussen JP, Chi EY, Fontenot J, Castelli L, Ye X et al (2008) Regulatory T cell-derived interleukin-10 limits inflammation at environmental interfaces. Immunity 28(4):546–558, Epub 2008/04/05
123. Huber S, Gagliani N, Esplugues E, O'Connor W Jr, Huber FJ, Chaudhry A et al (2011) Th17 cells express interleukin-10 receptor and are controlled by Foxp3(−) and Foxp3+ regulatory CD4+ T cells in an interleukin-10-dependent manner. Immunity 34(4):554–565, Epub 2011/04/23
124. Glocker EO, Kotlarz D, Boztug K, Gertz EM, Schaffer AA, Noyan F et al (2009) Inflammatory bowel disease and mutations affecting the interleukin-10 receptor. N Engl J Med 361(21):2033–2045, Epub 2009/11/06
125. Glocker EO, Frede N, Perro M, Sebire N, Elawad M, Shah N et al (2010) Infant colitis—it's in the genes. Lancet 376(9748):1272, Epub 2010/10/12
126. McGeachy MJ, Bak-Jensen KS, Chen Y, Tato CM, Blumenschein W, McClanahan T et al (2007) TGF-beta and IL-6 drive the production of IL-17 and IL-10 by T cells and restrain T(H)-17 cell-mediated pathology. Nat Immunol 8(12):1390–1397
127. Milner JD, Brenchley JM, Laurence A, Freeman AF, Hill BJ, Elias KM et al (2008) Impaired T(H)17 cell differentiation in subjects with autosomal dominant hyper-IgE syndrome. Nature 452(7188):773–776, Epub 2008/03/14
128. Glocker EO, Hennigs A, Nabavi M, Schaffer AA, Woellner C, Salzer U et al (2009) A homozygous CARD9 mutation in a family with susceptibility to fungal infections. N Engl J Med 361(18):1727–1735, Epub 2009/10/30
129. Ferwerda B, Ferwerda G, Plantinga TS, Willment JA, van Spriel AB, Venselaar H et al (2009) Human dectin-1 deficiency and mucocutaneous fungal infections. N Engl J Med 361(18): 1760–1767, Epub 2009/10/30
130. Kisand K, Boe Wolff AS, Podkrajsek KT, Tserel L, Link M, Kisand KV et al (2010) Chronic mucocutaneous candidiasis in APECED or thymoma patients correlates with autoimmunity to Th17-associated cytokines. J Exp Med 207(2):299–308, Epub 2010/02/04
131. Puel A, Doffinger R, Natividad A, Chrabieh M, Barcenas-Morales G, Picard C et al (2010) Autoantibodies against IL-17A, IL-17F, and IL-22 in patients with chronic mucocutaneous candidiasis and autoimmune polyendocrine syndrome type I. J Exp Med 207(2):291–297, Epub 2010/02/04
132. Puel A, Cypowyj S, Bustamante J, Wright JF, Liu L, Lim HK et al (2011) Chronic mucocutaneous candidiasis in humans with inborn errors of interleukin-17 immunity. Science 332(6025):65–68, Epub 2011/02/26

Chapter 12
Inflammatory Bowel Disease at the Intersection of Autophagy and Immunity: Insights from Human Genetics

Natalia B. Nedelsky, Petric Kuballa, Adam B. Castoreno, and Ramnik J. Xavier

Abstract Studies using human genetics have identified more than 160 loci that affect the risk of developing inflammatory bowel disease (IBD), including Crohn's disease (CD) and ulcerative colitis (UC). Several of these genes have been found to play key roles in the process of autophagy, a lysosome-based degradation pathway. Although historically considered to be a relatively nonselective process of degradation of cytosolic contents, autophagy has recently been revealed to have several selective and immune-specific functions that are relevant to the maintenance of intestinal homeostasis, including xenophagy, mitophagy, antigen presentation, secretion, and inflammasome regulation. In this chapter, we review the evidence that links autophagy-related genes, their immune-specific functions, and possible mechanisms of IBD pathogenesis. We summarize the basic molecular events underlying general and selective autophagy and present evidence suggesting possible pathogenic mechanisms revealed by studies of IBD-associated risk alleles of *ATG16L1* and *IRGM*. Finally, we review chemical biology-based experimental approaches for identifying autophagy regulatory pathways that may have implications for the development of therapeutics.

N.B. Nedelsky, Ph.D.
Gastrointestinal Unit, Center for the Study of Inflammatory Bowel Disease, and Center for Computational and Integrative Biology, Massachusetts General Hospital, 185 Cambridge Street, Simches 7, Boston, MA 02114, USA

P. Kuballa, Ph.D. • A.B. Castoreno, Ph.D. • R.J. Xavier, M.D., Ph.D. (✉)
Gastrointestinal Unit, Center for the Study of Inflammatory Bowel Disease, and Center for Computational and Integrative Biology, Massachusetts General Hospital, 185 Cambridge Street, Simches 7, Boston, MA 02114, USA

Broad Institute of Harvard University and Massachusetts Institute of Technology, Cambridge, MA, USA
e-mail: xavier@molbio.mgh.harvard.edu

M. D'Amato and J.D. Rioux (eds.), *Molecular Genetics of Inflammatory Bowel Disease*, DOI 10.1007/978-1-4614-8256-7_12, © Springer Science+Business Media New York 2013

Introduction

Genome-wide association studies (GWAS) of IBD have been a major success story for genome-level studies of complex human disease. One of the particularly exciting results from these studies was the identification of autophagy as an unanticipated player in determining risk of IBD. This pathway, long considered to be a relatively nonselective process of bulk cytoplasmic degradation by the lysosome, is thought to have evolutionary roots as a mechanism to maintain metabolic homeostasis in response to decreased nutrient levels [1]. However, studies over the last 10 years have expanded the functions of autophagy to include roles in organelle clearance, antigen presentation, cell death, development, and the degradation of intracellular pathogens, among others. The role of autophagy in immune-related processes [2–4] was particularly unexpected and has served as a guide in many studies seeking to connect human genetics to mechanisms of IBD pathogenesis [5–7].

These genetic studies have identified multiple genes within the autophagy network as playing a role in CD, including *ATG16L1*, *IRGM*, *NOD2*, and *LRRK2* [8–11]. In the case of UC, the autophagy-related genes *DAP1* and *SMURF1* have also been identified as determinants of risk [12, 13]. Further studies subsequently identified polymorphisms in *ULK1*, *ATG2A*, and *GABARAPL1* as harboring significant associations with CD [14, 15]. Although autophagy has been implicated in the disease process in a range of other disorders, including neurodegenerative disease [16, 17], atherosclerosis, and cancer, to date no compelling genetic association has been found between autophagy and many immune-related diseases, including multiple sclerosis, celiac disease, rheumatoid arthritis, and psoriasis. However, targeting sequencing studies have also begun to identify coding mutations in essential autophagy genes as playing roles in disease, raising the possibility that more disease-associated autophagy-related polymorphisms have yet to be uncovered [18, 19].

In this chapter, we review studies that have employed human genetics to understand the link between autophagy and IBD. We begin by examining the idea of the intestinal mucosa as a physiological system wherein homeostasis is particularly vulnerable to disruptions in autophagy. We then present a basic overview of the molecular events involved in autophagy, followed by discussions of how autophagy may function as a selective process with critical roles in IBD-related immunity. These roles include (1) degradation of intracellular pathogens (xenophagy), (2) mitophagy, (3) antigen presentation, (4) secretion, and (5) inflammasome regulation. In the cases of xenophagy, mitophagy, and antigen presentation, we present detailed evidence of how autophagy functions in these pathways. Next, we review results from studies of two particularly well-examined CD risk single-nucleotide polymorphisms (SNPs), rs2241880 (ATG16L1, coding variant T300A) and rs13361189 (IRGM). Finally, we discuss approaches from the field of chemical biology that may be used to identify autophagy regulatory pathways with possible implications for therapeutics development.

Autophagy in the Intestinal Mucosa

As mentioned above, autophagy has been implicated in several immune-related processes that influence IBD pathogenesis, including xenophagy, mitophagy, antigen presentation, secretion and vesicular trafficking, and cytokine-based regulation of inflammasome activity. One major point of intersection between these processes is their critical importance in the intestinal mucosa. This complex environment must maintain a delicate balance of immune responses, which are continuously stimulated by metabolic stresses and antigens from food. Microbes also play a major role in this milieu, which carries the highest bacterial load in the body. This tissue, therefore, may be especially susceptible to disruption in autophagy, with consequent alteration of homeostatic processes. For example, mice with conditional knockout of *Atg7* in intestinal epithelial cells show increased severity of colitis when exposed to the infectious pathogen *Citrobacter rodentium*, while mice with deficiency in the core autophagy gene *Atg16l1* in hematopoietic cells show increased susceptibility to chemically induced models of colitis [20, 21].

Studies in human genetics have provided evidence that microbial stimulation may exist as a node of functional interaction between CD susceptibility genes, including *ATG16L1* and *NOD2*. For example, expression of the ATG16L1 CD-associated risk allele (ATG16L1 T300A) results in impairment of autophagy and antigen presentation, both of which are enhanced by muramyl dipeptide (MDP), the bacterial ligand for the receptor NOD2 [22, 23]. Although no evidence has been found for an epistatic interaction between these two genes, this type of functional interaction suggests that autophagy may operate as a key point of intersection among multiple susceptibility genes.

Basic Molecular Events in Autophagy

The process of autophagy begins with an appearance of a flat membrane sheet termed the isolation membrane (also known as the phagophore assembly site in yeast), the origin of which remains unknown. The membrane elongates and expands to encompass its cargo, eventually fusing to form a double-membrane structure known as an autophagosome. The outer autophagosomal membrane then fuses with the lysosomal membrane, forming a degradative autolysosome. It should be noted that the source of the autophagosomal membrane is still a highly debated topic and the existence of multiple sources cannot be excluded. In the case of the ER, a novel Ω (omega)-shaped vesicle called the omegasome has been identified, which resides on the ER and might precede the appearance of the isolation membrane [24].

Many of the molecular events underlying the assembly and elongation of autophagy membranes were initially identified by genetic screens in *Saccharomyces cerevisiae* [25, 26]. These screens have identified approximately 35 core autophagy proteins, about half of which have clear homologues in mammals. For the purposes of this chapter, we divide these genes into five functional categories: (1) the ULK complex (ATG1 complex in yeast); (2) ATG9; (3) Vps34 complexes; and (4, 5) two ubiquitin-like conjugation systems anchored by ATG12 and ATG8, respectively.

The ULK complex (composed of ATG13, FIP200, and either ULK1 or ULK2) plays a central role in starvation-induced autophagy and is thought to be involved in the initiation of the autophagic cascade. The complex is recruited to sites of autophagosome formation, where it likely acts to recruit and disassemble other ATG protein complexes [27–30]. ATG9, a transmembrane protein, shuttles between autophagosomes, the trans-Golgi network, and late endosomes in a cycling process that may involve the ULK complex. ATG9 is of special interest to studies of xenophagy, since a recent report suggested that ATG9 is essential for the formation of the isolation membrane in anti-*Salmonella* autophagy in mouse embryonic fibroblasts (MEFs) [31]. In mammals, the Vps34 complex is likely required for the recruitment of several autophagy proteins to sites of autophagosome formation. This complex plays a critical role in supplying phosphatidylinositol 3-phosphate (PI3P) to the growing autophagosome, where it acts as an essential driving force in autophagosome assembly [32–36]. Furthermore, depending on the identity of its protein components, some alternative versions of the Vps34 complex may serve regulatory roles in autophagosome maturation and autophagosome-lysosome fusion [32–34, 37–39].

Two ubiquitin-like conjugation systems play important roles in autophagosome formation and maturation. The ATG12-based conjugation system is initiated by the E1-like ATG7, which activates and transfers ATG12 to the E2-like ATG10. ATG12 is then subsequently conjugated to ATG5 [40, 41]. ATG5 and ATG12 then bind with ATG16L1, forming a complex that is essential for autophagosome formation [42]. In addition to a key role in autophagosome formation, the second ubiquitin-like conjugation system may play additional roles in cargo recognition [43]. In this system, ATG8 is activated by the E1-like ATG7, is transferred to the E2-like ATG3, and is conjugated to phosphatidylethanolamine through the E3-like action of the ATG5-ATG12 complex. The conjugation of phosphatidylethanolamine to ATG8 allows subsequent incorporation of ATG8 into the inner and outer membranes of autophagosomes [44–46]. Mammals appear to have two families of ATG8 homologues, known as LC3s and GABARAPs. While all ATG8 homologues can be incorporated into the autophagosomal membrane, the incorporation of LC3s and GABARAPs might occur at different stages, with LC3s acting at the stage of biogenesis and GABARAPs playing a more important role in autophagosome maturation [47]. Given that ATG8 remains associated with autophagosomes even after their fusion with lysosomes, tracking ATG8 and its homologues allows for localization of autophagosomes. Furthermore, since the inner membranes of autophagosomes are degraded upon formation of autolysosomes, tracking ATG8 enables the measurement of autophagic flux as reflected by ATG8 degradation [44, 45].

Several layers of regulation exist in the initiation and progression through the steps of autophagic degradation. One important regulatory influence is the cytoskeleton. Autophagosome trafficking involves movement along microtubules, which facilitates fusion with lysosomes [48, 49]. Changes in lysosome positioning regulate mTORC1 signaling, since peripheral localization of lysosomes and mTORC1 increases mTORC1 activity, while perinuclear lysosomal clustering decreases activity [49]. Heat shock protein beta-1 (HSPB1), a protein that interacts with and mediates reorganization of the actin cytoskeleton, is also linked to mitophagy [50]. Disruption of HSPB1 signaling can lead to accumulation of dysfunctional mitochondria, mitochondrial fragmentation, and effects in mitochondrial respiration in MEFs [51].

Transcription also serves as a regulatory influence in the autophagic process. Autophagosome and lysosome biogenesis appear to be transcriptionally coregulated by the transcription factor EB (TFEB) [52]. Overexpression of TFEB induces autophagy, while a block of TFEB expression by RNAi decreases autophagy. Interestingly, TFEB action is implicated in basal as well as starvation-induced autophagy, and therefore TFEB might represent a transcriptional master regulator of the autophagy pathway. Furthermore, translocation of TFEB from the cytosol to the nucleus is strongly enhanced upon removal of nutrients, suggesting that TFEB function is sensitive to nutrients and growth factors. Under fed conditions, phosphorylation on serine 142 of TFEB can be mediated by the extracellular signal-regulated kinase 2 (ERK2) and might be sufficient to retain TFEB in the cytosol. A comprehensive analysis of 51 genes with known functions in the autophagy pathway sheds some light on transcriptional regulation of autophagy in the context of starvation and TFEB function. Strikingly, TFEB-dependent transcription patterns generally correlate with starvation effects, further emphasizing the role of TFEB as an autophagy regulator that can directly or indirectly sense nutritional status [52]. However, the transcription of the majority of genes is not significantly affected by starvation or TFEB expression levels [52], suggesting that transcription factors other than TFEB control the expression of most autophagy and autophagy-related genes, possibly in a nutrient-insensitive fashion.

Xenophagy

Autophagy has now been well described as playing a key role in the degradation of intracellular pathogens, including bacteria, in a specialized process known as xenophagy. However, the environment of the intestinal mucosa provides special challenges to this function. As this tissue is exposed to both commensal and pathogenic bacteria, the interactions between these flora and host immune responses must be tightly controlled.

Molecular Basis of Xenophagy

The molecular events surrounding the process of xenophagy follow the same basic pattern as those involved in autophagic degradation of other cellular contents, such as organelles or aggregated protein deposits. However, the mechanisms underlying the selection of cargo for degradation by xenophagy remain largely unknown. Components of the innate immune response can interact with autophagy proteins to target bacteria for autophagic degradation. Mechanisms involved in xenophagy of other species of bacteria may share some components with anti-*Salmonella* xenophagy, but to date, detailed molecular mechanisms have been best explored for *Salmonella*.

Upon infection of epithelial cells, *Salmonella enterica* serotype Typhimurium (*S*. Typhimurium) resides in *Salmonella*-containing vacuoles (SCVs). The environment of the SCV promotes bacterial survival and replication. The SCV protective niche is established by effectors of two type III secretion systems (T3SSs) encoded by *Salmonella* pathogenicity islands 1 and 2 (SP-1 and SP-2). The secretion and activity of T3SS components is highly regulated both temporally and spatially. For example, pH sensing regulates the transition from translocon protein secretion and pore formation to effector secretion [53]. Once secreted into the host cell, the activity of effectors can be modulated. For example, the T3SS SP-1 effectors SopA, SopE, SptP, and SopB are all ubiquitinated shortly after infection by the host cell machinery [54]. The phosphoinositide phosphatase activity of SopB regulates several processes that promote internalization and survival of *S*. Typhimurium, including cytoskeletal rearrangements that promote invasion, SCV biogenesis and maturation, and prosurvival Akt activation [54]. The ability of SopB to regulate these functions is dependent upon its ubiquitination status [54, 55]. A SopB mutant that cannot be ubiquitinated has a number of defects, including a failure to relocalize from the plasma membrane to SCVs, disrupted recruitment of Rab5 to SCVs, and perturbed *S*. Typhimurium replication [54]. Thus, ubiquitination-dependent relocalization of SopB to SCVs promotes the establishment of SCVs as a niche environment for replication. This is due in part to the ability of SopB to reduce the net negative charge of PI(4,5)P2 and phosphatidylserine lipids on SCV membranes, thereby disrupting maturation and fusion with lysosomes [55].

Although most *S*. Typhimurium resides in SCVs, a small but significant fraction is released into the cytosol from damaged SCVs. This cytosolic *S*. Typhimurium rapidly becomes ubiquitinated. Ubiquitinated bacteria may be recognized by selective cargo receptors, including p62 and NBR1, which act as bridging factors that bind to both ubiquitin and the autophagosome. p62 and NBR1 share a similar protein structure and domain composition and may show some level of redundancy, particularly in degradation of protein aggregates (aggrephagy) and mitochondria (mitophagy). However, while p62 is important for the autophagic clearance of *Salmonella*, NBR1 is not [56]. Therefore, the mechanism of xenophagy, which relies on ubiquitination events on or around intracellular bacteria, might differ from other ubiquitin-dependent selective autophagy events. Interestingly, nuclear dot protein 52 (NDP52), in addition to p62, has been implicated in the recognition of ubiquitin-coated invasive bacteria, but not in other forms of selective autophagy [57]. NDP52 contains a putative LIR motif at its N terminus, can directly bind ubiquitin through a C-terminal ubiquitin-binding zinc finger, and can form homo-oligomers via its coiled-coil domain [57–59]. p62 and NDP52 appear to act cooperatively in the recognition and killing of *S*. Typhimurium, *Listeria monocytogenes*, and *Shigella flexneri* [60, 61]. However, p62 and NDP52 are recruited to distinct microdomains of invading *Salmonella*, suggesting that ubiquitin coats on *Salmonella* might be heterogeneous [60].

Interestingly, in macrophages but not in epithelial cells, proteasomes localize to ubiquitin-coated *Salmonella* with potential effects on efficient rupture of damaged SCVs and major histocompatibility complex (MHC) class I antigen presentation [62, 63]. Taken together, ubiquitin-dependent autophagic targeting of intracellular pathogens and by-products has implications for pathogen growth and the regulation of inflammatory responses. However, in most cases, the E3 ubiquitin ligases, the nature of ubiquitin modifications, and the substrates involved in ubiquitin-dependent xenophagy are not known and may be pathogen- as well as cell type-specific. One important exception is LRSAM1, an LRR- and RING-domain protein that was recently identified as an E3 ubiquitin ligase crucial for ubiquitin-dependent autophagy of *S.* Typhimurium [64]. Interestingly, LRSAM1 was required for ubiquitination of intracellular bacteria but dispensable for ubiquitination of aggregated proteins, confirming that LRSAM1 serves as a selective and novel antibacterial sensor that mediates target selection for the xenophagy pathway. To date, however, LRSAM1 remains the only known ubiquitin ligase associated with antibacterial autophagy, and a detailed biochemical characterization of the aforementioned factors will yield critical insights into host–pathogen interactions and may boost the development of novel pharmaceuticals aimed to combat infectious diseases.

Xenophagy and IBD Risk Genes

As mentioned above, intestinal tissues are in constant interaction with both commensal and pathogenic bacteria. To prevent inappropriate responses to commensal bacteria, the cells of the intestinal mucosa must be educated to tolerate these microbes and their products [65]. Breakdown of this tolerance, with consequent inappropriate inflammation, has been proposed to play a major role in the pathogenesis of IBD. This model is supported by the finding that the IBD-associated genes *ATG16L1*, *IRGM*, and *ULK1* all play roles in the process of xenophagy. ATG16L1, IRGM, and ULK1 each function in the degradation of *S.* Typhimurium, while ATG16L1 and IRGM also play a role in antibacterial autophagy of a CD-relevant strain of *Escherichia coli* (adherent-invasive *E. coli*, or AIEC) [10, 66–69]. ULK1 may also function in the degradation of *L. monocytogenes*. Furthermore, the function of LRRK2 in reactive oxygen species (ROS)-driven bacterial killing, as well as its role in autophagy, raises the possibility that this protein may also be involved in xenophagy of gut bacteria [70–72]. Defects in these gene products might lead to impaired bacterial handling, promoting an environment of chronic inflammation and inappropriate host responses to commensal bacteria.

Some intracellular pathogens have also been highlighted as potential contributors to risk of CD, including mycobacteria. Genetic loci that include *IRGM*, *LRRK2*, and *NOD2* have been associated with increased susceptibility to leprosy and tuberculosis, suggesting an interesting point of interaction between autophagy genes,

CD, and mycobacteria. These studies, combined with reports that autophagy may play a key role in control of *Mycobacterium tuberculosis*, have led to the suggestion that control of mycobacteria might be altered in CD [73, 74]. Genetic studies appear to support this idea, especially in the context of granulomas, a histological hallmark of both CD and tuberculosis: polymorphisms in loci containing the autophagy genes *ATG2A* and *GABARAPL1* have been linked to granulomas in CD patients. To date, however, connections between CD and the mycobacterioses remain largely speculative, to be answered perhaps by future well-powered GWAS for tuberculosis [75].

The recent wealth of data from the human microbiome project has also highlighted how little we understand of the complex relationship between IBD and microbial stimulation. Certainly the diversity and composition of the gut microbiota are major factors influencing gut homeostasis, and particular dietary nutrients and metabolites are likely to interact with host genetics to influence host-microbiome interactions. An imbalance in the composition of the gut microbiome, termed dysbiosis, has been associated with IBD, including shifts in relative abundances of bacterial taxa, decreases in the diversity of the bacterial community, and alterations in the functional composition of the microbiome.

Mitophagy

The autophagic processing of mitochondria (mitophagy) was directly implicated in the context of IBD by a recent study that identified *SMURF1* as a susceptibility gene for UC [13]. This gene is an ubiquitin ligase that was recently identified, through an image-based genome-wide siRNA screen, as an important mediator of selective viral autophagy and mitophagy [76].

Molecular Basis of Mitophagy

In contrast to other substrates for selective autophagy, for example, surplus peroxisomes or ribosomes, mitophagy is well studied and underlying biochemical mechanisms for this process are emerging. Upon loss of mitochondrial membrane potential, PTEN-induced putative kinase 1 (PINK1) accumulates on the outer mitochondrial membrane [77, 78]. This step is accompanied by stabilization of PINK1 protein and may involve a membrane potential-dependent block of PINK1 proteolysis by the mitochondrial inner membrane protease presenilin-associated rhomboid-like protein (PARL). This protease is effective in healthy mitochondria but impaired in damaged mitochondria [77, 79]. PINK1 then recruits Parkin1 to damaged mitochondria, which results in Parkin1-dependent ubiquitination of some mitochondrial proteins such as voltage-dependent anion channel 1 (VDAC1) and mitofusins [80–83]. p62 accumulates around ubiquitin-decorated damaged mitochondria and

PB1-domain-mediated oligomerization of p62 results in mitochondrial clustering, eventually leading to autophagic degradation of mitochondria [80]. It should be noted that the mitochondrial ubiquitination event required for mitophagy may involve multiple substrates and/or may be cell type-specific, as p62 recruitment to damaged mitochondria and mitophagy is not impaired in VDAC1-deficient MEFs [78]. In addition, activating molecule in beclin 1-regulated autophagy (Ambra1), an autophagy-regulating protein initially identified in neuronal cells [84], is recruited to damaged mitochondria in a Parkin1-dependent fashion [85] and knockdown of Ambra1 expression results in impairment of mitophagy. Interestingly, Ambra1 recruitment most likely involves direct binding to Parkin1, but Ambra1 is not a substrate for Parkin1-mediated ubiquitination [85]. Thus, Parkin1-mediated mitophagy relies at least in part on effects other than ubiquitination.

Mitophagy and IBD Risk Genes

As mentioned above, the best evidence linking mitophagy to IBD comes from the identification of *SMURF1* as a gene associated with susceptibility to UC. SMURF1 is a HECT domain-containing E3 ubiquitin ligase that interacts with the selective autophagy factor p62 [76]. Smurf1$^{-/-}$ MEFs show defects in the degradation of Sindbis and herpes simplex virus as well as impaired clearance of damaged mitochondria. Supporting this observation, Smurf1-deficient mice display an accumulation of damaged mitochondria in the heart, brain, and liver. Interestingly, Smurf1$^{-/-}$ MEFs are competent for starvation-induced bulk autophagy, indicating that Smurf1 is a specific mediator of selective autophagy. However, a mechanistic link between *SMURF1* polymorphisms and IBD pathogenesis remains to be examined.

Antigen Presentation

Evidence for a link between IBD, antigen presentation, and autophagy originates from studies of CD-associated risk variants in NOD2 and ATG16L1. Although a role for autophagy in antigen presentation has been well described in basic studies, how this function may be altered in the context of IBD remains relatively unknown.

Molecular Basis of Autophagy in Antigen Presentation

MHC molecules on the cell surface present peptide antigens to T cells. MHC class I molecules interact with the T cell receptor (TCR) of CD8$^+$ cytotoxic T cells, while MHC class II molecules interact with the TCR of CD4$^+$ helper T cells. MHC class I antigens are derived from proteasomal degradation of cytosolic proteins. In contrast, MHC class II antigens are delivered to the lysosome before transfer to the MHC class II loading compartment and transport to the cell surface. Delivery of these antigens to the lysosome can occur either via endocytosis of protein antigens from the extracellular space or by autophagy of cytoplasmic material [86–88]. Autophagy also plays an important role in the presentation of pathogen antigens by dendritic cells (DCs). Mice lacking Atg5 specifically in DCs show impaired $CD4^+$ T cell priming following infection with herpes simplex virus, an impairment that is due to a decreased ability of Atg5-deficient DCs to process and present antigens for presentation on MHC class II molecules [89]. Furthermore, these cells show delayed fusion of phagosomes to lysosomes. Autophagic degradation may also combine with vesicular trafficking to facilitate presentation of citrullinated self-peptides by DCs to $CD4^+$ T cells. This finding is particularly relevant to IBD, since immune responses against citrullinated self-proteins are associated with autoimmunity [90, 91].

Antigen Presentation and IBD Risk Genes

Interestingly, DCs from patients with CD-associated risk variants of ATG16L1 or NOD2 show defects in autophagy induction, bacterial trafficking, and MHC class II antigen presentation to $CD4^+$ T cells [23]. Furthermore, autophagy was recently reported to destabilize the immune synapse between DCs and T cells; in this report, decreased expression of ATG16L1 or IRGM resulted in hyperstable interactions between DCs and T cells as well as increased activation of T cells, suggesting a mechanism by which adaptive immunity might be increased in patients with CD who carry ATG16L1 risk alleles [92].

Vesicular Trafficking and Secretion

The suggestion that the secretory pathway might be modulated by autophagy in the context of CD arose from the GWAS-based identification of *ATG16L1* as a common CD-associated risk gene. Patients carrying the CD-associated variant of this gene (ATG16L1 T300A), as well as Atg16l1 hypomorphic ($Atg16l1^{HM}$) mice, were found to show abnormalities in specialized epithelial cells called Paneth cells. These cells play a central role in innate immunity and are an important source of antimicrobial peptides in the small intestine, serving to prevent microbial invasion and control the composition of the gut microflora [93]. ATG16L1 T300A patients and $Atg16l1^{HM}$ mice exhibit decreased numbers of granules and diffuse staining for lysozyme in their Paneth cells, raising the possibility that autophagy plays a key role in secretion in these cells [94, 95]. Supporting this hypothesis, mice lacking Atg5 or Atg7 in intestinal epithelial cells also show Paneth cell defects [96]. Interestingly, the Paneth cell phenotype in Atg16l1^{HM} mice can be influenced by exposure to a pathogen, since the cellular defects can be eliminated by maintaining mice in a virus-free environment [94]. Paneth cell defects have also been observed in mice lacking other CD-associated genes, including *Nod2* and *Xbp1* [97, 98]. More recently, a role was reported for Paneth cells in sensing nutrient availability, demonstrating that caloric restriction reduces Paneth cell-specific signaling by mTORC1, a regulator of autophagy [99]. Taken together, these reports suggest that Paneth cell homeostasis and function are intimately linked to autophagy. However, whether autophagy directly regulates vesicular trafficking and secretion remains unknown, since the molecular mechanisms by which ATG16L1 might regulate these cellular functions are poorly understood.

Inflammasome Hyperactivation and Cytokine Secretion

Inflammasomes are molecular scaffolds that activate caspase 1 and maturation of the proinflammatory cytokines interleukin (IL)-1β and IL-18. Several types of inflammasomes have been described, which are activated by a number of endogenous and exogenous signals.

Recent studies have shown that autophagy defects can lead to inflammasome hyperactivation. For example, stimulation of Atg16l1$^{-/-}$ fetal macrophages with lipopolysaccharide (LPS) induces elevated levels of activate caspase 1 and enhanced secretion of IL-1β and IL-18 [21]. Mice with specific deletion of Atg16l1 in hematopoietic cells also show higher serum levels of IL-1β and IL-18 in response to dextran sulfate sodium (DSS), accompanied by increased inflammation and mortality.

The underlying mechanism linking autophagy to regulation of the inflammasome remains unknown. One possibility is that defects specifically in mitophagy result in increased levels of ROS, which are known to cause hyperactivation of inflammatory activity [100–102]. Supporting this idea, Atg16l1$^{-/-}$ macrophages treated with a ROS scavenger did not show the increased IL-1β secretion described above [21].

Samples from patients have also shown ATG16L1 coding variant-dependent alterations in IL-1β secretion. In these studies, stimulation of peripheral blood mononuclear cells with MDP, but not LPS, resulted in a relative increase of IL-1β secretion from ATG16L1 T300A-expressing cells. Although no difference was found in levels of activated caspase 1, protein levels of pro-IL-1β and mRNA levels of IL-1β were increased in the context of the ATG16L1 risk allele [103].

The autophagy adapter p62 may also provide a connection between IL-1β signaling and ATG16L1 [104]. In a recent study, we showed that levels of p62 are normally regulated by ATG16L1-based activation of the ubiquitin ligase Cullin-3, which promotes proteasomal degradation of p62. Loss of ATG16L1, however, can result in decreased levels of Cullin-3 and decreased degradation of p62. As p62 can act as a scaffold in the IL-1β signaling pathway, elevated levels of p62 may result in amplified IL-1β signaling [104].

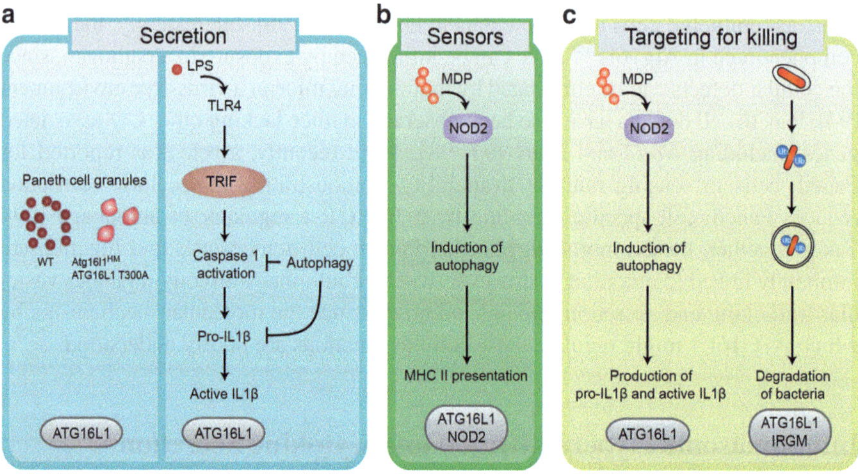

Fig. 12.1 Immune-related autophagy functions. (**a**) Paneth cells from CD patients expressing the CD-associated ATG16L1 T300A allele, as well as mice expressing a hypomorphic allele of Atg16l1 (Atg16l1^{HM}), show altered granule morphology, suggesting a role for autophagy in secretion of antimicrobial peptides. Another role for autophagy in secretion is suggested by observations in macrophages, in which secretion of active IL-1β (as well as active IL-18, not shown) upon LPS stimulation is regulated by autophagy at two levels. In this pathway, loss of ATG16L1 expression is associated with increased cytokine secretion. (**b**) Autophagy can also act as a sensor for bacterial products. In dendritic cells, activation of NOD2 by MDP induces autophagy that leads to MHC class II antigen presentation. This pathway appears to be altered in the context of CD-associated SNPs in *ATG16L1* and *NOD2*. (**c**) Autophagy plays a role in targeting bacteria and bacterial products for degradation/killing. In one pathway, observed in primary blood mononuclear cells, activation of NOD2 by MDP induces autophagy that leads to production of pro-IL-1β and secretion of active IL-1β. The CD-associated variant ATG16L1 T300A is associated with increased amounts of active IL-1β upon MDP stimulation. Autophagy can also directly degrade bacteria. In epithelial cells, bacteria become ubiquitinated and targeted for autophagic degradation in a process that requires the CD-associated genes *ATG16L1* and IRGM

ATG16L1 T300A

One of the best-studied CD-associated risk alleles is the T300A coding polymorphism in *ATG16L1*. Atg16l1$^{-/-}$ mice are not viable [21], but an alternative hypomorphic model (Atg16l1^{HM}) has yielded important clues to how the T300A coding polymorphism may result in altered autophagy and increased inflammation [94, 95]. The site of the T300A polymorphism is near a WD-repeat domain, which is present in mammalian *ATG16L1* but absent in yeast Atg16. This domain is not required for autophagy, consistent with the finding that the T300A polymorphism does not affect classical autophagy [42, 67].

As a core autophagy protein, *ATG16L1* has been shown to play multiple roles in IBD-relevant processes, including xenophagy, antigen presentation, IL-1β production, and secretion (see Fig. 12.1). Interestingly, the CD-associated risk allele appears to have different effects on these functions depending on the cell type examined [42, 67]. It is perhaps not surprising, then, that T300A has also been associated with defects in antigen presentation increases in IL-1β production and abnormalities in Paneth cell secretion in patient samples [23, 95, 103]. In addition, antibacterial autophagy of *Salmonella* is affected by the presence of the ATG16L1 CD risk allele [67], and patients with the T300A allele show increased susceptibility to infection by *Helicobacter pylori* [105]. These findings illustrate that disease genes are likely to have specific functions in specific cell types, and researchers must consider the cell type- and stimulation-specific contexts used to examine disease-associated phenotypes.

IRGM rs13361189

IRGM (immunity-related GTPase family M) is a human gene recently identified as playing an important role in antibacterial and antiviral autophagy. To date, xenophagy of pathogens including *S.* Typhimurium, *M. tuberculosis*, and CD-associated AIEC has been reported to be dependent on IRGM [66, 68, 69, 106].

A synonymous SNP within the human *IRGM* locus (rs13361189) is associated with CD, and a 20 kb deletion polymorphism within the 5′ untranslated region of *IRGM* is in perfect linkage disequilibrium with this SNP [68, 107]. However, both rs13361189 and the deletion polymorphism have also been reported to be in perfect linkage disequilibrium with a synonymous exonic SNP (rs10065172). This SNP is associated with decreased binding of the microRNA miR-196 to the 3′ UTR of *IRGM*, an event that downregulates expression of IRGM [69]. This finding is consistent with the observation that expression of miR-196 is increased in epithelial cells of the inflamed ileum and colon of CD patients compared with controls [69]. This result may be relevant to autophagic targeting of pathogenic bacteria, since knockdown of IRGM or overexpression of miR-196 affects the autophagic targeting of AEIC. Furthermore, an additional *IRGM* polymorphism (−261TT) is associated with an increase of IRGM expression and protection against *M. tuberculosis* [74].

Autophagy Interaction Network

Despite the approaches described in the studies above, in which individual autophagy genes are examined in relative isolation, it is important to note that a complex network of autophagy-related proteins exists in the cell and that cellular functions captured within this network likely extend beyond the strict bounds of lysosomal degradation.

As mentioned above, autophagy is implicated in many cellular pathways, including adaptations to changes in environmental factors such as nutrient availability. Many autophagy and autophagy-related proteins have been characterized, and distinct autophagic complexes with functions in autophagosome assembly and maturation have been identified and analyzed in great detail. The current picture suggests a stepwise process, but the precise regulation of autophagy in changing environmental settings—including fluctuations in growth factor supply, cytokine signaling, and the presence of pathogens—likely involves complex cross talk between pathway components. Behrends and colleagues recently used an approach which might be described as "pathway proteomics" to investigate the network organization of the human autophagy system [108]. In doing so, they provided a basis for understanding the autophagy network on a protein level. In brief, interaction partners for 32 epitope-tagged stably expressed proteins, representing core autophagy as well as autophagy-related processes, were identified using mass spectrometry following immunoprecipitation of the respective tagged protein. A subset of proteins interacting with primary baits were used as baits in secondary screens, leading to the identification of 751 high-confidence candidate interactions between a total of 65 baits and an additional 409 proteins. A high coverage of known interactions, a high level of validation (reciprocal immunoprecipitations and in vitro assays), and validation of autophagy functions of network components via RNAi allow for the designation of these results as a true autophagy protein interaction network. Interestingly, different autophagy subnetworks such as the ULK complex, Vps34 complex, ATG8 conjugation system, and PI3P signaling members were revealed to have an underappreciated high level of interconnectivity. Novel interaction partners for the mammalian ATG8 homologues were also identified. These interactors may be shared among all ATG8 homologues, specific for LC3 or GABARAP subfamily members or specific for individual ATG8 homologues. However, differential cellular abundance of distinct ATG8 homologues could in part account for different efficiencies with respect to the co-immunoprecipitation of ATG8 interaction partners [108]. Nevertheless, differences in LC3-interacting region (LIR) motif composition among different ATG8 homologues can affect binding to p62, NBR1 [109], and possibly control differential requirements of LC3 and GABARAP subfamily members in early or late phases of autophagosome biogenesis [47]. In addition, it will be interesting to see if some of the novel ATG8 interactors might function as receptors for certain types of selective autophagy. Good candidates might be recovered by a comparison of ATG8 interactors identified by these proteomics studies and a list of proteins predicted to combine LIR and ubiquitin-binding motifs (Kay Hofmann, unpublished data in [110]).

Interestingly, pharmacological induction of autophagy led to alterations of protein–protein interactions within some, but not all, subnetworks [108]. It will be interesting to see if such clusters might be targets of microRNAs, which generally modify the translation of multiple genes at the same time. In addition, it is possible that different subnetworks are altered by different stimuli, possibly in a cell type-specific fashion, which could ultimately help in the design of disease-specific autophagy-enhancing or autophagy-inhibiting drugs.

Chemical Modulators of Autophagy

As highlighted throughout this chapter, many unanswered questions remain regarding the mechanistic basis of multiple autophagy functions. We believe that the field of chemical biology is poised to help answer these questions, providing the tools to dissect the regulatory nodes that exist at the intersections of autophagy and immunity. Furthermore, studies in this field may help to identify druggable targets within the autophagy pathway and its regulatory influences.

Small molecules have been proven to be useful tools in autophagy research, demonstrating success by increasing our knowledge of autophagy regulation while simultaneously identifying potential therapeutic compounds. Small-molecule screens are particularly important in identifying probes for dissecting complex biological processes such as autophagy, as they allow the dissection of multiple steps in a pathway, provide temporal control over target function, and are often reversible. These screens are also particularly powerful in enabling researchers to identify molecules that parallel gene activity. Furthermore, by pursuing the mechanism of action (MOA) of active compounds, novel regulatory pathways can be identified.

Target identification and MOA determination is a rate-limiting step to small-molecule discovery in phenotype-based screens and relies on the integration of multiple complementary approaches. One such approach is candidate based, which can be used to determine whether novel small molecules target known MOAs. Several unbiased approaches to determine small-molecule MOAs can also be used, including (1) combining quantitative proteomics (SILAC) with affinity enrichment to identify proteins that interact with the small molecule and yield the observed phenotype, (2) compound profiling and connectivity analysis using gene expression signatures, and (3) genetic complementation of small-molecule effects by RNAi or overexpression screening to identify genes that function in the same pathway as the small molecule [111, 112].

Several forward chemical screens have been used to identify compounds with relevance to autophagy. These cell-based phenotypic screens employed libraries of FDA-approved drugs and known bioactive compounds to identify chemical modulators of autophagy [113–116]. Various readouts were used to measure autophagic activity, including GFP-LC3 puncta formation, the clearance of mutant huntingtin and A53T α (alpha)-synuclein aggregates, and degradation of luciferase-fused LC3. Autophagy-related small molecules discovered in these screens can be broadly classified into two groups: (1) mTOR-dependent molecules, which consist of compounds that induce autophagy, and (2) mTOR-independent molecules, which can be either inducers or suppressors of autophagy. The first group includes inhibitors of class I PI3K/Akt/mTOR signaling, such as allosteric and ATP-competitive mTOR kinase, Akt, PI3K, and dual mTOR/PI3K inhibitors. The second (mTOR-independent) group includes inhibitors of cAMP-Epac-PLC-ε (epsilon)-IP3 and Ca^{2+}-calpain-Gs-α (alpha) signaling [117].

Results from such chemical screens have been carried forward into cell and animal models of autophagy-related diseases. For example, (SMER) compounds were originally identified in a screen for enhancers of rapamycin-induced growth defects in *S. cerevisiae* [115]. SMERs 10, 18, or 28 were found to be active in mammalian cell models of Huntington's and Parkinson's disease, promoting autophagic clearance of mutant huntingtin and A53T α (alpha)-synuclein aggregates [115]. In vivo, SMERs protect against neurodegeneration in a *Drosophila* model of Huntington's disease [115].

Like the SMERs, the pan-nitric oxide synthase (NOS) inhibitor Nω (omega)-nitro-L-arginine methyl ester hydrochloride (L-NAME) can block neurodegeneration in the same *Drosophila* Huntington's disease model and, similar to rapamycin, can clear mutant huntingtin aggregates in a zebrafish Huntington's disease model, suggesting that NOS inhibitors have therapeutic potential [118]. L-NAME triggers functional autophagy in a variety of cell types as measured by LC3 processing, the accumulation of RFP-LC3 puncta, and the ATG5-dependent clearance of mutant huntingtin aggregates [118]. While NOS inhibitors such as L-NAME decrease NO, complementary experiments to increase NO levels using NO donors and NO synthase overexpression revealed an inhibitory role for NO in autophagy [118]. The MOA of NO was found to involve two different mechanisms of inhibiting autophagy. First, S-nitrosylation of JNK1 impairs phosphorylation of BCL2, leading to increased BCL2-belin 1 interaction. Second, S-nitrosylation of IKKβ promotes the TSC2-dependent activation of mTORC1 [118].

The role of HMGB1 in regulating mitophagy following rotenone-induced disruption of oxidative phosphorylation is a further example that highlights the utility of small-molecule–gene interactions in identifying key autophagy pathways [51]. Mitochondrial stress mediated by rotenone, a small molecule that inhibits complex I in the electron transport chain, led to mitophagy that was facilitated by HMGB1 and its transcriptional target HSPB1. HMGB1 and HSPB1 were found to control mitochondrial homeostasis, as knockdown of either HMGB1 or HSPB1 resulted in perturbed morphology, glycolysis, ATP production, and mitochondrial fragmentation [51, 119]. Furthermore, rotenone-induced mitophagy was disrupted by cytochalasin D, suggesting a role for the actin cytoskeleton in HMGB1-HSPB1-mediated mitophagy [51].

To enhance the therapeutic utility of small molecules, the relationships between compound structure and biological activity (structure-activity relationships or SARs) can be studied to determine which chemical groups present in the compound are important for activity and how modifications of the structure can improve selectivity and potency. In one example, Chen et al. synthesized a novel set of diphenylbutylpiperidines that demonstrated tenfold improved potency over lead compounds fluspirilene and penfluridol as measured by LC3-GFP puncta [120].

In addition to autophagy activators, autophagy inhibitors can also be useful as probes in determining whether a given state is dependent on classical autophagy or whether a compound truly induces autophagic flux. PI3K inhibitors 3-MA and wortmannin and the vacuolar H^+ ATPase inhibitor bafilomycin A1 are typically used for these purposes. Autophagy inhibitors can also have therapeutic potential. Chou and colleagues identified DBeQ, a reversible ATP-competitive inhibitor of the p97 ATPase [121]. Inhibition of p97 following DBeQ treatments triggers diverse phenotypes, including a blockage of autophagosome maturation, activation of caspase 3 and 7, and inhibition of cancer cell proliferation, suggesting a potential therapeutic utility for p97 inhibitors in cancer [121].

Unlike the PI3K class I inhibitors discussed above, the discovery of potent and selective inhibitors of class III VPS34 has proven difficult. Class III VPS34 contains a smaller, more rigid ATP binding pocket compared to class I p110γ, possibly adding to the difficulty in finding inhibitors [122]. 3-MA, a widely used inhibitor of VPS34, is typically used at millimolar concentrations, thus likely triggering many off-target effects that complicate results. To gain insight into developing VPS34 selective inhibitors, the structures of VPS34 in complexes with 3-MA, PIK-90, PIK-93, and PI103 were solved [122]. By employing structure-based design, Miller and colleagues were able to synthesize PT210, an analog of PIK-93, with tenfold more selectivity for VPS34 than for class I p110γ [122]. Improved VPS34 inhibitors will be useful to probe the role of VPS34 in autophagy with more selectivity and without confounding effects of class I p110γ inhibition.

The success of small-molecule screens in identifying and dissecting autophagy regulatory pathways is notable for its repercussions for therapeutics. The identification of new druggable targets may lead, after extensive optimization, to potential therapeutics in human diseases with autophagy-dependent and/or autophagy-modulatory features. Experimental results such as those obtained using SMERs and NOS inhibitors suggest that small molecules may also be employed to target antibacterial autophagy at multiple stages.

Concluding Remarks

Recent research has provided significant insight into autophagic functions in immunity, including xenophagy, mitophagy, antigen presentation, vesicular trafficking and secretion, and cytokine activity. However, substantial questions remain regarding the mechanistic basis of such functions as well as how individual disease risk alleles modulate these functions. In particular, studies of the microbiome may be critically important in understanding the relationship between xenophagy, immunity, and IBD. We anticipate that microbiome-wide studies (MWAS), in combination with detailed insights into the function of autophagy in health and disease, will help inform the development of novel biological and chemical entities for the treatment of patients suffering from these inflammatory disorders.

Acknowledgments This work was supported by funding from the Crohn's and Colitis Foundation of America and NIH grants DK043351, DK083756, and DK086502 to R.J.X.

References

1. Mizushima N, Yoshimori T, Ohsumi Y (2011) The role of Atg proteins in autophagosome formation. Annu Rev Cell Dev Biol 27:107–132, Epub 2011/08/02
2. Li W, Zou W, Yang Y, Chai Y, Chen B, Cheng S et al (2012) Autophagy genes function sequentially to promote apoptotic cell corpse degradation in the engulfing cell. J Cell Biol 197(1):27–35, Epub 2012/03/28

3. Qu X, Zou Z, Sun Q, Luby-Phelps K, Cheng P, Hogan RN et al (2007) Autophagy gene-dependent clearance of apoptotic cells during embryonic development. Cell 128(5):931–946, Epub 2007/03/14
4. Pierdominici M, Vomero M, Barbati C, Colasanti T, Maselli A, Vacirca D et al (2012) Role of autophagy in immunity and autoimmunity, with a special focus on systemic lupus erythematosus. FASEB J 26(4):1400–1412, Epub 2012/01/17
5. Levine B, Mizushima N, Virgin HW (2011) Autophagy in immunity and inflammation. Nature 469(7330):323–335, Epub 2011/01/21
6. Kuballa P, Nolte WM, Castoreno AB, Xavier RJ (2012) Autophagy and the immune system. Annu Rev Immunol 30:611–646, Epub 2012/03/28
7. Deretic V (2011) Autophagy in immunity and cell-autonomous defense against intracellular microbes. Immunol Rev 240(1):92–104, Epub 2011/02/26
8. Franke A, McGovern DP, Barrett JC, Wang K, Radford-Smith GL, Ahmad T et al (2010) Genome-wide meta-analysis increases to 71 the number of confirmed Crohn's disease susceptibility loci. Nat Genet 42(12):1118–1125, Epub 2010/11/26
9. Barrett JC, Hansoul S, Nicolae DL, Cho JH, Duerr RH, Rioux JD et al (2008) Genome-wide association defines more than 30 distinct susceptibility loci for Crohn's disease. Nat Genet 40(8):955–962, Epub 2008/07/01
10. Rioux JD, Xavier RJ, Taylor KD, Silverberg MS, Goyette P, Huett A et al (2007) Genome-wide association study identifies new susceptibility loci for Crohn disease and implicates autophagy in disease pathogenesis. Nat Genet 39(5):596–604, Epub 2007/04/17
11. Hampe J, Franke A, Rosenstiel P, Till A, Teuber M, Huse K et al (2007) A genome-wide association scan of nonsynonymous SNPs identifies a susceptibility variant for Crohn disease in ATG16L1. Nat Genet 39(2):207–211, Epub 2007/01/04
12. Anderson CA, Boucher G, Lees CW, Franke A, D'Amato M, Taylor KD et al (2011) Meta-analysis identifies 29 additional ulcerative colitis risk loci, increasing the number of confirmed associations to 47. Nat Genet 43(3):246–252, Epub 2011/02/08
13. Jostins L, Ripke S, Weersma RK, Duerr RH, McGovern DP, Hui KY et al (2012) Host-microbe interactions have shaped the genetic architecture of inflammatory bowel disease. Nature 491(7422):119–124, Epub 2012/11/07
14. Henckaerts L, Cleynen I, Brinar M, John JM, Van Steen K, Rutgeerts P et al (2011) Genetic variation in the autophagy gene ULK1 and risk of Crohn's disease. Inflamm Bowel Dis 17(6):1392–1397, Epub 2011/05/12
15. Brinar M, Vermeire S, Cleynen I, Lemmens B, Sagaert X, Henckaerts L et al (2012) Genetic variants in autophagy-related genes and granuloma formation in a cohort of surgically treated Crohn's disease patients. J Crohns Colitis 6(1):43–50, Epub 2012/01/21
16. Correia SC, Santos RX, Perry G, Zhu X, Moreira PI, Smith MA (2012) Mitochondrial importance in Alzheimer's, Huntington's and Parkinson's diseases. Adv Exp Med Biol 724:205–221, Epub 2012/03/14
17. Youle RJ, Narendra DP (2011) Mechanisms of mitophagy. Nat Rev Mol Cell Biol 12(1): 9–14, Epub 2010/12/24
18. Saitsu H, Nishimura T, Muramatsu K, Kodera H, Kumada S, Sugai K et al (2013) De novo mutations in the autophagy gene WDR45 cause static encephalopathy of childhood with neurodegeneration in adulthood. Nat Genet 45(4):445–449, Epub 2013/02/26
19. Cullup T, Kho AL, Dionisi-Vici C, Brandmeier B, Smith F, Urry Z et al (2013) Recessive mutations in EPG5 cause Vici syndrome, a multisystem disorder with defective autophagy. Nat Genet 45(1):83–87, Epub 2012/12/12
20. Inoue J, Nishiumi S, Fujishima Y, Masuda A, Shiomi H, Yamamoto K et al (2012) Autophagy in the intestinal epithelium regulates Citrobacter rodentium infection. Arch Biochem Biophys 521(1–2):95–101, Epub 2012/04/06
21. Saitoh T, Fujita N, Jang MH, Uematsu S, Yang BG, Satoh T et al (2008) Loss of the autophagy protein Atg16L1 enhances endotoxin-induced IL-1beta production. Nature 456(7219): 264–268, Epub 2008/10/14

22. Travassos LH, Carneiro LA, Ramjeet M, Hussey S, Kim YG, Magalhaes JG et al (2010) Nod1 and Nod2 direct autophagy by recruiting ATG16L1 to the plasma membrane at the site of bacterial entry. Nat Immunol 11(1):55–62, Epub 2009/11/10
23. Cooney R, Baker J, Brain O, Danis B, Pichulik T, Allan P et al (2010) NOD2 stimulation induces autophagy in dendritic cells influencing bacterial handling and antigen presentation. Nat Med 16(1):90–97, Epub 2009/12/08
24. Tooze SA, Yoshimori T (2010) The origin of the autophagosomal membrane. Nat Cell Biol 12(9):831–835, Epub 2010/09/03
25. Yang Z, Klionsky DJ (2010) Eaten alive: a history of macroautophagy. Nat Cell Biol 12(9):814–822, Epub 2010/09/03
26. Tsukada M, Ohsumi Y (1993) Isolation and characterization of autophagy-defective mutants of Saccharomyces cerevisiae. FEBS Lett 333(1–2):169–174, Epub 1993/10/25
27. Jung CH, Jun CB, Ro SH, Kim YM, Otto NM, Cao J et al (2009) ULK-Atg13-FIP200 complexes mediate mTOR signaling to the autophagy machinery. Mol Biol Cell 20(7):1992–2003, Epub 2009/02/20
28. Kamada Y, Yoshino K, Kondo C, Kawamata T, Oshiro N, Yonezawa K et al (2010) Tor directly controls the Atg1 kinase complex to regulate autophagy. Mol Cell Biol 30(4):1049–1058, Epub 2009/12/10
29. Cheong H, Nair U, Geng J, Klionsky DJ (2008) The Atg1 kinase complex is involved in the regulation of protein recruitment to initiate sequestering vesicle formation for nonspecific autophagy in Saccharomyces cerevisiae. Mol Biol Cell 19(2):668–681, Epub 2007/12/14
30. Chan EY, Longatti A, McKnight NC, Tooze SA (2009) Kinase-inactivated ULK proteins inhibit autophagy via their conserved C-terminal domains using an Atg13-independent mechanism. Mol Cell Biol 29(1):157–171, Epub 2008/10/22
31. Kageyama S, Omori H, Saitoh T, Sone T, Guan JL, Akira S et al (2011) The LC3 recruitment mechanism is separate from Atg9L1-dependent membrane formation in the autophagic response against Salmonella. Mol Biol Cell 22(13):2290–2300, Epub 2011/04/29
32. Itakura E, Kishi C, Inoue K, Mizushima N (2008) Beclin 1 forms two distinct phosphatidylinositol 3-kinase complexes with mammalian Atg14 and UVRAG. Mol Biol Cell 19(12):5360–5372, Epub 2008/10/10
33. Matsunaga K, Saitoh T, Tabata K, Omori H, Satoh T, Kurotori N et al (2009) Two Beclin 1-binding proteins, Atg14L and Rubicon, reciprocally regulate autophagy at different stages. Nat Cell Biol 11(4):385–396, Epub 2009/03/10
34. Zhong Y, Wang QJ, Li X, Yan Y, Backer JM, Chait BT et al (2009) Distinct regulation of autophagic activity by Atg14L and Rubicon associated with Beclin 1-phosphatidylinositol-3-kinase complex. Nat Cell Biol 11(4):468–476, Epub 2009/03/10
35. Sun Q, Fan W, Chen K, Ding X, Chen S, Zhong Q (2008) Identification of Barkor as a mammalian autophagy-specific factor for Beclin 1 and class III phosphatidylinositol 3-kinase. Proc Natl Acad Sci U S A 105(49):19211–19216, Epub 2008/12/04
36. Matsunaga K, Morita E, Saitoh T, Akira S, Ktistakis NT, Izumi T et al (2010) Autophagy requires endoplasmic reticulum targeting of the PI3-kinase complex via Atg14L. J Cell Biol 190(4):511–521, Epub 2010/08/18
37. Takahashi Y, Coppola D, Matsushita N, Cualing HD, Sun M, Sato Y et al (2007) Bif-1 interacts with Beclin 1 through UVRAG and regulates autophagy and tumorigenesis. Nat Cell Biol 9(10):1142–1151, Epub 2007/09/25
38. Liang C, Feng P, Ku B, Dotan I, Canaani D, Oh BH et al (2006) Autophagic and tumour suppressor activity of a novel Beclin1-binding protein UVRAG. Nat Cell Biol 8(7):688–699, Epub 2006/06/27
39. Liang C, Lee JS, Inn KS, Gack MU, Li Q, Roberts EA et al (2008) Beclin1-binding UVRAG targets the class C Vps complex to coordinate autophagosome maturation and endocytic trafficking. Nat Cell Biol 10(7):776–787, Epub 2008/06/17
40. Mizushima N, Noda T, Yoshimori T, Tanaka Y, Ishii T, George MD et al (1998) A protein conjugation system essential for autophagy. Nature 395(6700):395–398, Epub 1998/10/06

41. Mizushima N, Sugita H, Yoshimori T, Ohsumi Y (1998) A new protein conjugation system in human. The counterpart of the yeast Apg12p conjugation system essential for autophagy. J Biol Chem 273(51):33889–33892, Epub 1998/12/16
42. Fujita N, Saitoh T, Kageyama S, Akira S, Noda T, Yoshimori T (2009) Differential involvement of Atg16L1 in Crohn disease and canonical autophagy: analysis of the organization of the Atg16L1 complex in fibroblasts. J Biol Chem 284(47):32602–32609, Epub 2009/09/29
43. Noda NN, Ohsumi Y, Inagaki F (2010) Atg8-family interacting motif crucial for selective autophagy. FEBS Lett 584(7):1379–1385, Epub 2010/01/20
44. Kabeya Y, Mizushima N, Ueno T, Yamamoto A, Kirisako T, Noda T et al (2000) LC3, a mammalian homologue of yeast Apg8p, is localized in autophagosome membranes after processing. EMBO J 19(21):5720–5728, Epub 2000/11/04
45. Kabeya Y, Mizushima N, Yamamoto A, Oshitani-Okamoto S, Ohsumi Y, Yoshimori T (2004) LC3, GABARAP and GATE16 localize to autophagosomal membrane depending on form-II formation. J Cell Sci 117(pt 13):2805–2812, Epub 2004/06/01
46. Satoo K, Noda NN, Kumeta H, Fujioka Y, Mizushima N, Ohsumi Y et al (2009) The structure of Atg4B-LC3 complex reveals the mechanism of LC3 processing and delipidation during autophagy. EMBO J 28(9):1341–1350, Epub 2009/03/27
47. Weidberg H, Shvets E, Shpilka T, Shimron F, Shinder V, Elazar Z (2010) LC3 and GATE-16/GABARAP subfamilies are both essential yet act differently in autophagosome biogenesis. EMBO J 29(11):1792–1802, Epub 2010/04/27
48. Kimura S, Noda T, Yoshimori T (2008) Dynein-dependent movement of autophagosomes mediates efficient encounters with lysosomes. Cell Struct Funct 33(1):109–122, Epub 2008/04/05
49. Korolchuk VI, Saiki S, Lichtenberg M, Siddiqi FH, Roberts EA, Imarisio S et al (2011) Lysosomal positioning coordinates cellular nutrient responses. Nat Cell Biol 13(4):453–460, Epub 2011/03/12
50. Lavoie JN, Hickey E, Weber LA, Landry J (1993) Modulation of actin microfilament dynamics and fluid phase pinocytosis by phosphorylation of heat shock protein 27. J Biol Chem 268(32):24210–24214, Epub 1993/11/15
51. Tang D, Kang R, Livesey KM, Kroemer G, Billiar TR, Van Houten B et al (2011) High-mobility group box 1 is essential for mitochondrial quality control. Cell Metab 13(6):701–711, Epub 2011/06/07
52. Settembre C, Di Malta C, Polito VA, Garcia Arencibia M, Vetrini F, Erdin S et al (2011) TFEB links autophagy to lysosomal biogenesis. Science 332(6036):1429–1433, Epub 2011/05/28
53. Yu XJ, McGourty K, Liu M, Unsworth KE, Holden DW (2010) pH sensing by intracellular Salmonella induces effector translocation. Science 328(5981):1040–1043, Epub 2010/04/17
54. Patel JC, Hueffer K, Lam TT, Galan JE (2009) Diversification of a Salmonella virulence protein function by ubiquitin-dependent differential localization. Cell 137(2):283–294, Epub 2009/04/22
55. Bakowski MA, Braun V, Lam GY, Yeung T, Heo WD, Meyer T et al (2010) The phosphoinositide phosphatase SopB manipulates membrane surface charge and trafficking of the Salmonella-containing vacuole. Cell Host Microbe 7(6):453–462, Epub 2010/06/15
56. Zheng YT, Shahnazari S, Brech A, Lamark T, Johansen T, Brumell JH (2009) The adaptor protein p62/SQSTM1 targets invading bacteria to the autophagy pathway. J Immunol 183(9):5909–5916, Epub 2009/10/09
57. Thurston TL, Ryzhakov G, Bloor S, von Muhlinen N, Randow F (2009) The TBK1 adaptor and autophagy receptor NDP52 restricts the proliferation of ubiquitin-coated bacteria. Nat Immunol 10(11):1215–1221, Epub 2009/10/13
58. Korioth F, Gieffers C, Maul GG, Frey J (1995) Molecular characterization of NDP52, a novel protein of the nuclear domain 10, which is redistributed upon virus infection and interferon treatment. J Cell Biol 130(1):1–13, Epub 1995/07/01
59. Sternsdorf T, Jensen K, Zuchner D, Will H (1997) Cellular localization, expression, and structure of the nuclear dot protein 52. J Cell Biol 138(2):435–448, Epub 1997/07/28

60. Cemma M, Kim PK, Brumell JH (2011) The ubiquitin-binding adaptor proteins p62/SQSTM1 and NDP52 are recruited independently to bacteria-associated microdomains to target Salmonella to the autophagy pathway. Autophagy 7(3):341–345, Epub 2010/11/17
61. Mostowy S, Sancho-Shimizu V, Hamon M, Simeone R, Brosch R, Johansen T et al (2011) p62 and NDP52 target intracytosolic Shigella and Listeria to different autophagy pathways. J Biol Chem 286(30):26987–26995, Epub 2011/06/08
62. Veiga E, Cossart P (2005) Ubiquitination of intracellular bacteria: a new bacteria-sensing system? Trends Cell Biol 15(1):2–5, Epub 2005/01/18
63. Birmingham CL, Smith AC, Bakowski MA, Yoshimori T, Brumell JH (2006) Autophagy controls Salmonella infection in response to damage to the Salmonella-containing vacuole. J Biol Chem 281(16):11374–11383, Epub 2006/02/24
64. Huett A, Heath RJ, Begun J, Sassi SO, Baxt LA, Vyas JM et al (2012) The LRR and RING domain protein LRSAM1 is an E3 ligase crucial for ubiquitin-dependent autophagy of intracellular Salmonella typhimurium. Cell Host Microbe 12(6):778–790, Epub 2012/12/19
65. Reis BS, Mucida D (2012) The role of the intestinal context in the generation of tolerance and inflammation. Clin Dev Immunol 2012:157948, Epub 2011/09/29
66. Lapaquette P, Glasser AL, Huett A, Xavier RJ, Darfeuille-Michaud A (2010) Crohn's disease-associated adherent-invasive E. coli are selectively favoured by impaired autophagy to replicate intracellularly. Cell Microbiol 12(1):99–113, Epub 2009/09/15
67. Kuballa P, Huett A, Rioux JD, Daly MJ, Xavier RJ (2008) Impaired autophagy of an intracellular pathogen induced by a Crohn's disease associated ATG16L1 variant. PLoS One 3(10):e3391, Epub 2008/10/15
68. McCarroll SA, Huett A, Kuballa P, Chilewski SD, Landry A, Goyette P et al (2008) Deletion polymorphism upstream of IRGM associated with altered IRGM expression and Crohn's disease. Nat Genet 40(9):1107–1112, Epub 2009/01/24
69. Brest P, Lapaquette P, Souidi M, Lebrigand K, Cesaro A, Vouret-Craviari V et al (2011) A synonymous variant in IRGM alters a binding site for miR-196 and causes deregulation of IRGM-dependent xenophagy in Crohn's disease. Nat Genet 43(3):242–245, Epub 2011/02/01
70. Tong Y, Giaime E, Yamaguchi H, Ichimura T, Liu Y, Si H et al (2012) Loss of leucine-rich repeat kinase 2 causes age-dependent bi-phasic alterations of the autophagy pathway. Mol Neurodegener 7:2, Epub 2012/01/11
71. Plowey ED, Cherra SJ III, Liu YJ, Chu CT (2008) Role of autophagy in G2019S-LRRK2-associated neurite shortening in differentiated SH-SY5Y cells. J Neurochem 105(3):1048–1056, Epub 2008/01/10
72. Gardet A, Benita Y, Li C, Sands BE, Ballester I, Stevens C et al (2010) LRRK2 is involved in the IFN-gamma response and host response to pathogens. J Immunol 185(9):5577–5585, Epub 2010/10/06
73. Kumar D, Nath L, Kamal MA, Varshney A, Jain A, Singh S et al (2010) Genome-wide analysis of the host intracellular network that regulates survival of Mycobacterium tuberculosis. Cell 140(5):731–743, Epub 2010/03/10
74. Intemann CD, Thye T, Niemann S, Browne EN, Amanua Chinbuah M, Enimil A et al (2009) Autophagy gene variant IRGM -261T contributes to protection from tuberculosis caused by Mycobacterium tuberculosis but not by M. africanum strains. PLoS Pathog 5(9):e1000577, Epub 2009/09/15
75. Chiodini RJ, Chamberlin WM, Sarosiek J, McCallum RW (2012) Crohn's disease and the mycobacterioses: a quarter century later. Causation or simple association? Crit Rev Microbiol 38(1):52–93, Epub 2012/01/17
76. Orvedahl A, Sumpter R Jr, Xiao G, Ng A, Zou Z, Tang Y et al (2011) Image-based genome-wide siRNA screen identifies selective autophagy factors. Nature 480(7375):113–117, Epub 2011/10/25
77. Matsuda N, Sato S, Shiba K, Okatsu K, Saisho K, Gautier CA et al (2010) PINK1 stabilized by mitochondrial depolarization recruits Parkin to damaged mitochondria and activates latent Parkin for mitophagy. J Cell Biol 189(2):211–221, Epub 2010/04/21

78. Narendra D, Kane LA, Hauser DN, Fearnley IM, Youle RJ (2010) p62/SQSTM1 is required for Parkin-induced mitochondrial clustering but not mitophagy; VDAC1 is dispensable for both. Autophagy 6(8):1090–1106, Epub 2010/10/05
79. Jin SM, Lazarou M, Wang C, Kane LA, Narendra DP, Youle RJ (2010) Mitochondrial membrane potential regulates PINK1 import and proteolytic destabilization by PARL. J Cell Biol 191(5):933–942, Epub 2010/12/01
80. Geisler S, Holmstrom KM, Skujat D, Fiesel FC, Rothfuss OC, Kahle PJ et al (2010) PINK1/Parkin-mediated mitophagy is dependent on VDAC1 and p62/SQSTM1. Nat Cell Biol 12(2):119–131, Epub 2010/01/26
81. Tanaka A, Cleland MM, Xu S, Narendra DP, Suen DF, Karbowski M et al (2010) Proteasome and p97 mediate mitophagy and degradation of mitofusins induced by Parkin. J Cell Biol 191(7):1367–1380, Epub 2010/12/22
82. Gegg ME, Cooper JM, Chau KY, Rojo M, Schapira AH, Taanman JW (2010) Mitofusin 1 and mitofusin 2 are ubiquitinated in a PINK1/parkin-dependent manner upon induction of mitophagy. Hum Mol Genet 19(24):4861–4870, Epub 2010/09/28
83. Ziviani E, Tao RN, Whitworth AJ (2010) Drosophila parkin requires PINK1 for mitochondrial translocation and ubiquitinates mitofusin. Proc Natl Acad Sci U S A 107(11):5018–5023, Epub 2010/03/03
84. Fimia GM, Stoykova A, Romagnoli A, Giunta L, Di Bartolomeo S, Nardacci R et al (2007) Ambra1 regulates autophagy and development of the nervous system. Nature 447(7148): 1121–1125, Epub 2007/06/26
85. Van Humbeeck C, Cornelissen T, Hofkens H, Mandemakers W, Gevaert K, De Strooper B et al (2011) Parkin interacts with ambra1 to induce mitophagy. J Neurosci 31(28): 10249–10261, Epub 2011/07/15
86. Dengjel J, Schoor O, Fischer R, Reich M, Kraus M, Muller M et al (2005) Autophagy promotes MHC class II presentation of peptides from intracellular source proteins. Proc Natl Acad Sci U S A 102(22):7922–7927, Epub 2005/05/17
87. Kasai M, Tanida I, Ueno T, Kominami E, Seki S, Ikeda T et al (2009) Autophagic compartments gain access to the MHC class II compartments in thymic epithelium. J Immunol 183(11):7278–7285, Epub 2009/11/17
88. Schmid D, Pypaert M, Munz C (2007) Antigen-loading compartments for major histocompatibility complex class II molecules continuously receive input from autophagosomes. Immunity 26(1):79–92, Epub 2006/12/22
89. Lee HK, Mattei LM, Steinberg BE, Alberts P, Lee YH, Chervonsky A et al (2010) In vivo requirement for Atg5 in antigen presentation by dendritic cells. Immunity 32(2):227–239, Epub 2010/02/23
90. Ireland JM, Unanue ER (2011) Autophagy in antigen-presenting cells results in presentation of citrullinated peptides to CD4 T cells. J Exp Med 208(13):2625–2632, Epub 2011/12/14
91. Paludan C, Schmid D, Landthaler M, Vockerodt M, Kube D, Tuschl T et al (2005) Endogenous MHC class II processing of a viral nuclear antigen after autophagy. Science 307(5709): 593–596, Epub 2004/12/14
92. Wildenberg ME, Vos AC, Wolfkamp SC, Duijvestein M, Verhaar AP, Te Velde AA et al (2012) Autophagy attenuates the adaptive immune response by destabilizing the immunologic synapse. Gastroenterology 142(7):1493–503.e6, Epub 2012/03/01
93. Bevins CL, Salzman NH (2011) Paneth cells, antimicrobial peptides and maintenance of intestinal homeostasis. Nat Rev Microbiol 9(5):356–368, Epub 2011/03/23
94. Cadwell K, Patel KK, Maloney NS, Liu TC, Ng AC, Storer CE et al (2010) Virus-plus-susceptibility gene interaction determines Crohn's disease gene Atg16L1 phenotypes in intestine. Cell 141(7):1135–1145, Epub 2010/07/07
95. Cadwell K, Liu JY, Brown SL, Miyoshi H, Loh J, Lennerz JK et al (2008) A key role for autophagy and the autophagy gene Atg16l1 in mouse and human intestinal Paneth cells. Nature 456(7219):259–262
96. Cadwell K, Patel KK, Komatsu M, Virgin HW, Stappenbeck TS (2009) A common role for Atg16L1, Atg5 and Atg7 in small intestinal Paneth cells and Crohn disease. Autophagy 5(2):250–252

97. Biswas A, Liu YJ, Hao L, Mizoguchi A, Salzman NH, Bevins CL et al (2010) Induction and rescue of Nod2-dependent Th1-driven granulomatous inflammation of the ileum. Proc Natl Acad Sci U S A 107(33):14739–14744, Epub 2010/08/04
98. Kaser A, Lee AH, Franke A, Glickman JN, Zeissig S, Tilg H et al (2008) XBP1 links ER stress to intestinal inflammation and confers genetic risk for human inflammatory bowel disease. Cell 134(5):743–756, Epub 2008/09/09
99. Yilmaz OH, Katajisto P, Lamming DW, Gultekin Y, Bauer-Rowe KE, Sengupta S et al (2012) mTORC1 in the Paneth cell niche couples intestinal stem-cell function to calorie intake. Nature 486(7404):490–495, Epub 2012/06/23
100. Zhou R, Yazdi AS, Menu P, Tschopp J (2011) A role for mitochondria in NLRP3 inflammasome activation. Nature 469(7329):221–225, Epub 2010/12/03
101. Bulua AC, Simon A, Maddipati R, Pelletier M, Park H, Kim KY et al (2011) Mitochondrial reactive oxygen species promote production of proinflammatory cytokines and are elevated in TNFR1-associated periodic syndrome (TRAPS). J Exp Med 208(3):519–533, Epub 2011/02/02
102. Wen H, Gris D, Lei Y, Jha S, Zhang L, Huang MT et al (2011) Fatty acid-induced NLRP3-ASC inflammasome activation interferes with insulin signaling. Nat Immunol 12(5):408–415, Epub 2011/04/12
103. Plantinga TS, Crisan TO, Oosting M, van de Veerdonk FL, de Jong DJ, Philpott DJ et al (2011) Crohn's disease-associated ATG16L1 polymorphism modulates pro-inflammatory cytokine responses selectively upon activation of NOD2. Gut 60(9):1229–1235, Epub 2011/03/17
104. Lee J, Kim HR, Quinley C, Kim J, Gonzalez-Navajas J, Xavier R et al (2012) Autophagy suppresses interleukin-1beta (IL-1beta) signaling by activation of p62 degradation via lysosomal and proteasomal pathways. J Biol Chem 287(6):4033–4040, Epub 2011/12/15
105. Raju D, Hussey S, Ang M, Terebiznik MR, Sibony M, Galindo-Mata E et al (2012) Vacuolating cytotoxin and variants in Atg16L1 that disrupt autophagy promote Helicobacter pylori infection in humans. Gastroenterology 142(5):1160–1171, Epub 2012/02/16
106. Singh SB, Ornatowski W, Vergne I, Naylor J, Delgado M, Roberts E et al (2010) Human IRGM regulates autophagy and cell-autonomous immunity functions through mitochondria. Nat Cell Biol 12(12):1154–1165, Epub 2010/11/26
107. Parkes M, Barrett JC, Prescott NJ, Tremelling M, Anderson CA, Fisher SA et al (2007) Sequence variants in the autophagy gene IRGM and multiple other replicating loci contribute to Crohn's disease susceptibility. Nat Genet 39(7):830–832, Epub 2007/06/08
108. Behrends C, Sowa ME, Gygi SP, Harper JW (2010) Network organization of the human autophagy system. Nature 466(7302):68–76, Epub 2010/06/22
109. Rozenknop A, Rogov VV, Rogova NY, Lohr F, Guntert P, Dikic I et al (2011) Characterization of the Interaction of GABARAPL-1 with the LIR Motif of NBR1. J Mol Biol 410(3):477–487, Epub 2011/05/31
110. Kraft C, Peter M, Hofmann K (2010) Selective autophagy: ubiquitin-mediated recognition and beyond. Nat Cell Biol 12(9):836–841, Epub 2010/09/03
111. Ong SE, Schenone M, Margolin AA, Li X, Do K, Doud MK et al (2009) Identifying the proteins to which small-molecule probes and drugs bind in cells. Proc Natl Acad Sci U S A 106(12):4617–4622, Epub 2009/03/04
112. Lamb J, Crawford ED, Peck D, Modell JW, Blat IC, Wrobel MJ et al (2006) The Connectivity Map: using gene-expression signatures to connect small molecules, genes, and disease. Science 313(5795):1929–1935, Epub 2006/09/30
113. Zhang L, Yu J, Pan H, Hu P, Hao Y, Cai W et al (2007) Small molecule regulators of autophagy identified by an image-based high-throughput screen. Proc Natl Acad Sci U S A 104(48): 19023–19028, Epub 2007/11/21
114. Balgi AD, Fonseca BD, Donohue E, Tsang TC, Lajoie P, Proud CG et al (2009) Screen for chemical modulators of autophagy reveals novel therapeutic inhibitors of mTORC1 signaling. PLoS One 4(9):e7124, Epub 2009/09/23
115. Sarkar S, Perlstein EO, Imarisio S, Pineau S, Cordenier A, Maglathlin RL et al (2007) Small molecules enhance autophagy and reduce toxicity in Huntington's disease models. Nat Chem Biol 3(6):331–338, Epub 2007/05/09

116. Farkas T, Hoyer-Hansen M, Jaattela M (2009) Identification of novel autophagy regulators by a luciferase-based assay for the kinetics of autophagic flux. Autophagy 5(7):1018–1025, Epub 2009/08/05
117. Fleming A, Noda T, Yoshimori T, Rubinsztein DC (2011) Chemical modulators of autophagy as biological probes and potential therapeutics. Nat Chem Biol 7(1):9–17, Epub 2010/12/18
118. Sarkar S, Korolchuk VI, Renna M, Imarisio S, Fleming A, Williams A et al (2011) Complex inhibitory effects of nitric oxide on autophagy. Mol Cell 43(1):19–32, Epub 2011/07/06
119. Kang R, Livesey KM, Zeh HJ 3rd, Loze MT, Tang D (2011) Metabolic regulation by HMGB1-mediated autophagy and mitophagy. Autophagy 7(10):1256–1258, Epub 2011/06/22
120. Chen G, Xia H, Cai Y, Ma D, Yuan J, Yuan C (2011) Synthesis and SAR study of diphenylbutylpiperidines as cell autophagy inducers. Bioorg Med Chem Lett 21(1):234–239, Epub 2010/12/04
121. Chou TF, Brown SJ, Minond D, Nordin BE, Li K, Jones AC et al (2011) Reversible inhibitor of p97, DBeQ, impairs both ubiquitin-dependent and autophagic protein clearance pathways. Proc Natl Acad Sci U S A 108(12):4834–4839, Epub 2011/03/09
122. Miller S, Tavshanjian B, Oleksy A, Perisic O, Houseman BT, Shokat KM et al (2010) Shaping development of autophagy inhibitors with the structure of the lipid kinase Vps34. Science 327(5973):1638–1642, Epub 2010/03/27

Chapter 13
The Epithelial Barrier

Stefan Koch, Asma Nusrat, and Charles A. Parkos

Abstract The intestinal epithelium is the body's first line of defense against harmful contents of the gut, and defects in the epithelial barrier are thought to contribute to the initiation and perpetuation of inflammatory bowel diseases. Recent genome-wide association studies have identified a number of mutations in genes implicated in the regulation of the intestinal epithelium, which may result in barrier dysfunction and thereby predispose to the development of IBD. In this chapter, we will review the role of the epithelial barrier in the pathogenesis of intestinal inflammation and introduce relevant animal models that link epithelial barrier defects to increased colitis susceptibility and IBD susceptibility genes that are associated with epithelial barrier regulation.

The intestinal epithelium consists of a cohesive monolayer of epithelial cells that separate the content of the gut lumen from underlying tissues. The epithelium has two major functions; it absorbs nutrients and water from the digestive tract, while at the same time acting as an impermeable barrier for potentially harmful foreign materials, such as bacteria and viruses. To maintain stringent barrier function despite continuous antigen exposure and mechanical stress, intestinal epithelial cells (IEC) are constantly being replenished by a small pool of highly proliferative stem cells at the base of epithelial crypts. The progeny of these intestinal stem cells differentiates

S. Koch, Ph.D.
Department of Pathology and Laboratory Medicine, Epithelial Pathobiology Unit, Emory University, Atlanta, GA, USA

Division of Molecular Embryology, German Cancer Research Center (DKFZ), Heidelberg, Germany

A. Nusrat, M.D. • C.A. Parkos, M.D., Ph.D. (✉)
Department of Pathology and Laboratory Medicine, Epithelial Pathobiology and Mucosal Inflammation Research Unit, Emory University, 615 Michael Street, Room 105B, Atlanta, GA 30322, USA
e-mail: cparkos@emory.edu

M. D'Amato and J.D. Rioux (eds.), *Molecular Genetics of Inflammatory Bowel Disease*,
DOI 10.1007/978-1-4614-8256-7_13, © Springer Science+Business Media New York 2013

into specialized epithelial cells as cells migrate along the crypt-surface axis, and senescent IEC at the surface tip are ultimately shed into the gut lumen. Thus, the intestinal epithelium undergoes complete renewal every 5–7 days, without compromising barrier integrity.

Structure and Function of the Epithelial Tight Junction Complex

IEC are anchored to one another and to the surrounding connective tissue through various transmembrane proteins, which are clustered in distinct cell–cell and cell-matrix adhesion structures and connect to the cytoskeleton of the cell (see Fig. 13.1). Among these, the apical junctional complex—consisting of the tight junction at the apex of the lateral cell membrane and the more basally located adherens junction—is of critical importance for the regulation of paracellular solute flux and cell migration [1, 2]. In particular, claudin proteins in the epithelial tight junction form a belt-like continuous barrier around the cell, which restricts water and small molecule movement from the tissue into the lumen and vice versa. To date, 27 claudin protein family members have been identified [3]. Based on their preference to promote or restrict paracellular permeability, several claudins have been separated into "leaky" or "tight" functional groups, and their respective expression pattern is thought to confer tissue-specific barrier properties (e.g., low permeability in the epidermis and high permeability in the kidney). Consequently, genetic deletion of individual claudins in transgenic mice is typically associated with epithelial barrier dysfunction and—in some cases—chronic inflammation, although functionally related claudins can often compensate the loss of just one family member [4].

Epithelial Barrier Defects in IBD Pathogenesis

Considering the crucial importance of the epithelial barrier in the regulation of intestinal homeostasis, it would be expected that preexisting barrier dysfunction can result in a pronounced inflammatory response caused by increased antigen translocation across the epithelium. Indeed, numerous studies have demonstrated increased intestinal permeability in IBD patients [5–8], which suggests that epithelial barrier function is severely compromised in intestinal inflammation. The results present a chicken-and-egg problem, however: is intestinal inflammation triggered by increased transepithelial antigen translocation caused by a congenital barrier defect, or is barrier function in IBD patients secondary to mucosal leukocyte infiltration and cytokine secretion (see Fig. 13.2)? The answer, more likely than not, is "both." On the one hand, it is well known that various inflammatory mediators increase transepithelial permeability and disrupt epithelial homeostasis. For example, the prominent pro-inflammatory cytokines interleukin IL-4, interferon IFN-γ, and tumor necrosis factor TNF-α, whose expression is upregulated in the intestinal mucosa of

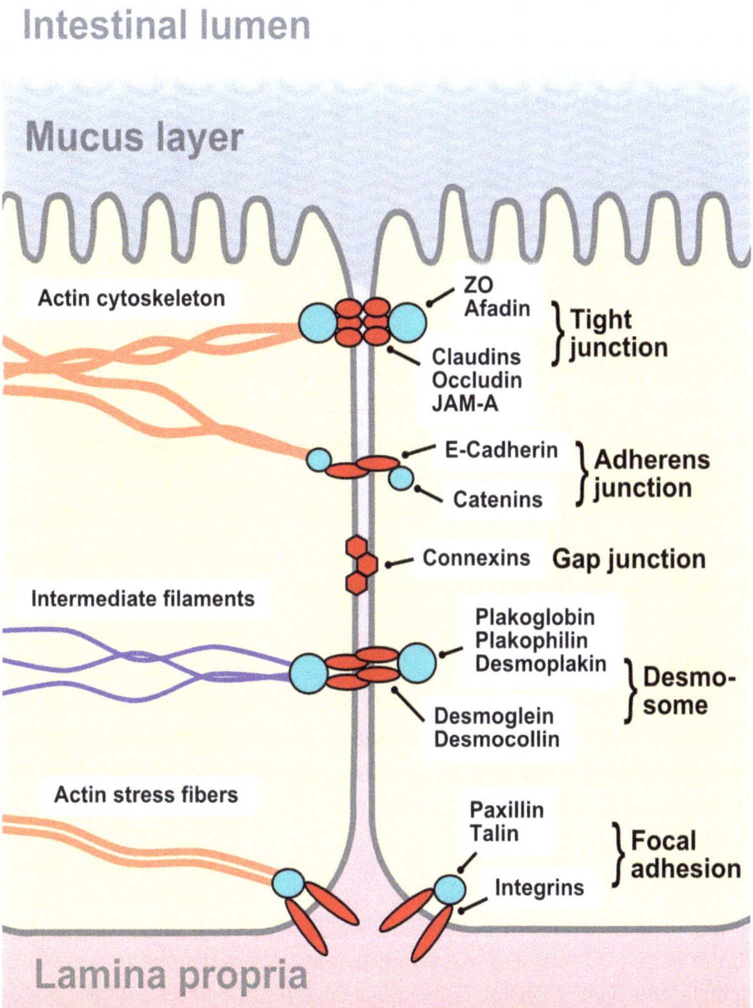

Fig. 13.1 Schematic representation of epithelial junctional complexes. Transmembrane adhesion molecules are shown in *red*; intracellular scaffold proteins are represented in *blue*

persons with IBD, impair barrier function both by inducing aberrant IEC apoptosis and by disassembling the apical junctional complex [9]. Studies from our laboratory and others have shown that key tight junction and adherens junction molecules, including E-cadherin, JAM-A, and occludin, are internalized or degraded during active intestinal inflammation [2]. In addition, it is thought that chronic IBD is associated with a switch from a "tight" to a "leaky" claudin expression profile in epithelial cells, which exacerbates barrier dysfunction and perpetuates inflammation. For example, it has been shown that in active Crohn's disease (CD) mucosa the pore-forming claudin-2 is upregulated, whereas the sealing claudin-5 and claudin-8 are downregulated and removed from epithelial tight junctions [10].

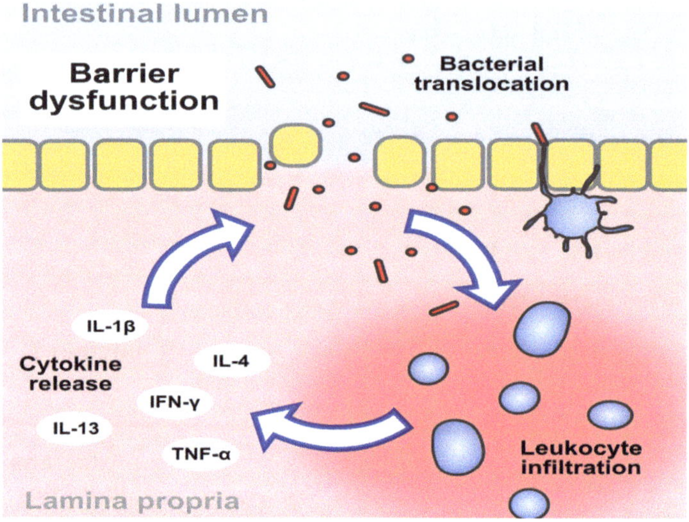

Fig. 13.2 Epithelial barrier defects contribute to the pathogenesis of intestinal inflammation. Breaches in the epithelial barrier result in bacterial translocation into the lamina propria, which causes leukocytes infiltration and the release of inflammatory cytokines. Many cytokines, including IL-1β, IL-4, and IFN-γ increase epithelial permeability, thereby perpetuating the cycle of inflammation

On the other hand, there is evidence that at least in some cases, barrier defects may precede leukocyte infiltration and overt inflammation. Daniel Hollander and colleagues [11] found that healthy relatives of CD patients had increased transepithelial small molecule permeability, which was much higher than in controls. Although this observation raises the possibility that an inherited barrier defect may increase IBD susceptibility, the findings have remained controversial because no such effect was observed in subsequent studies [12–14]. Interestingly, however, Soderholm et al. [15] reported that baseline intestinal permeability was elevated in CD patients and their spouses, rather than their first-degree relatives. In contrast, after acetylsalicylic acid administration, small molecule flux was strongly increased in patients and relatives, whereas spouses responded like healthy controls. It is thus possible that normal transepithelial permeability is determined by environmental factors such as nutrition and that genetic predisposition alters the response of the intestinal epithelium to luminal or mucosal stimuli. However, this intriguing hypothesis remains to be further investigated in future studies.

Animal Models of Intestinal Epithelial Barrier Dysfunction

The remarkable clinical heterogeneity of chronic intestinal inflammation and the limited tools to study epithelial permeability defects in the human gut make it challenging to evaluate the contribution of barrier dysregulation to IBD etiology.

Table 13.1 Select transgenic mouse models of epithelial barrier dysfunction and intestinal inflammation

Model	Intestinal phenotype	AJC molecular changes	Reference(s)
SAMP1/YitFc	Spontaneous chronic ileitis preceded by increased intestinal permeability	Claudin-2 ↑; occludin ↓	[16, 17]
Il10$^{-/-}$	Spontaneous severe enterocolitis preceded by increased intestinal permeability	Not determined	[19, 20]
p120-catenin$^{-/-\,a}$	Spontaneous severe enterocolitis caused by epithelial barrier disruption	E-cadherin ↓; α-catenin ↓; β-catenin ↓	[21]
JAM-A$^{-/-}$	Enlarged lymphoid follicles and enhanced susceptibility to experimental colitis associated with intestinal barrier defect	Claudin-10 ↑; claudin-15 ↑	[24]
Muc2$^{-/-}$	Spontaneous chronic colitis associated with epithelial barrier dysfunction	Claudin-10 ↑; claudin-1 ↓; claudin-5 ↓	[29]
Hnf4α$^{-/-\,a}$	Enhanced susceptibility to experimental colitis associated with intestinal barrier defect	Claudin-2 ↑; claudin-7 ↓; ZO-1 ↓	[35, 36]
Mlck transgene[a]	Enhanced susceptibility to experimental colitis associated with intestinal barrier defect	None detected	[38]

AJC apical junctional complex
[a]Intestine-specific

Considerable interest has been paid to animal models of barrier dysfunction, and there is growing evidence that innate defects in the epithelial barrier increase susceptibility to intestinal inflammation. Although no perfect in vivo model for human IBD has been described to date, results from multiple lines of investigation (summarized in Table 13.1 and discussed below) point toward a common pathological phenotype resulting from diverse genetic defects in IEC homeostatic pathways.

SAMP1/YitFc Mice

One of the most intriguing rodent models of intestinal inflammation is the SAMP1/YitFc (SAMP) mouse, which was derived by repeated brother–sister mating of senescence-accelerated mice (SAM). Matsumoto and colleagues [16] observed that in addition to frequent skin ulceration described in earlier studies, SAMP mice develop spontaneous inflammatory skip lesions in the ileum and cecum, with striking similarities to CD. Disease onset was dependent on the enteric microflora, as germfree mice showed no signs of inflammation, whereas enteritis was inducible by colonization with commensals. More recently, work from Theresa Pizzaro's laboratory showed that SAMP mice exhibit a pronounced intestinal epithelial barrier defect, which precedes the development of overt intestinal inflammation [17].

When compared to the founder AKR mouse strain, SAMP mice have a significantly higher mRNA expression of the "leaky" claudin-2 and, conversely, reduced expression of occludin. Interestingly, barrier dysfunction was also observed in germfree mice, which strongly suggests that the primary epithelial barrier defect is causally linked to the development of intestinal inflammation in these animals. Although the underlying reason for barrier dysfunction in SAMP mice remains to be determined, genetic analyses have implicated mutations in several apical junctional complex-associated genes, including *Cldn2* (claudin-2), *Cdh1* (E-cadherin), and *Mllt4* (afadin), which alone or in combination may compromise junction integrity [18].

Il10-Deficient Mice

Similar to SAMP mice, mice with targeted deletion of the *Il10* gene spontaneously develop intestinal inflammation at around 10 weeks after birth [19]. However, inflammation in these animals extends throughout most of the intestinal tract, and local inflammation restricted to the proximal colon can also be observed in germfree mice, indicative of a more severe disease phenotype. Importantly, mutations in the IL10 pathway are strongly associated with increased susceptibility to IBD in humans, consistent with reduced anti-inflammatory signaling. Although IL10 deficiency primarily affects the function of the innate immune system, there is evidence that loss of IL10 also results in a pronounced epithelial barrier defect. Madsen et al. [20] reported that *Il10*-deficient mice have microbiota-dependent, increased intestinal permeability which was observed weeks before the onset of inflammation. Taken together, these data suggest that barrier dysfunction contributes to disease development in animals with compromised immune homeostasis.

p120-Catenin-Deficient Mice

The stability of epithelial adherens junctions is in part regulated by the catenin protein family, which are intracellular binding partners of E-cadherin. As reported recently, mice deficient for p120-catenin in IEC suffer from early-onset, wasting enterocolitis, with notable fragility of the epithelial monolayer [21]. Structural analysis revealed increased intestinal permeability and neutrophil recruitment in transgenic mice, which was caused by a loss of E-cadherin, as well as α-catenin and β-catenin from the IEC lateral membrane. Thus, genetic impairment of adherens junction integrity results in a breakdown of the epithelial barrier and catastrophic inflammation.

JAM-A-Deficient Mice

Junctional adhesion molecule JAM-A is a tight junction-associated adhesion molecule with important roles in mediating cell–cell contacts, cell migration, and epithelial cell proliferation [22, 23]. Studies from our laboratory have shown that mice with genomic deletion of the JAM-A coding gene (*F11r*) exhibit increased intestinal permeability and bacterial translocation across the epithelium [24]. Unlike the transgenic animals introduced above, *F11r*$^{-/-}$ mice do not develop spontaneous colitis; however, we observed that JAM-A-deficient mice show signs of subclinical mucosal immune activation, as indicated by enlarged lymphoid follicles and higher myeloperoxidase activity resulting from an increased number of lamina propria neutrophils. In addition, JAM-A-deficient mice were found to be more susceptible to chemically induced colitis, with an earlier disease onset, more severe tissue damage, and increased mortality compared to wild-type controls. Structurally, we observed that IEC from JAM-A-deficient mice had increased expression of claudin-10 and claudin-15, indicative of a leaky epithelial barrier.

Of importance to human pathology, although no JAM-A (*F11R*) defect has been observed in IBD patients, it has been shown that mutations in the *MAGI2* gene are associated with the development of ulcerative colitis [25]. MAGI proteins are PDZ domain-containing scaffold proteins that localize to the tight junction and promote the formation of intracellular signaling complexes, which are at least in part stabilized by JAM family members. It is thus feasible that loss of JAM-associated proteins may compromise the epithelial barrier and increase IBD susceptibility.

Mouse Models of Mucus Layer Defects

The intestinal epithelial barrier is itself protected by a layer of mucus secreted by a specialized type of IEC, the goblet cell. The mucus gel physically limits access of bacteria to surface IEC and stores antimicrobial peptides mainly derived from small intestinal Paneth cells, thereby creating a comparatively germfree environment in the immediate vicinity of the intestinal epithelium. It is thus feasible that alterations in the mucus layer may contribute to IBD pathogenesis, and indeed, changes in mucin deposition and glycosylation have been observed in CD and ulcerative colitis (UC) patients, as well as in some of their healthy relatives [26]. Importantly, a recent study additionally identified goblet cells as an interface for immune cell education [27]. McDole et al. observed that goblet cells in the small intestine act as passages for luminal antigens, which are processed by tolerogenic dendritic cells in the lamina propria.

In agreement with these reports, several recent studies have shown that mice with various genetic defects in goblet cell function and mucus secretion or assembly are prone to intestinal inflammation. For example, mice deficient for mucin 2, the major component of the intestinal mucus layer, spontaneously develop mild colitis and are exceptionally susceptible to chemically induced inflammation [28]. Loss of mucus integrity in these animals was found to be associated with a pronounced epithelial barrier defect, as evidenced by increased claudin-10 mRNA expression and decreased claudin-1 and claudin-5 message levels [29]. Interestingly, aberrant mucin 2 production was also observed in *Il10*$^{-/-}$ mice, suggesting that loss of IL-10 signaling results in a multifactorial imbalance in mucosal homeostasis [30]. Similar to *Muc2* gene-deficient animals, mice lacking core-1- and core-3-derived O-glycans (i.e., mucin-bound oligosaccharide side chains), or the mucin 2-modifying enzyme AGR2, exhibit varying degrees of colitis susceptibility [31–33]. Although the exact nature of the barrier defect in these animal models remains to be determined, collectively the studies suggest that multiple genetic defects converging on mucus secretion and modification can compromise epithelial barrier function, resulting in enhanced bacterial translocation and increased intestinal inflammation.

HNF4α-Deficient Mice

A somewhat unexpected candidate gene for the development of IBD, *HNF4A*, coding for the transcription factor hepatocyte nuclear factor (HNF)-4α, was recently identified in genome-wide association scans (GWASs) [34]. HNF-4α is strongly expressed in the intestinal epithelium and has been implicated in downregulation of epithelial cell proliferation pathways controlled by Wnt/β-catenin signaling. In particular, it has been shown that IEC-specific deletion of HNF-4α increases epithelial turnover and IEC differentiation into goblet cells [35, 36]. In parallel, Cattin et al. [36] reported that *Hnf4α*-deficient mice exhibit increased intestinal permeability, which is associated with increased claudin-2 expression and reduced claudin-7 and zonula occludens [37] -1 levels. HNF-4α may thus indirectly control intestinal permeability, through transcriptional regulation of epithelial differentiation pathways.

MLCK Transgenic Mice

Additional support for the hypothesis that an innate barrier defect promotes intestinal inflammation comes from a report on mice with IEC-specific over-expression of a constitutively active myosin light chain kinase construct (CA-MLCK) [38]. Myosin-dependent contractility of the perijunctional actin belt is an important mechanism regulating tight junction integrity and, consequently, activation of MLCK results in partial disassembly of the apical junctional complex [2]. Similar to *F11r* and *Hnf4α*-deficient animals, CA-MLCK transgenic mice do not spontaneously develop colitis, but show signs of heightened immune activation in the mucosa and are susceptible to experimental colitis.

IBD Susceptibility Genes Implicated in Epithelial Barrier Regulation

As we have seen, a multitude of studies on transgenic animals suggest that single genetic mutations are sufficient to upset fine-tuning of homeostatic regulation of the intestinal epithelial barrier that may promote development and perpetuation of

Table 13.2 Select susceptibility loci and candidate genes for IBD with a possible role in epithelial barrier regulation, identified in genome-wide association scans (GWAS)

Locus	Candidate gene (protein)	Function in intestinal epithelial cells	Reference(s)
16q22	*CDH1* (E-cadherin)	Principal cell adhesion molecule of the epithelial adherens junction	[34]
1q21	*ECM1* (extracellular matrix protein 1)	Secreted glycoprotein that contributes to cell proliferation	[41, 43]
7p22	*GNA12* (guanine nucleotide-binding protein α12)	Inhibits tight junction assembly by phosphorylation of ZO proteins through Src and HSP90	[37, 44]
20q13	*HNF4A* (hepatocyte nuclear factor 4α)	Transcription factor that regulates differentiation along the intestinal crypt-surface axis	[34, 36]
1q23	*ITLN1* (intelectin-1)	Lectin thought to protect the brush border membrane	[54, 55]
7q31	*LAMB1* (laminin β1)	Secreted protein involved in enterocyte differentiation	[34, 59]
18p11	*PTPN2* (protein tyrosine phosphatase N2)	Inhibits IFN-γ-induced expression of claudin-2	[56, 57]
7q22	*MUC3A* (mucin 3A)	Essential glycoprotein in the protective mucus layer	[52, 60]
12q12	*MUC19* (mucin 19)	Essential glycoprotein in the protective mucus layer	[51]

chronic intestinal inflammation. The obvious questions, of course, are if and how these findings translate to human disease and what lessons we can learn from the animal models. We further discuss here possible roles of recently identified IBD susceptibility genes—summarized in Table 13.2—in the regulation of intestinal epithelial homeostasis and barrier function. Notably, many of the mutations found in humans result in similar phenotypes observed in transgenic mice and suggest that animal models are valuable aids in studies of IBD pathobiology.

CDH1 (E-cadherin)

E-cadherin is the main adhesion molecule in the epithelial adherens junction. Loss of E-cadherin expression results in catastrophic failure of the epithelial barrier, and downregulation of E-cadherin can be seen in active IBD mucosa [39]. A recent GWAS identified a new UC susceptibility region harboring, among others, the *CDH1* gene encoding E-cadherin [34]. Although complete loss of function is unlikely because deletion of E-cadherin is incompatible with life, it is possible that genomic mutations of *CDH1* result in a functionally restricted form of the protein that compromises epithelial barrier function. Indeed, Muise and colleagues [40] reported that a common polymorphism in *CDH1* results in a truncated E-cadherin isoform that does not correctly localize to the plasma membrane. Importantly, these authors additionally found that this mutation is a risk allele for CD, but interestingly, it was not observed for UC. Nevertheless, these findings indicate that functional changes in E-cadherin impair the epithelial barrier and promote intestinal inflammation.

Extracellular Matrix Protein 1

Extracellular matrix protein 1 (ECM1) is a secreted glycoprotein implicated in number of biological processes, including endothelial cell proliferation and angiogenesis; however, its function in epithelial homeostasis is less well understood. It has been reported that expression of ECM1 is enhanced in epithelial cancers and that it promotes tumor growth through as yet unknown mechanisms [41, 42]. The genomic cluster containing the *ECM1* gene has been identified as an UC susceptibility region, and it has been suggested that mutations in *ECM1* may impair epithelial homeostasis, potentially by reduced activation of NF-κB signaling pathways [43]. Considering that NF-κB is a major regulator of the mucosal immune response and cell survival pathways, it is thus possible that loss of ECM1 may increase cytokine-induced epithelial cell apoptosis and, conversely, reduce proliferation; however, this hypothesis remains to be addressed.

GNA12 (Guanine Nucleotide-Binding Protein α12)

Another recently discovered UC-risk gene is *GNA12*, encoding the heterotrimeric G protein α12 (Gα12) [37]. G proteins are crucial signaling mediators, which relay signals from cell surface receptors to intracellular molecules, but little is known about how Gα12 may regulate epithelial cell homeostasis. It has been shown that Gα12 associates with the tight junction scaffold protein ZO-1 and that it inhibits tight junction integrity in kidney epithelial cells by activation of tyrosine kinase Src, which in turn disrupts the interaction of ZO-1 with claudin-1 and occludin [44]. Interestingly, the authors found that Gα12 itself is activated by heat shock protein (HSP) 90, a stress response chaperone whose expression is increased during inflammation. Recently, these authors reported that reactive oxygen species, which have been shown to accelerate epithelial regeneration [45], activate Gα12 and disrupt the barrier function of kidney epithelial cells [46]. In light of these observations, it appears counterintuitive that *GNA12* mutations might confer susceptibility for chronic colitis. However, it is possible that Gα12 may have different functions in IEC or that the polymorphisms in UC patients increase its activity.

HNF4A (Hepatocyte Nuclear Factor-4α)

As we have discussed in the previous section, HNF-4α is a transcription factor that regulates epithelial homeostasis and barrier function in the intestine. The specific function of HNF-4α in the human remains enigmatic at this time; however, studies using inducible gene-deficient mice revealed that HNF-4α regulates IEC homeostasis by Wnt/β-catenin signaling-dependent modulation of cell proliferation and differentiation [36], which may—directly or indirectly—have potent effects on regulation of barrier function.

MUC3A and MUC19 (Mucin 3A and 19)

Considering the strong evidence from animal models that defective goblet cell differentiation and epithelial mucin secretion may predispose to the development of chronic intestinal inflammation, it would be expected that impaired goblet cell function may similarly contribute to human IBD. Indeed, it has been noted that mucin composition changes during inflammation, with increased secretion of mucins 1, 2, and 5 and reduced expression of mucins 9 and 17 [47–50]. Recent genome-wide association studies identified rare coding mutation in *MUC19* as a risk factor for Crohn's disease and ulcerative colitis [51]. In addition, Kyo et al. discovered multiple single-nucleotide polymorphisms in the *MUC3A* gene that are weakly associated with the development of Crohn's disease and ulcerative colitis [52, 53]. The specific function and expression of these mucins has not been investigated to date; however, considering the critical role of the mucus layer in the protection of the intestinal epithelium, it is not surprising that even minor alterations in goblet cell function may contribute to IBD pathobiology.

Intelectin-1, Laminin β1, and Protein Tyrosine Phosphatase N2

Multiple additional risk factors identified in genome-wide association studies are thought to protect the intestinal epithelial barrier; however, their specific roles are somewhat enigmatic. Intelectin-1 (ITLN1) is a D-galactosyl-specific lectin expressed in Paneth and goblet cells [54] that was recently found to increase susceptibility for Crohn's disease [55]. Although intelectin-1 has been suggested to promote epithelial barrier function, how this is achieved is poorly understood. Most likely intelectin-1 binds bacteria at the brush border membrane and thus restricts access of potential pathogens to IEC. Alternatively, it is possible that intelectin-1 physically stabilizes the apical membrane of enterocytes, which prevents the release of digestive enzymes into the gut lumen. Laminins are integral structural proteins in the extracellular matrix, and *LAMB1* mutations are associated with increased risk of ulcerative colitis [34]. The effect of loss of function of laminin β1 (LAMB1) in the intestine has not been investigated at this point. Considering the function of laminins as anchoring proteins for epithelial cells, it is conceivable that *LAMB1* mutations weaken the attachment of the epithelium to the basal lamina and thereby drive sloughing off of the mucosa in inflamed tissue. Finally, *PTPN2* has been identified as a risk allele for Crohn's disease [56]. Scharl et al. [57] found that protein tyrosine phosphatase N2 (PTPN2) is induced by IFN-γ and that it attenuates IFN-γ-mediated activation of STAT 1 and 2. Consequently, loss of *PTPN2* results in increased transepithelial permeability and may be involved in the perpetuation of intestinal inflammation.

Concluding Remarks

Taken together, the above observations strongly support a model in which genetic alterations of proteins involved in the regulation of the epithelial barrier promote the activation of the mucosal immune system, most likely by allowing more luminal antigens to cross into the lamina propria. It is interesting to note that these proteins encompass a wide spectrum of functions, ranging from cell–cell adhesion molecules to cytokines and transcription factors, and that their loss results in a remarkably similar intestinal phenotype (see Fig. 13.3). Not all of the mutations investigated in transgenic animal models are associated with overt inflammation, but rather

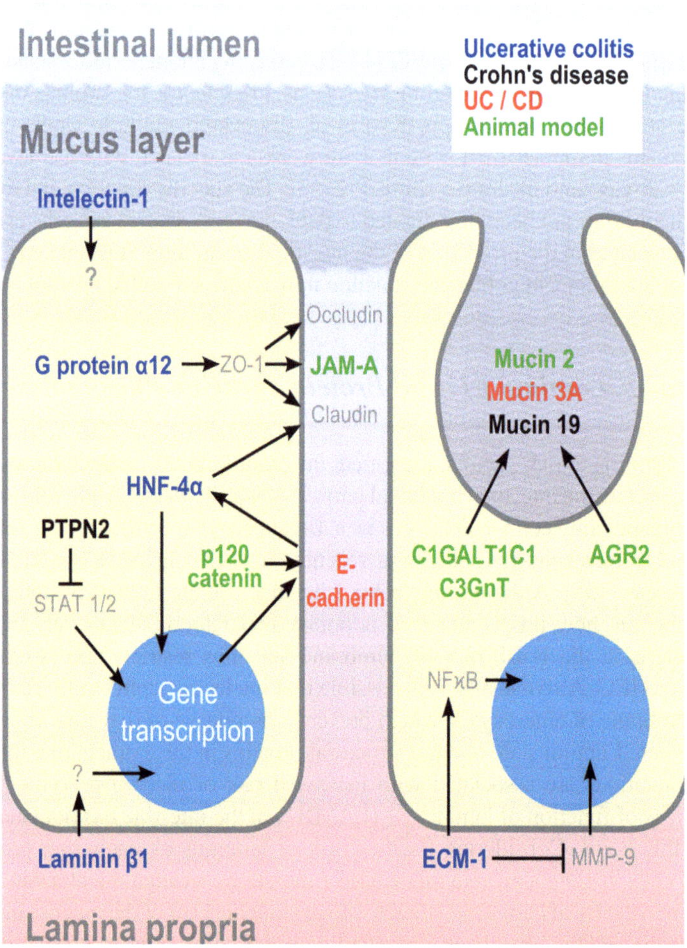

Fig. 13.3 Schematic representation of IBD susceptibility genes, implicated in regulating the epithelial barrier. *Arrows* indicate potential functional connections to intracellular signaling pathways and other structural proteins

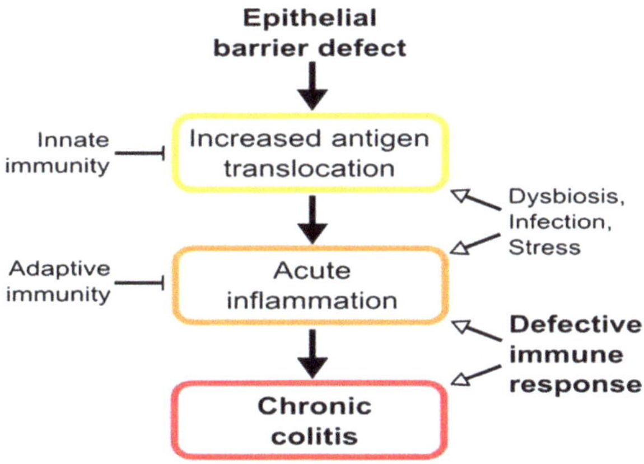

Fig. 13.4 Proposed model of the progression from primary intestinal barrier dysfunction to chronic colitis. Innate and adaptive immunity dampen the pathological response to increased mucosal antigen translocation. Additional defects in the immune system result in chronic inflammation

appear to universally increase the susceptibility to colitis. This observation is compatible with the hypothesis that in most cases, the development of IBD requires two "hits," for example, a host immune defect in conjunction with intestinal dysbiosis [58]. It appears that even when the integrity of the epithelial barrier is impaired, both innate and adaptive immunities are able to suppress inflammation (see Fig. 13.4). Like man, mice may thus be able to compensate to some extent for the loss of barrier-promoting proteins, but will be afflicted with intestinal inflammation if mucosal homeostasis is compromised further.

Also of note is the fact that most of the known susceptibility genes related to epithelial barrier function appear to increase the risk for developing ulcerative colitis, but are not associated with Crohn's disease. It is therefore feasible to speculate that a primary barrier defect can initiate the extensive crypt erosion and superficial inflammation seen in UC, but may not contribute to the striking transmural inflammatory phenotype found in CD. If this is the case, treatment aimed specifically at restoring epithelial barrier function may emerge as a valuable tool for inducing and sustaining remission in ulcerative colitis patients.

References

1. Chin AC, Parkos CA (2007) Pathobiology of neutrophil transepithelial migration: implications in mediating epithelial injury. Annu Rev Pathol 2:111–143, Epub 2007/11/28
2. Ivanov AI, Parkos CA, Nusrat A (2010) Cytoskeletal regulation of epithelial barrier function during inflammation. Am J Pathol 177(2):512–524, Epub 2010/06/29
3. Mineta K, Yamamoto Y, Yamazaki Y, Tanaka H, Tada Y, Saito K et al (2011) Predicted expansion of the claudin multigene family. FEBS Lett 585(4):606–612, Epub 2011/02/01
4. Van Itallie CM, Anderson JM (2006) Claudins and epithelial paracellular transport. Annu Rev Physiol 68:403–429, Epub 2006/02/08
5. Bjarnason I, O'Morain C, Levi AJ, Peters TJ (1983) Absorption of 51chromium-labeled ethylenediaminetetraacetate in inflammatory bowel disease. Gastroenterology 85(2):318–322, Epub 1983/08/01

6. Magnusson KE, Sundqvist T, Sjodahl R, Tagesson C (1983) Altered intestinal permeability to low-molecular-weight polyethyleneglycols (PEG 400) in patients with Crohn's disease. Acta Chir Scand 149(3):323–327, Epub 1983/01/01
7. Pearson AD, Eastham EJ, Laker MF, Craft AW, Nelson R (1982) Intestinal permeability in children with Crohn's disease and coeliac disease. Br Med J (Clin Res Ed) 285(6334):20–21, Epub 1982/07/03
8. Ukabam SO, Clamp JR, Cooper BT (1983) Abnormal small intestinal permeability to sugars in patients with Crohn's disease of the terminal ileum and colon. Digestion 27(2):70–74, Epub 1983/01/01
9. Koch S, Nusrat A (2012) The life and death of epithelia during inflammation: lessons learned from the gut. Annu Rev Pathol 7:35–60, Epub 2011/08/16
10. Zeissig S, Burgel N, Gunzel D, Richter J, Mankertz J, Wahnschaffe U et al (2007) Changes in expression and distribution of claudin 2, 5 and 8 lead to discontinuous tight junctions and barrier dysfunction in active Crohn's disease. Gut 56(1):61–72, Epub 2006/07/11
11. Hollander D, Vadheim CM, Brettholz E, Petersen GM, Delahunty T, Rotter JI (1986) Increased intestinal permeability in patients with Crohn's disease and their relatives. A possible etiologic factor. Ann Intern Med 105(6):883–885, Epub 1986/12/01
12. Teahon K, Smethurst P, Levi AJ, Menzies IS, Bjarnason I (1992) Intestinal permeability in patients with Crohn's disease and their first degree relatives. Gut 33(3):320–323, Epub 1992/03/01
13. Munkholm P, Langholz E, Hollander D, Thornberg K, Orholm M, Katz KD et al (1994) Intestinal permeability in patients with Crohn's disease and ulcerative colitis and their first degree relatives. Gut 35(1):68–72, Epub 1994/01/01
14. Ainsworth M, Eriksen J, Rasmussen JW, Schaffalitzky de Muckadell OB (1989) Intestinal permeability of 51Cr-labelled ethylenediaminetetraacetic acid in patients with Crohn's disease and their healthy relatives. Scand J Gastroenterol 24(8):993–998
15. Soderholm JD, Olaison G, Lindberg E, Hannestad U, Vindels A, Tysk C et al (1999) Different intestinal permeability patterns in relatives and spouses of patients with Crohn's disease: an inherited defect in mucosal defence? Gut 44(1):96–100, Epub 1998/12/24
16. Matsumoto S, Okabe Y, Setoyama H, Takayama K, Ohtsuka J, Funahashi H et al (1998) Inflammatory bowel disease-like enteritis and caecitis in a senescence accelerated mouse P1/Yit strain. Gut 43(1):71–78, Epub 1998/10/15
17. Olson TS, Reuter BK, Scott KG, Morris MA, Wang XM, Hancock LN et al (2006) The primary defect in experimental ileitis originates from a nonhematopoietic source. J Exp Med 203(3): 541–552, Epub 2006/03/01
18. Reuter BK, Pizarro TT (2009) Mechanisms of tight junction dysregulation in the SAMP1/YitFc model of Crohn's disease-like ileitis. Ann N Y Acad Sci 1165:301–307, Epub 2009/06/23
19. Kuhn R, Lohler J, Rennick D, Rajewsky K, Muller W (1993) Interleukin-10-deficient mice develop chronic enterocolitis. Cell 75(2):263–274, Epub 1993/10/22
20. Madsen KL, Malfair D, Gray D, Doyle JS, Jewell LD, Fedorak RN (1999) Interleukin-10 gene-deficient mice develop a primary intestinal permeability defect in response to enteric microflora. Inflamm Bowel Dis 5(4):262–270, Epub 1999/12/01
21. Smalley-Freed WG, Efimov A, Burnett PE, Short SP, Davis MA, Gumucio DL et al (2010) p120-catenin is essential for maintenance of barrier function and intestinal homeostasis in mice. J Clin Invest 120(6):1824–1835, Epub 2010/05/21
22. Nava P, Capaldo CT, Koch S, Kolegraff K, Rankin CR, Farkas AE et al (2011) JAM-A regulates epithelial proliferation through Akt/beta-catenin signalling. EMBO Rep 12(4):314–320, Epub 2011/03/05
23. Severson EA, Parkos CA (2009) Structural determinants of Junctional Adhesion Molecule A (JAM-A) function and mechanisms of intracellular signaling. Curr Opin Cell Biol 21(5): 701–707, Epub 2009/07/18
24. Laukoetter MG, Nava P, Lee WY, Severson EA, Capaldo CT, Babbin BA et al (2007) JAM-A regulates permeability and inflammation in the intestine in vivo. J Exp Med 204(13): 3067–3076, Epub 2007/11/28

25. Wapenaar MC, Monsuur AJ, van Bodegraven AA, Weersma RK, Bevova MR, Linskens RK et al (2008) Associations with tight junction genes PARD3 and MAGI2 in Dutch patients point to a common barrier defect for coeliac disease and ulcerative colitis. Gut 57(4):463–467, Epub 2007/11/09
26. Rhodes JM (1996) Unifying hypothesis for inflammatory bowel disease and associated colon cancer: sticking the pieces together with sugar. Lancet 347(8993):40–44, Epub 1996/01/06
27. McDole JR, Wheeler LW, McDonald KG, Wang B, Konjufca V, Knoop KA et al (2012) Goblet cells deliver luminal antigen to CD103+ dendritic cells in the small intestine. Nature 483(7389):345–349, Epub 2012/03/17
28. Van der Sluis M, De Koning BA, De Bruijn AC, Velcich A, Meijerink JP, Van Goudoever JB et al (2006) Muc2-deficient mice spontaneously develop colitis, indicating that MUC2 is critical for colonic protection. Gastroenterology 131(1):117–129, Epub 2006/07/13
29. Lu P, Burger-van Paassen N, van der Sluis M, Witte-Bouma J, Kerckaert JP, van Goudoever JB et al (2011) Colonic gene expression patterns of mucin muc2 knockout mice reveal various phases in colitis development. Inflamm Bowel Dis 17(10):2047–2057, Epub 2011/01/08
30. Schwerbrock NM, Makkink MK, van der Sluis M, Buller HA, Einerhand AW, Sartor RB et al (2004) Interleukin 10-deficient mice exhibit defective colonic Muc2 synthesis before and after induction of colitis by commensal bacteria. Inflamm Bowel Dis 10(6):811–823, Epub 2005/01/01
31. Fu J, Wei B, Wen T, Johansson ME, Liu X, Bradford E et al (2011) Loss of intestinal core 1-derived O-glycans causes spontaneous colitis in mice. J Clin Invest 121(4):1657–1666, Epub 2011/03/09
32. An G, Wei B, Xia B, McDaniel JM, Ju T, Cummings RD et al (2007) Increased susceptibility to colitis and colorectal tumors in mice lacking core 3-derived O-glycans. J Exp Med 204(6): 1417–1429, Epub 2007/05/23
33. Park SW, Zhen G, Verhaeghe C, Nakagami Y, Nguyenvu LT, Barczak AJ et al (2009) The protein disulfide isomerase AGR2 is essential for production of intestinal mucus. Proc Natl Acad Sci U S A 106(17):6950–6955, Epub 2009/04/11
34. Barrett JC, Lee JC, Lees CW, Prescott NJ, Anderson CA, Phillips A et al (2009) Genome-wide association study of ulcerative colitis identifies three new susceptibility loci, including the HNF4A region. Nat Genet 41(12):1330–1334, Epub 2009/11/17
35. Ahn SH, Shah YM, Inoue J, Morimura K, Kim I, Yim S et al (2008) Hepatocyte nuclear factor 4alpha in the intestinal epithelial cells protects against inflammatory bowel disease. Inflamm Bowel Dis 14(7):908–920, Epub 2008/03/15
36. Cattin AL, Le Beyec J, Barreau F, Saint-Just S, Houllier A, Gonzalez FJ et al (2009) Hepatocyte nuclear factor 4alpha, a key factor for homeostasis, cell architecture, and barrier function of the adult intestinal epithelium. Mol Cell Biol 29(23):6294–6308, Epub 2009/10/07
37. Anderson CA, Boucher G, Lees CW, Franke A, D'Amato M, Taylor KD et al (2011) Meta-analysis identifies 29 additional ulcerative colitis risk loci, increasing the number of confirmed associations to 47. Nat Genet 43(3):246–252, Epub 2011/02/08
38. Su L, Shen L, Clayburgh DR, Nalle SC, Sullivan EA, Meddings JB et al (2009) Targeted epithelial tight junction dysfunction causes immune activation and contributes to development of experimental colitis. Gastroenterology 136(2):551–563, Epub 2008/11/26
39. Kucharzik T, Walsh SV, Chen J, Parkos CA, Nusrat A (2001) Neutrophil transmigration in inflammatory bowel disease is associated with differential expression of epithelial intercellular junction proteins. Am J Pathol 159(6):2001–2009, Epub 2001/12/06
40. Muise AM, Walters TD, Glowacka WK, Griffiths AM, Ngan BY, Lan H et al (2009) Polymorphisms in E-cadherin (CDH1) result in a mis-localised cytoplasmic protein that is associated with Crohn's disease. Gut 58(8):1121–1127, Epub 2009/04/29
41. Wang L, Yu J, Ni J, Xu XM, Wang J, Ning H et al (2003) Extracellular matrix protein 1 (ECM1) is over-expressed in malignant epithelial tumors. Cancer Lett 200(1):57–67, Epub 2003/10/11
42. Matsuda A, Suzuki Y, Honda G, Muramatsu S, Matsuzaki O, Nagano Y et al (2003) Large-scale identification and characterization of human genes that activate NF-kappaB and MAPK signaling pathways. Oncogene 22(21):3307–3318, Epub 2003/05/23

43. Fisher SA, Tremelling M, Anderson CA, Gwilliam R, Bumpstead S, Prescott NJ et al (2008) Genetic determinants of ulcerative colitis include the ECM1 locus and five loci implicated in Crohn's disease. Nat Genet 40(6):710–712, Epub 2008/04/29
44. Sabath E, Negoro H, Beaudry S, Paniagua M, Angelow S, Shah J et al (2008) Galpha12 regulates protein interactions within the MDCK cell tight junction and inhibits tight-junction assembly. J Cell Sci 121(Pt 6):814–824, Epub 2008/02/21
45. Swanson PA 2nd, Kumar A, Samarin S, Vijay-Kumar M, Kundu K, Murthy N et al (2011) Enteric commensal bacteria potentiate epithelial restitution via reactive oxygen species-mediated inactivation of focal adhesion kinase phosphatases. Proc Natl Acad Sci U S A 108(21): 8803–8808, Epub 2011/05/11
46. Yu W, Beaudry S, Negoro H, Boucher I, Tran M, Kong T et al (2012) H2O2 activates G protein, alpha 12 to disrupt the junctional complex and enhance ischemia reperfusion injury. Proc Natl Acad Sci U S A 109(17):6680–6685, Epub 2012/04/12
47. Senapati S, Ho SB, Sharma P, Das S, Chakraborty S, Kaur S et al (2010) Expression of intestinal MUC17 membrane-bound mucin in inflammatory and neoplastic diseases of the colon. J Clin Pathol 63(8):702–707, Epub 2010/08/13
48. Yamamoto-Furusho JK, Mendivil-Rangel EJ, Fonseca-Camarillo G (2011) Reduced expression of mucin 9 (MUC9) in patients with ulcerative colitis. Inflamm Bowel Dis 18(3):E601, Epub 2011/10/27
49. Shaoul R, Okada Y, Cutz E, Marcon MA (2004) Colonic expression of MUC2, MUC5AC, and TFF1 in inflammatory bowel disease in children. J Pediatr Gastroenterol Nutr 38(5):488–493, Epub 2004/04/21
50. Furr AE, Ranganathan S, Finn OJ (2010) Aberrant expression of MUC1 mucin in pediatric inflammatory bowel disease. Pediatr Dev Pathol 13(1):24–31, Epub 2008/11/26
51. Rivas MA, Beaudoin M, Gardet A, Stevens C, Sharma Y, Zhang CK et al (2011) Deep resequencing of GWAS loci identifies independent rare variants associated with inflammatory bowel disease. Nat Genet 43(11):1066–1073, Epub 2011/10/11
52. Kyo K, Muto T, Nagawa H, Lathrop GM, Nakamura Y (2001) Associations of distinct variants of the intestinal mucin gene MUC3A with ulcerative colitis and Crohn's disease. J Hum Genet 46(1):5–20, Epub 2001/04/06
53. Kyo K, Parkes M, Takei Y, Nishimori H, Vyas P, Satsangi J et al (1999) Association of ulcerative colitis with rare VNTR alleles of the human intestinal mucin gene, MUC3. Hum Mol Genet 8(2):307–311, Epub 1999/02/05
54. Wrackmeyer U, Hansen GH, Seya T, Danielsen EM (2006) Intelectin: a novel lipid raft-associated protein in the enterocyte brush border. Biochemistry 45(30):9188–9197, Epub 2006/07/27
55. Barrett JC, Hansoul S, Nicolae DL, Cho JH, Duerr RH, Rioux JD et al (2008) Genome-wide association defines more than 30 distinct susceptibility loci for Crohn's disease. Nat Genet 40(8):955–962, Epub 2008/07/01
56. Parkes M, Barrett JC, Prescott NJ, Tremelling M, Anderson CA, Fisher SA et al (2007) Sequence variants in the autophagy gene IRGM and multiple other replicating loci contribute to Crohn's disease susceptibility. Nat Genet 39(7):830–832, Epub 2007/06/08
57. Scharl M, Paul G, Weber A, Jung BC, Docherty MJ, Hausmann M et al (2009) Protection of epithelial barrier function by the Crohn's disease associated gene protein tyrosine phosphatase n2. Gastroenterology 137(6):2030–2040.e5, Epub 2009/10/13
58. Kaser A, Zeissig S, Blumberg RS (2010) Inflammatory bowel disease. Annu Rev Immunol 28:573–621, Epub 2010/03/03
59. Vachon PH, Beaulieu JF (1995) Extracellular heterotrimeric laminin promotes differentiation in human enterocytes. Am J Physiol 268(5 Pt 1):G857–G867, Epub 1995/05/01
60. Satsangi J, Parkes M, Louis E, Hashimoto L, Kato N, Welsh K et al (1996) Two stage genome-wide search in inflammatory bowel disease provides evidence for susceptibility loci on chromosomes 3, 7 and 12. Nat Genet 14(2):199–202, Epub 1996/10/01

Chapter 14
Host Interactions with Bacteria: From "Entente Cordiale" to "Casus Belli"

Nouara Lhocine and Philippe J. Sansonetti

Abstract Host–bacteria interactions are highly diverse in humans and animals in general. In the gastrointestinal tract they can range from mutualistic to pathogenic interactions. Host and intestinal symbionts form a superorganism where co-evolution has set up a dynamic but fragile homeostasis. Symbiotic bacteria are contained in the intestinal lumen by tightly controlled innate immune mechanisms referred as "physiological inflammation", which is tightly regulated by sustained mechanisms of innate immune tolerance responding to host–bacteria cross talks that remain to be fully deciphered. Conversely, pathogenic bacteria need to be quickly perceived and discriminated from symbiotic bacteria in order for the host to develop rapid and efficient bactericidal responses referred as "pathologic inflammation". Recognition of pathogen/microbe-associated molecular patterns (P/MAMP) by pathogen recognition receptors (PRR) can hardly account for discriminating bacterial symbionts from bacterial pathogens, which largely share similar PAMPs. Unlike bacterial symbionts, the pathogens engage the host epithelial surface by tightly adhering to cells, possibly invading them, multiplying intracellularly, introducing massive amounts of PAMPs in their cytosol and altering their membranes by the secretion of pore-forming toxins and various secretory translocators. A large part of these "aggressive" events is recognized by dedicated systems, for instance, the toll-like receptors (TLR) or the cytosolic NOD-like receptors (NLR) that activate major pro-inflammatory pathways such as the NF-κB cascade and the inflammasome, leading to the release of the potent inflammatory cytokine IL-1β. These danger signals are amplified by endogenous signals, the damage-associated molecular patterns (DAMPs), derived from the damage induced to the host and mediated by PRRs

N. Lhocine, Ph.D. • P.J. Sansonetti, M.D. (✉)
Institut Pasteur Paris, Unité de Pathogénie Microbienne Moléculaire,
28 rue du Docteur Roux, Bâtiment Duclau, 75724 Paris Cedex 15, France
e-mail: nouara.lhocine@pasteur.fr

M. D'Amato and J.D. Rioux (eds.), *Molecular Genetics of Inflammatory Bowel Disease*,
DOI 10.1007/978-1-4614-8256-7_14,

and other receptors such as purinergic receptors for ATP. On the top of the PRR-associated first wave of signalling, this second wave of signalling that is strongly related to pathogen-associated dangers will complete the discrimination and innate inflammatory response to pathogens. However, bacterial classification in symbionts versus pathogens is far too simple and may not reflect the reality of host–bacteria interactions. The existence of pathobionts as well as the emergence of inflammatory bowel diseases associated with a loss of bacterial mutualism indicates that there is a spectrum of situations between the mutualistic and pathogenic poles.

Introduction

The human gastrointestinal tract is continuously exposed to microorganisms, ranging from symbionts to pathogens. These bacteria establish complex and dynamic interactions with the intestinal mucosa. Symbiotic bacteria have adapted to selective pressure exerted by the host through evolution and activate innate immune responses, leading to "physiological inflammation" and contributing actively to intestinal immune homeostasis. Pathogens can also interact with the intestinal mucosa to promote invasion, thus compromising intestinal homeostasis. They penetrate intestinal host tissues, proliferate and disseminate to other hosts, inducing "pathological inflammation" and resulting in damage or death. Thus the mammalian innate immune system has to deal with symbiotic and pathogenic bacteria, in sickness and in health, in order to maintain or re-establish intestinal homeostasis. As proposed by Janeway [1], microbial structures called pathogen/microbe-associated molecular patterns (P/MAMP) are recognized by host pathogen recognition receptors (PRRs) of the innate immune system. This recognition can lead to activation or regulation of the innate and adaptive immune systems. However both symbionts and pathogens produce these molecular structures such as lipopolysaccharide (LPS) or peptidoglycan (PGN). This recognition raises the question of the discrimination between symbionts and pathogens by the host immune system. How does the intestinal mucosa face the challenge to be simultaneously tolerant to symbiotic bacteria that populate the gut lumen and release PAMPs and antigenic molecules but also highly responsive to the occurrence of pathogenic bacteria that have evolved strategies for attaching to and invading mucosal surfaces?

In this chapter we will mainly focus on the discrimination between symbionts and pathogens by the host immune system. We first discuss how host and bacteria interact to enable the establishment of a stable intestinal microbiota that participates in host fitness. This microbiota is beneficial to the host as it constitutes a physical barrier against pathogenic infection but also stimulates the development of the innate immune system. However this physiological inflammation that contributes to the containment of the microbiota in the intestinal lumen is not sufficient to counteract the attack by true pathogens. These bacteria indeed developed virulence strategies to cross the intestinal epithelial barrier, breach innate immune host defences, manipulate host signalling and invade deeper host tissues. Thus the innate immune system has to mount rapid and efficient responses to limit the infection. We will then focus on the recognition of this pathogenic threat by the host innate immune system and we will try to decipher the mechanisms that permit the discrimination between symbionts and pathogens necessary to set up a specific and adapted response. We will consider that, in addition to recognition of PAMPs, the immune system responds to other signals associated with infection or inflammation, allowing the discrimination between harmless and virulent bacteria. The host immune system would then adapt its response to the level of encountered threat. In particular, damaged cells can release in the extracellular milieu endogenous molecules that signal the danger such as molecules associated to cell death [2, 3]. These molecules, called DAMPs (damage-associated molecular patterns), are thus not strictly specific to pathogenic infection but translate the emergence of a danger coming from either a pathogen or a "harmful symbiont", namely, pathobionts, but also from a sterile inflammation. Indeed these DAMPs are not necessarily due to the pathogens themselves but can be released after the induction of damage by the host innate immune response and the activation of pathologic inflammation.

Host–Symbionts Interactions: Adapted to Live with Our Best Enemies

Until recently, microbes associated with humans were largely described as aggressors engaging host surfaces, controlling host immune defences, expanding their population and propagating to other hosts. Microorganisms were essentially considered as pathogens and interactions between host and bacteria were mostly associated with infection [4]. Studies of the host immune system were thus based on the dichotomy between recognition and elimination of microorganisms and tolerance of self-molecules to maintain host homeostasis. However it is increasingly recognized that interactions between microorganisms and their hosts are not exclusively detrimental. Indeed these interactions can range from a symbiotic association to a deadly infection. "Illness is the exception rather than the rule", state Scott Merrell and Stanley Falkow [5]. As a matter of fact, the dominant forms of human–bacteria interactions are those in which microorganisms—now collectively called microbiota—do not cause harm (commensal interactions) or even benefit to the host (mutualistic interactions). Obligate and facultative symbionts associate with eukaryotes and both partners can take advantage from these interactions, forming superorganisms in which homeostasis is preserved. We thus share a complex and subtle relationship with our microbiota.

Humans acquire a resident microbiota at birth. A complex and dense microbial flora colonizes the adult intestinal tract, with its highest density in the terminal ileum and in the colon. The human gastrointestinal tract harbours from 10 to 100 trillion microorganisms, most of which are anaerobic bacteria. Many studies have focused on determining the core elements of the microbiota that are far from being fully defined. It was estimated that more than 500 bacterial species are present in the intestine [6]. The number and composition change along the gastrointestinal tract. Predominant communities belong to two major groups, the Firmicutes (Gram-positive anaerobes) and the Bacteroidetes (Gram-negative anaerobes), suggesting that selective pressure may participate to this diversity. Despite the high variability of bacterial abundance and variety, metagenomic sequencing points out the existence of a common core of microbial genes, shared among at least 50 % of individuals [7, 8]. Moreover other vertebrates are colonized by related but distinct microbiota from those described in humans. These symbionts contribute to many functions that are beneficial to the host, especially by their metabolic capacities [9]. They achieve fermentation of non-digestible substrates, produce essential vitamins, generate short-chain fatty acids from glycans and contribute to ionic absorption of calcium or magnesium. Furthermore they shape the gut immune system, affect components of the enteric nervous system, contribute to oral tolerance to food antigens and play a role in wound repair of the intestinal mucosa after epithelial damage. Finally, these symbionts participate in immune homeostasis and protect the host against colonization and invasion of pathogenic bacteria by secreting bactericidal substances and competing for nutrients and niche colonization [10].

Since the most ancient time, the bacterial communities have been selected, meaning that competing organisms underwent biological co-evolution and co-adaptation to persist in specific host niches. Each host has co-evolved with its own microbiota. Mechanisms have been selected to promote and maintain mutualistic interactions between bacteria and eukaryotes, particularly in the gastrointestinal tract [11]. In the intestinal lumen, symbiotic bacteria benefit from a situation of active host tolerance. Indeed co-evolution has selected host immune mechanisms at mucosal surfaces that control immune overresponse to the microbiota and help maintaining a low level of responsiveness that is nevertheless sufficient to contain the symbionts intraluminally (i.e. physiological inflammation), while keeping the ability to recognize and fight pathogens. Compelling evidences show that the microbiota can influence the host immune response and elicit immune mechanisms that modify the balance between proinflammatory and regulatory responses [12]. In case of dysbiosis, the microbiota composition is altered, leading to inappropriate host immune response and possibly to inflammatory disorders [13, 14]. The "hygiene hypothesis" [15] or the "disappearing microbiota" hypothesis [16] state that alterations in the human microbiota, mainly due to hygiene and antibiotics, could be an important factor increasing the incidence of some diseases such as allergy, asthma, inflammatory bowel diseases, obesity and diabetes. A major issue is now to decipher the role of indigenous microbial communities in human health and disease. A cellular microbiology of symbiosis is quickly emerging which necessitates the development of a novel array of cell and animal models.

The Law of the Strongest: The Pathogenic Attack

Pathogenic bacteria constitute only a small proportion of bacterial species. Historically however, these bacteria have been the cause of deadly and widespread epidemics like the medieval "Great Plague", or "Black Death", and pox that marked the spirit of mankind. Work by Louis Pasteur, Robert Koch and their schools, at the end of the nineteenth century, enlightened the beginning of host–pathogen interaction studies. In spite of the implementation of hygiene, vaccination and antibiotic treatment in the last century that reduced the global morbidity and mortality of infectious diseases, pathogens remain an important public health threat in low-income countries, and more than 300 new infectious diseases emerged since 1940 [17] with 87 "novel" pathogens identified since 1980 [5, 18]. When a pathogen breaches a host anatomic barrier, the innate immune system provides immediate defence against infection to limit bacterial entry, proliferation and propagation. Activation of innate immunity by invading microorganisms needs to be very rapid and bacterial recognition is central to innate immunity [19, 20]. Invading pathogens are rapidly sensed in a non-specific manner by the host innate immune system using a limited set of receptors. Pathogen recognition leads to the immediate activation of humoral and cellular components of the innate immune system. This participates in bacterial clearance and favours immune cell recruitment to sites of infection but also crucial activation of the adaptive immune system, which confers final eradication capacity and long-lasting immune memory [21].

The host immune system and pathogens have co-evolved, leading to an evolutionary arm race [22] that can be deleterious for both players. A tempting hypothesis has been proposed by May and Anderson [23], according to which virulence was the first step on the way to a symbiotic interaction. It was also assumed that immune processes evolved to avoid overreaction to pathogens to eventually achieve mutualism [24]. The best example of this ultimate symbiosis could be the mitochondria. According to the endosymbiotic theory, this organelle may derive from primitive bacteria, contributing to oxidative metabolism when oxygen appeared on earth [25].

Pathogenic and symbiotic bacteria may belong to the same genus or species, but are different by the relationships they maintain with the host. Symbiotic bacteria provide a first line of defence against pathogenic bacteria as they compete with pathogens for nutrients and constitute a colonization barrier [22]. However, most enteric pathogens are not as well metabolically equipped as symbionts [26]. Indeed their genomes contain a limited number of genes involved in saccharide uptake and hydrolysis compared to symbiotic bacteria. Unnecessary or detrimental metabolic pathways may have been lost to confer advantages in a selective niche. Therefore they may be less adapted to compete with symbionts for nutrients and the virulence strategies they developed allow them to gain access to host tissues, thus possibly resolving these nutritional issues and replicating [27, 28]. In parallel, symbionts are devoid of genes that promote invasion and subversion of host tissues [29].

Pathogenic bacteria can cross the intestinal epithelial barrier, breach innate immune host defences and invade deeper host tissues. The main differences between symbionts and pathogenic bacteria thus reside in the latter producing effectors mediating adherence or penetration of the intestinal epithelium, but also innate immune response evasion (Box 14.1).

Pathogens inherited specific pathogenicity genes usually organized in pathogenicity islands [30]. The gene clusters present on pathogenicity islands encode for adhesins, invasins, secretory apparatus such as the type-three secretion system (TTSS) and their dedicated effectors, enzymes, toxins and hemolysins. These bacterial factors are necessary for the pathogens to adhere to and colonize the epithelial surface or to breach the epithelial barrier. Further they participate in host-cell manipulation and subversion of immune responses to promote bacterial survival, colonization, proliferation and dissemination to other sites. Pathogens are able to dampen

Box 14.1 Virulence Factors of Pathogenic Bacteria

- Virulence factors are produced by pathogenic bacteria to elicit adhesion, colonization, immunoevasion and proliferation into host tissues
- They are mainly encoded by pathogenicity islands
- They include:
  - Adhesion factors to promote adherence to host cells and invasion
  - Fimbriae or pili
  - Non-pilus adhesins and invasins
  - Flagella
  - Invasion factors for digestion of the host extracellular matrix and polysaccharides to invade deeper host tissues after bacterial adhesion
  - Hyaluronidase, collagenase, neuraminidase, lecithinase, hemolysin and phospholipase
  - Toxins that modify the host–cell environment, subvert host-cell processes and modulate host defence. They can be at the surface of the bacteria or secreted
  - Endotoxins
  - Pore-forming toxins
  - Exotoxins with enzymatic activity such as mucinases. These effectors can also inhibit the immune response, affect the host-cell cytoskeleton or mimic host-cell proteins to hijack host signalling
  - Secretion systems that participate in the transport of bacterial toxins from the bacterial cytoplasm into host cells
  - Type I–VI secretion systems, and especially the TTSS from enteropathogenic bacteria
  - Outer membrane vesicles

innate immune defences at any stage of their progression, such as inhibiting expression of antimicrobial peptides and mucins, evading or suppressing phagocytic killing. This ability to overcome and/or manipulate host innate immune responses also defines the identity of being a pathogen. Intracellular colonization by pathogens contributes to mucosal inflammation as invaded epithelial cells produce proinflammatory molecules that recruit immune cells, and ultimately to epithelial destruction.

Microbial Sensing by Host Receptors: The Crossroad Between Inflammation and Tolerance

Innate immunity is the initial step of defence against bacterial infection. The front-line target cells need to recognize microorganisms with prokaryote-specific receptors to induce antimicrobial innate immune responses. Invariant microbial structures called pathogen-associated molecular patterns (PAMP) are recognized by a variety of germ line-encoded pattern recognition receptors (PRR) [31]. Receptor activation induces a response characterized by a burst of inflammation in the infection site, tissue destruction and recruitment of immune cells such as phagocytes and antigen-presenting cells. PRRs play an important role in costimulation of the adaptive immune system. They recognize foreign organisms, i.e. viruses, bacteria, parasites and fungi. Different types of ligands activate these receptors such as LPS, PGN, non-methylated DNA, RNA, flagellins, lipopeptides, toxins and fimbriae. Four families of PRRs cooperate to recognize microorganisms: the toll-like receptors (TLR), the nucleotide oligomerization domain-like (NOD) receptors (NLR), the RIG-I-like receptors (RLR) and the C-type lectin receptors (CLR). They may be secreted in the extracellular fluid (such as mannan-binding lectin, C-reactive protein or serum amyloid protein), intracellular (such as members of the NLR family, MDA-5 or RIG-1) or membrane-anchored such as the TLRs that are expressed either at the cell surface or associated with endosomes.

TLRs represent a family of highly conserved transmembrane molecules with an intracellular domain similar to the cytoplasmic domain of the Interleukin-1 receptor (IL-1R) [32]. The extracellular domain is the recognition site. It is characterized by leucine-rich repeats (LRR) and determines the ligand specificity. Some TLRs are present at the surface membrane (TLR1, 2, 6, 4 and 5) whereas others are defined to intracellular endosomal compartments (TLR3, 7, 8 and 9). The ligand specificity also depends on TLRs association, as they may homo- or heterodimerize. TLR specificities have been widely studied. One of the most described is TLR4 that recognizes LPS of the outer cell membrane of Gram-negative bacteria. TLR expression was first detected on blood monocytes, but intestinal epithelial cells (IECs) also express at least a subset of these TLRs.

Bacterial ligands are moreover recognized intracellularly by NLRs [33]. These multidomain proteins usually contain an LRR domain that determines the recognition specificity, a nucleotide-binding and self-oligomerization domain called NACHT domain and effector-binding domains that can be either caspase recruitment domains (CARD) for the NLRC family that comprises NOD1 and NOD2 or pyrin effector domains (PYR) for the NLRP subclass. This binding domain interacts with adaptor molecules to initiate signalling cascades following receptor activation [34]. NOD1 and NOD2 are key cytosolic sensors. They recognize muropeptides that are fragments of the PGN, a component of the bacterial cell wall. PGN consists of carbohydrate chains of β (1–4) linked, alternating *N*-acetylglucosamine and *N*-acetylmuramic acid sugars, cross-linked by short peptide chains. NOD1 is found in all cell types and binds to a diaminopimelate containing the GlcNAc-MurNAc tripeptide (GM-TriDAP or DAP) found in mainly Gram-negative bacterial PGN, whereas NOD2 detects muramyl dipeptide (MDP), the minimal bioactive PGN motif shared by all bacteria, in myelomonocytic and IECs [35–37].

Activation of most of these PRRs initiates signal transduction pathways that converge on the key transcription factor Nuclear Factor-κB (NF-κB) and leads to antimicrobial and proinflammatory gene expression and also recruitment of immune cells to the site of infection [34]. NOD1 and NOD2 are the only NLR members able to activate the NF-κB pathway upon stimulation. Other members of the NLR family, the

NLRPs, participate in the activation of caspase-1 by the inflammasome (described in chapter "Genetic Overlap Between Inflammatory Bowel Disease and Other Diseases").

Symbiotic and pathogenic bacteria share many striking similarities. Notably, the microbial structures that are recognized by the host innate immune system are common. The molecular motifs recognized by NLRs or TLRs are ubiquitously present in intestinal bacteria and are therefore important not only for mucosal host defence but also for intestinal homeostasis. As these immune activators are found in symbiotic and pathogenic bacteria, it was proposed to rename these components MAMPs for microbial-associated molecular patterns. These similarities make them, in general, unlikely candidates for discrimination between symbiotic and pathogenic microorganisms. How does the intestinal mucosa avoid the continuous activation of inflammatory responses by symbionts that are a source of proinflammatory PAMPs? The association found between mutations affecting NOD2 and Crohn's disease illustrates this question [38, 39]. PRRs do not generally efficiently discriminate between pathogenic and symbiotic microorganisms because none of the PAMPs is specific to one category, and they do not recognize pathogenicity-specific components. TLRs expressed by IECs can sense mucosal surfaces to monitor bacterial densities reflected by MAMPs concentration [40, 41]; thus, they play important functions to maintain intestinal epithelial homeostasis [42]. Innate immune recognition of symbiotic bacteria by TLRs at the mucosal surface is also necessary under steady-state conditions to limit inflammation within the intestine [43]. Similarly, NOD1 and NOD2 participate to host defence but also to intestinal homeostasis [44, 45].

Symbionts and Pathogens: Persona Non Grata Beyond the Mucosal Barrier

Controlling bacterial interactions at the intestinal mucosal surfaces is an important host strategy to limit bacterial contacts and prevent bacterial invasion. Symbiotic bacteria are contained in the intestinal lumen and tolerated. A general view is that most of the microbiota is maintained, as a complex population, away from the epithelial surface, by a process earlier qualified as "physiological inflammation" (see Fig. 14.1a). It is a complex mixture of a physical barrier, largely achieved by the

Fig. 14.1 (continued) Tight-junction opening, secretion of pore-forming toxins (PFT), insertion of secretory translocators such as the type-three secretion system (TTSS) and secretion of toxins and effectors (mucinases, adhesins, invasins) are necessary for host tissue invasion. Activation of PRRs initiates signal transduction pathways that lead to proinflammatory gene expression but also recruitment and activation of immune cells to the site of infection such as DCs, macrophages, T cells and polymorphonuclear neutrophils (PMN). Endogenous signals amplify this response. Indeed, damage-associated molecular patterns (DAMPs) derived from the damage induced to the host such as cell lysis or cytoskeleton modifications and mediated by PRRs and other receptors such as P2X7, the receptor for ATP, participate in the formation of molecular scaffolds named inflammasomes that induce a caspase-1-dependent cellular death called pyroptosis, but also cytokine activation such as IL-1β (beta) and IL-18

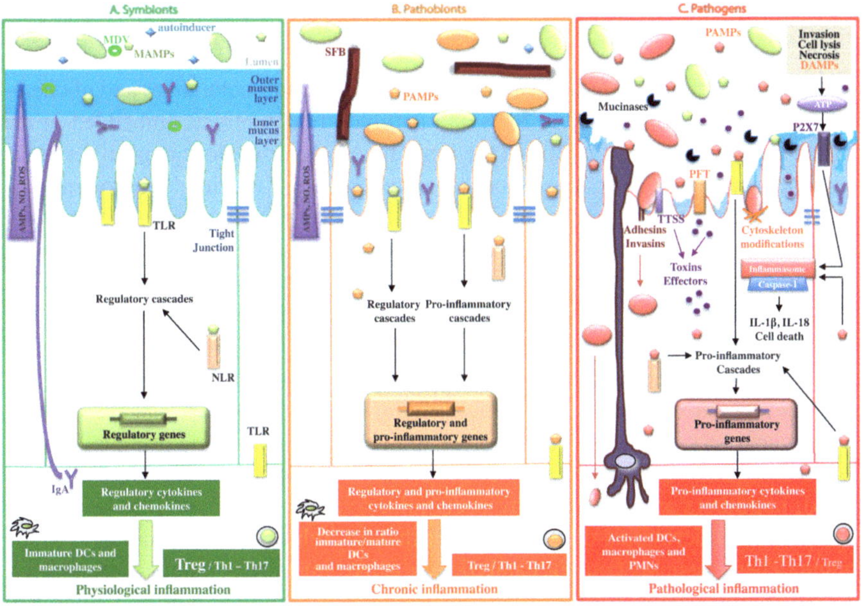

Fig. 14.1 Host–bacteria interactions in health and disease. The human gastrointestinal tract is continuously exposed to microorganisms, ranging from symbionts to pathogens, including pathobionts. These bacteria establish complex and dynamic interactions with the intestinal mucosa. (**a**) Symbiotic bacteria participate in immune homeostasis and protect the host against colonization and invasion of pathogenic bacteria by secreting bactericidal substances and competing for nutrients and niche colonization. Symbionts communicate among themselves by releasing small molecules such as autoinducers. They are contained in the intestinal lumen by tightly controlled innate immune mechanisms referred as "physiological inflammation", contributing to intestinal immune homeostasis. Intestinal epithelial cells (IECs) are attached together by tight junctions that form a sealed barrier to the luminal environment, preventing bacterial penetration. The production of mucus by goblet cells, NOS, ROS, antimicrobial peptides and microvillus-derived vesicles (MDV) containing catalytically active intestinal alkaline phosphatase by IECs and IgA by subepithelial B cells prevents the overt stimulation of intestinal innate immunity by symbionts and limit mucosal inflammation. Microbial structures called PAMP are recognized by host PRRs of the innate immune system such as toll-like receptors (TLR) and NOD-like receptors (NLR), inducing regulatory signalling cascades that participate in intestinal homeostasis by inducing the production of regulatory cytokines and chemokines. Several subtypes of differentiated T cells are associated with the epithelial layer and participate to the mucosal immune response, the same as dendritic cells [117]. They produce effector cytokines necessary to contain bacteria to the intestine and stimulate macrophages located in the subepithelial area of the lamina propria to quickly phagocytose and kill the symbionts that would cross the epithelial barrier. To avoid overt immune responses, PRRs expression is restricted and limited to certain cell populations and the localization of the receptors on the cells is restricted. Host cells may also modify PAMPs to limit their agonist function on PRRs. Symbiotic bacteria may directly participate to the tolerogenic process by producing weakly stimulatory PAMPs. (**b**) Bacteria venturing towards the epithelial surface are quickly killed unless they can resist surface defence molecules. This is the case of a subcategory of symbionts called the pathobionts, illustrated by the segmented filamentous bacteria (SFB) *Clostridium*. Pathobionts are contained by innate immune mechanisms. They may be considered "good bacteria" as they achieve maturation of the mucosal immune system, inducing the maturation of naïve T cells into inflammatory (i.e. Th17) lymphocytes that are essential to the maintenance of mucosal innate immune protection. However, if the density of pathobionts is not balanced by sufficient density and diversity of symbionts, they may express their pathogenic potential and account for chronic inflammation of the gut. (**c**) Pathogenic bacteria engage the host epithelial surface by tightly adhering to cells, possibly invading them, multiplying intracellularly, introducing massive amounts of PAMPs in their cytosol and altering their membranes.

mucus, a great majority of the bacteria residing in its softer outer layer and of chemical compounds, such as NOS, ROS and antimicrobial peptides, embedded in the denser lattice of mucins that forms the inner mucus layer in close apposition to the epithelial surface [12, 34]. The mucus layer, composed of mucin glycoproteins produced by goblet cells, covers the intestinal epithelium and protects the mucosal surface from invasion, defining a relatively "germ-free" zone [46]. Bacteria are thus prevented to adhere directly to the intestinal epithelium [47]. Mucin 2-deficient mice do not exhibit this bacteria-free area and develop spontaneous colonic mucosal inflammation [48, 49]. Similarly, mucin upregulation is observed after infection with enteropathogenic bacteria such as *Salmonella*, *Yersinia* or *Shigella* [50, 51].

IECs play an important role in maintaining intestinal homeostasis. These cells produce and secrete microbicidal molecules such as antimicrobial proteins, mainly defensins or cathelicidins [50–53]. Some of these molecules are constitutively produced, but others are controlled by bacterial activation. IECs lining the gut lumen also produce microvillus-derived vesicles that are released and accumulate in the lumen. These vesicles contain catalytically active intestinal alkaline phosphatase, an enzyme responsible for LPS dephosphorylation, preventing intestinal inflammation [54–57]. These vesicles cluster on the luminal bacteria, inhibiting bacterial attachment to the host cells and limiting bacterial growth in the intestinal lumen [58]. In addition to their production of bactericidal molecules that keep the cellular apical environment relatively clear of symbiotic bacteria, IECs are attached together by tight junctions that form a sealed barrier to the luminal environment, preventing bacterial penetration. Another epithelial cell, the Paneth cell, secretes antimicrobial proteins at the base of the small intestinal crypts. Paneth cells contain microbicidal granules that they discharge when they sense bacteria, controlling the invasion by symbiotic and pathogenic bacteria.

Adaptive immunological responses are also induced to protect the intestinal surface and limit bacterial interactions with the mucosal surface. It is largely based upon IgA production by subepithelial B cells located in the lamina propria. Mice lacking IgA exhibit an increase in mucosa-associated bacteria [59]. Dendritic cells located in intestinal lymphoid structures called Peyer's patches or solitary nodules in the colon sample bacteria translocated through epithelial M cells. They interact with B cells that differentiate into plasma cells to produce IgA that transcytose across the epithelium [60]. In parallel, several subtypes of differentiated T cells are associated with the epithelial layer and participate to the mucosal immune response. They produce effector cytokines necessary to contain bacteria to the intestine [61] and stimulate macrophages located in the subepithelial area of the lamina propria to quickly phagocytose and kill the symbionts that would cross the epithelial barrier.

Bacteria venturing towards the epithelial surface are therefore quickly killed unless they can resist surface defence molecules. This is the case of a subcategory of symbionts called the pathobionts, illustrated by the segmented filamentous bacteria (SFB) *Clostridium* in the murine intestine [62], and probably Enterobacteriaceae and Enterococci in humans (see Fig. 14.1b). Pathobionts are contained by these innate immune mechanisms but may quickly trespass the mucosal barrier in case of immune failure such as immunosuppressive chemotherapies for cancer, leukemias and organ transplantation [63]. From their "distant" site of residence, symbionts communicate among themselves as any complex microbial population in various environments, releasing small molecules such as autoinducers [64] and PAMPs. This collection of molecules is sensed by the front line of IECs, which are likely, on this basis, to gauge the bacterial density and accordingly adjust its response that encompasses the dual necessity to fine tune its bactericidal response to the exact level of threat, and to elicit the signals maintaining the tolerogenic process. Microbiota detection therefore activates host mechanisms involved in mucosal homeostasis [65].

All the strategies that control the symbiotic luminal content are also involved in the fight against pathogenic infection, particularly the sensing of PAMPs, thus reemphasizing the need to better understand how the host can discriminate between symbionts, including their more adventurous companions, the pathobionts and the true pathogens. A characteristics of the pathogens is their capacity to engage the epithelial barrier and possibly to achieve its subversion and invasion [66]. One could therefore operationally discriminate symbionts and pathogens on the capacity of the latter to trespass the epithelial barrier, thus generating a systemic immune response, instead of the controlled mucosally localized response that is characteristic of the microbiota [67].

Mesenteric Lymph Nodes as Immune Firewalls Between Tolerance and Ignorance

Symbionts do not induce a strong systemic inflammatory reaction, indicating that the host immune system is highly adapted to the microbiota. On the contrary host–pathogen interaction leads to a strong systemic response. A major problem is again to understand how the immune system can distinguish pathogenic bacteria from symbiotic bacteria. The development of the mucosal immune system depends on colonization by the microbiota. Comparisons between germ-free and specific pathogen-free (SPF) animals indicate that the mucosal immune system is underdeveloped in germ-free animals. They exhibit a strong decrease in IgA-producing plasma cells in the lamina propria, a reduction in some subclasses of intraepithelial lymphocytes and hypoplastic lymphoid follicles. Interestingly, expansion of the structures of the spleen and lymph nodes is also dependent upon the presence of the microbiota, even if these organs are not in direct contact with the microbiota [68]. While the microbiota is restricted to the intestinal lumen, some symbionts can be sampled by mucosal lymphoid organs such as Peyer's patches where they interact with dendritic cells. A local immune response is induced but these primed dendritic cells are not found farther than the draining mesenteric lymph nodes (MLN). This indicates that these primed dendritic cells are limited to the mucosal immune system, thus restricting symbiont dissemination. Symbiotic bacteria normally do not penetrate beyond the draining MLNs that form a "firewall" between the mucosal and the systemic immune system. As symbionts do not go further, they are poorly recognized by the systemic immune system. The intravenous injection of symbiotic bacteria in mice induces a systemic immune response [69]. In parallel their ignorance by the adaptive systemic immune response disappears when MLNs are surgically removed [59]. This indicates that symbiotic bacteria normally do not prime the systemic immune response, mucosal immunity being sufficient to their containment. This highly compartmentalized immune response preserves the host capacity to mount an efficient systemic response against bacteria breaching the epithelial barriers. Conversely, pathogenic bacteria are able to trespass the draining MLN "barrier". Infection is not confined to mucosal tissues; therefore, pathogenic bacteria prime the adaptive systemic immune system.

The TLR-MYD88 (myeloid differentiation primary-response protein 88) signalling pathway can sense symbiotic bacteria and is important for the establishment and the maintenance of host–symbionts homeostasis. Classically the activation of this pathway leads to the induction of an inflammatory response that participates in bacterial clearance. This cascade is involved in many diverse processes such as the production of antimicrobial proteins like RegIIIγ, enlightening the importance of symbiotic bacteria in the activation of the host innate immune response [70]. Interestingly, innate immune defects such as a deficiency in TLR signalling adaptor molecules can result in the priming of adaptive immune system by symbiotic bacteria. Mice lacking MYD88 or TRIF (TIR-domain-containing adaptor protein inducing IFN-β) adaptors exhibit a complete loss of host–symbionts compartmentalization and produce IgG responses against symbionts, probably because many symbiotic bacteria cross the epithelial barrier and are not efficiently eliminated by phagocytes [71]. The systemic adaptive immune system can thus compensate a loss of the mucosal innate immunity. Therefore the activation of the systemic immune response is essentially a matter of balance. This balance is disrupted in the presence of pathogens or in the absence of an efficient mucosal immunity necessary to contain the symbionts upstream the MLNs.

Active Immune Tolerance Towards Symbiotic Bacteria

To avoid the overt stimulation of intestinal innate immune receptors by symbiotic PAMPs and to limit mucosal inflammation, several sophisticated strategies have been developed by symbiotic bacteria and host tissues.

PRRs expression is restricted and limited to certain cell populations. For example, intestinal macrophages show reduced PRRs expression contrary to macrophages present in other tissues [72–74]. Also TLR4 expression is restricted to crypt epithelial cells, limiting its activation by bacterial ligands. Paneth cells located at the base of these intestinal crypts express high levels of TLR4 and NOD2, protecting the crypts and probably the stem cells from bacteria [75, 76].

Second, the localization of the receptors on the cells can be restricted. Indeed intracellular localization of some receptors such as NOD2 requires the transport of the ligands into IECs or microbial invasion of the cytosol, thus likely avoiding major receptor activation by extracellular muropeptides and non-invasive bacteria [44, 77]. Similarly, TLRs distribution is restricted at the apical surface of IECs. TLR5, responsible for bacterial flagellin recognition, is only exposed at the basolateral side of differentiated IECs [78, 79].

Third, signalling regulations may prevent or modify immune activation. Indeed many negative regulators of the TLR signalling pathways have been described. For example, the negative regulator Tollip is expressed in IECs and directly interacts with IRAK-1, preventing its autophosphorylation and inhibiting any further signalling [80, 81]. Host cells may also modify PAMPs to limit their agonist function on PRRs as exemplified by IECs producing hydrolases, cleaving acyl chains from lipid A, the endotoxin moiety of LPS in Gram-negative bacteria [82, 83] or producing an alkaline phosphatase that dephosphorylates the lipid A [54], thus in both cases attenuating endotoxin activity on TLR4.

In parallel, symbiotic bacteria may directly participate to the tolerogenic process by producing weakly stimulatory PAMPs such as hypoacylated lipid A in *Bacteroides fragilis* [84, 85]. Symbiotic bacteria may also actively suppress epithelial proinflammatory signalling by interfering with NF-κB activation. For example, *Bacteroides thetaiotaomicron* promotes nuclear export of the NF-κB p65 subunit in a peroxisome proliferator-activated receptor γ-dependent manner [86]. Also the probiotic species *Lactobacillus casei* inhibits the degradation of the inhibitor I-κB in order to avoid proinflammatory gene induction [87].

Invasion, Damage and Inflammation Are Hallmarks of Pathogenic Bacteria

Intestinal homeostasis that prevails between the intestinal mucosa and symbiotic bacteria is fragile and can be subverted by pathogens (see Fig. 14.1c). Indeed enteropathogens have the capacity to disturb the intestinal epithelium, invade host cells and induce the inflammatory destruction of the intestinal mucosa with the help of virulence factors that are specific to pathogenic bacteria [66]. Pathogen recognition by host cells mainly occurs through PAMPs and PRRs. However this recognition may not be straightforward due to the recognition of harmless bacteria by PRRs or to PAMPs modifications by pathogens to subvert immune recognition, making difficult the discrimination between symbionts and pathogens. As a matter of fact, other signals are clearly necessary to activate a full immune response in case of true pathogens. In 1994, Polly Matzinger pioneered the concept of "danger signal" [3], suggesting that the host perceives the presence of pathogens or pathological conditions as much as the cellular damage they cause. The molecules that signal tissue and host damage (i.e. "danger molecules") may have two different origins: (1) host molecules released during the infectious process and called DAMPs or alarmins and (2) cellular structures affected by pathogenic factors [88]. Consequently tissues that undergo destruction and loss of integrity would trigger the immune system. These danger signals recruit and activate the innate immune system and participate in the restoration of the destroyed tissue. Notably some alarmins can signal through TLRs and NLRs to induce inflammatory and immune responses, suggesting that these receptors can also sense "self-ligands" [89].

Cells dying by necrosis release DAMPs like ATP, uric acid, DNA and DNA-binding proteins such as high-mobility group box-1 protein (HMGB1) into the extracellular milieu [88]. These secondary stimulatory molecules have a high proinflammatory impact. Necrotic cell lysates are a source of endogenous factors such as HSPs or uric acid that induce dendritic cell activation. Extracellular HSPs interact with several receptors including TLRs, leading to secretion of proinflammatory cytokines. HMGB-1 is a nuclear protein that binds to the nucleosome. When cells die by necrosis, this protein is released extracellularly. HMGB-1 has chemotactic activities on immune cells such as monocytes, macrophages, neutrophils and dendritic cells, but also proangiogenic and immunostimulatory activities. Similarly an increase of extracellular ATP concentration reflects the presence of dying cells and leads to its binding to the ligand-gated ion channel P2X purinergic receptor 7 (P2X7). This receptor is important for ion transport and allows potassium efflux after activation [90, 91]. Tissue architecture disorder during bacterial infection may also send signals to the immune system [92]. Blood vessel rupture and inflammation induce the extravascular relocation of fibrinogen that can activate macrophages through TLR4. Disruption of the extracellular matrix and basement membranes by bacterial proteases are also sensed like danger signals. Soluble fragments of heparin-sulphate proteoglycans released from cell surface and basement membranes can also activate TLR4 of dendritic cells. Host membrane recruitment is also considered as a danger signal, the same as membrane integrity disruption by the insertion of secretory systems such as the TTSS of enteropathogens. Indeed it may lead to expression of proinflammatory cytokines [93].

The innate immune system recognizes invading microbes, tissue damage or stress through conserved receptors such as the TLRs and NLRs that are activated by PAMPs or DAMPs. These PRRs activate signalling cascades that converge in the transcription of cytokines, chemokines and proteins involved in bacterial clearance [94]. Among the NLRs, NOD1 and NOD2 play major roles in the intestinal epithelium as they detect intracellular ligands and activate NF-κB signalling leading to the transcription of proinflammatory genes. However other NLRs are also important for the innate immune response to activate inflammation and limit microbial invasion. Especially some NLRs are involved in the post-translational activation of inflammatory caspases and participate in the formation of molecular scaffolds named inflammasomes [93]. These large multiprotein complexes are activated by pathogen-associated signatures or endogenous molecules of similar structure produced after tissue damage. Inflammasome activation leads to autocatalytic cleavage and activation of caspase-1, and processing and secretion of proinflammatory cytokines such as IL-1β and IL-18 [95]. Whereas IL-1β induces IL-17 release from Th17 cells to amplify early effector responses, IL-18 stimulates $CD8^+$ T cells and Th1 cells to secrete IFN-γ [96, 97]. Inflammasome activation induces a caspase-1-dependent cellular death called pyroptosis, but also autophagy and bacterial degradation [98]. Several types of inflammasome may be distinguished according to the

NLR that is involved and the ligand that activates this platform. For example, the NLRP1 inflammasome is activated after MDP or anthrax lethal toxin recognition and the NLRC4 inflammasome after flagellin sensing. The most studied inflammasome is the one containing NLRP3. This complex is activated by different stimuli such as crystals, pore-forming toxins, bacteria and viruses. Some cytosolic proteins like AIM2 and RIG-I, which do not belong to the NLR family, also form inflammasomes in response to cytosolic DNA and virus respectively. Bacterial infection can activate several inflammasomes [99]. For example, *Listeria monocytogenes* induces via listeriolysin O, flagellin and bacterial DNA the NLRP3, NLRC4 and AIM2 inflammasomes, whereas *Shigella flexneri* and *Salmonella typhimurium* activate the NLRP3 and NLRC4 platforms [95]. However, the contribution or redundancy of these different complexes is not clear. Interestingly, deregulated activation of inflammasomes is associated with autoinflammatory syndromes and other pathologies [99].

Host Responses Manipulation by Pathogens

Enteric bacteria have developed sophisticated strategies to overcome the innate immune system and successfully hijack host signalling. Manipulation is the hallmark of pathogens. Pathogenic bacteria secrete factors that first favour infection and bacterial survival inside the cells. These virulence factors target cytoskeletal components, host cell receptors and signalling molecules. For instance, *Shigella flexneri* secretes IpaB that prevents rapid epithelial turnover to maintain its replicative niche but also inhibits IEC detachment by delivering the OspE effector through the TTSS, reinforcing epithelial adhesion to the basal lamina [100, 101]. Some bacterial factors manipulate host membrane and regulate actin cytoskeleton remodelling to permit bacterial adhesion, invasion and propagation. Rho GTPases regulation enables tight junction opening, barrier function reduction and bacterial entry, for example. The bacterial effectors SopB, SopE, SopE2 and SipA participate in tight junction disruption during *Salmonella typhimurium* infection, for example, but also induce an inflammatory response [102–104]. Bacterial effectors can affect post-translational modifications on host cells by mimicking the corresponding activities such as ubiquitination or sumoylation to hijack signalling pathways. Some intracellular bacteria have also the capacities to escape from the early vacuole after invasion in order to avoid the unfriendly environment present in this compartment. Then they hijack host cytoskeleton to move intracellularly and disseminate.

In parallel, pathogenic bacteria have expanded mechanisms to avoid recognition by the host innate immune response. First they developed strategies to modify their PGN and thus avoid recognition by the innate immune receptors. Many modifications of the PGN are possible to ensure resistance to antibiotics or host degradative enzymes that target the cell wall but also impairment of detection by PRRs to avoid innate immune signalling. During the course of a bacterial infection, the structure and composition of PGN is likely to be modified through the action of bacterial enzymes. For instance, SltY, a bacterial lytic transglycosylase involved in PGN processing, is highly up-regulated during the infection of IECs by *Shigella flexneri* [105]. Many modifications are observed in pathogenic species, mainly GlcNAc N-deacetylation and MurNAc O-acetylation. *Listeria monocytogenes* is a gram-positive bacteria that plays on PGN structure. This pathogen produces and secretes autolysins important for the virulence such as the endopeptidase p60 and the *N*-acetylmuramidase NamA. These two enzymes cleave the PGN to generate the NOD2 agonist, thus modulating the inflammatory response and the microbial recognition [106].

Several mechanisms to circumvent and overcome innate immune responses also exist. Among these strategies, the secretion of bacterial effectors through the TTSS enables bacteria to manipulate a broad array of host pathways by regulating or mimicking host proteins in order to survive and colonize host tissues [107]. Some pathogens have evolved strategies to avoid phagocytosis by delivering effectors that impair the signalling downstream of phagocytic receptors such as YopH, YopE and YopT effectors secreted by *Yersinia* to inhibit phagocytosis [108]. Many bacterial effectors can directly modulate host proinflammatory pathways to suppress detrimental inflammation during invasion steps. Given the role of the NF-κB pathway in the immune responses, enteric bacteria have developed mechanisms to interfere with this cascade [109]. First pathogenic factors contribute to the inhibition of the cascade at the level of the TLRs by host protein mimicking. Indeed pathogenic effectors can share sequence or structural homology with the TLR domain necessary for the signalling, which is the TIR domain responsible for adaptation recruitment, interfering with TLR signalling and preventing thus further innate host defence. An example of bacterial mimicry is *Salmonella* effector TIR-like protein A (TlpA) that impairs TLR4-mediated NF-κB activation [110]. Secondly bacterial effectors act at different levels of the pathway to prevent its activation by regulating phosphorylation, acetylation, ubiquitination and neddylation of host proteins required for NF-κB activation [111]. This modulation can be focused on a specific protein or affect a more general mechanism such as the ubiquitin pathway. A major strategy to regulate inflammation and maintain this control for long periods may be epigenetic regulation by bacterial effectors. OspF, an effector produced by *Shigella*, directly inhibits in the nucleus the activation of NF-κB target genes. Indeed OspF is a phosphothreonine lyase that dephosphorylates irreversibly and inactivates the enzymes responsible for histone H3 phosphorylation in the nucleus, the MAPKs p38 and ERK2 [112]. Finally, enteric pathogens have evolved stratagems to block deleterious inflammation and to evade inflammasome activation. Enteropathogenic *Yersinia enterocolitica* bacteria exploit several processes to prevent caspase-1 activation and secretion of IL-1β and IL-18 after inflammasome activation [113]. Symbiotic bacteria have also developed means to control their recognition by the host and the activation of the host immune system. However these mechanisms are much less developed and do not enable such bacteria to invade tissues and proliferate in the case of immune homeostatic balance.

Symbionts, Pathobionts and Pathogens: Evolutionary Distant Relatives?

Host–bacteria interactions have evolved thanks to their high capacities for genomic modifications. Compared to multicellular eukaryotes, bacteria have higher generation times and highly efficient properties to increase their genetic variability, mainly by horizontal gene transfer. These characteristics allow bacteria to evolve and continuously adapt to the selective pressure imposed by environmental changes, particularly in the host. A Manichean view of the microbial world often classifies bacteria into two different types: the "good bacteria" versus the "bad bacteria" and symbionts versus pathogens. However this classification is far too simplistic. There is increasing evidence for a continuous and linear evolution between the "bona fide" symbionts and the "bona fide" pathogens. Between these two ends of the spectrum, a vast "grey zone" comprises species like *Escherichia coli* that affect a large number of intermediate isolates between strictly symbiotic or pathogenic strains, depending upon the number, variety and complementarity of pathogenic or metabolic traits present in their respective genomes, in addition to their common core genome [114].

In this "grey zone", emerge a category of microorganisms called the pathobionts [115]. Pathobionts may still be considered "good bacteria" as they achieve maturation of the mucosal immune system. For instance, in mice, the segmented filamentous bacterium clostridial strain SFB adheres to Peyer's patches in the healthy terminal ileum where it induces IgA production and activates B cells [116]. It also induces the maturation of naïve T cells into inflammatory (i.e. Th17) lymphocytes that are essential to the maintenance of mucosal innate immune protection [14, 117]. However, if the density of pathobionts is not balanced by sufficient density and diversity of symbionts, they may express their pathogenic potential and account for chronic inflammation of the gut [118]. In this context, the immune status of the host is essential to determine whether bacteria will be harmful or harmless to the host. Immunodeficient mice (SCID mice) reconstituted with CD4+CD45Rb^{high} T cells and colonized with SFB develop severe colitis and intestinal inflammation [119]. SFB does not only affect the gastrointestinal tract, it may affect the whole fitness of the host as SFB reconstitution in germ-free mice also increases rheumatoid arthritis and multiple sclerosis susceptibility [62, 120]. Similarly, *Helicobacter hepaticus* colonizes mucosal surfaces of the gastrointestinal tract and the liver. These bacteria trigger in wild-type animals inflammatory and tolerogenic responses that participate in physiological inflammation. However immunocompromised mice such as SCID and *Il10*$^{-/-}$ mice reconstituted with CD4+CD45Rb^{high} T cells or *Rag2*$^{-/-}$ mice exhibit pathogenic inflammation after *H. hepaticus* infection and can develop rapid colitis or even colon cancer [121–123]. In immunocompromised animals, *H. hepaticus* induces a deregulated inflammatory response that leads to pathological inflammation. The type VI secretion system of *H. hepaticus* mediates these effects as deletion of this apparatus leads to higher colonization and elevated inflammatory response in immunocompromised hosts, suggesting that it is involved in the tolerogenic responses [63]. However the secreted effectors involved in this process have not been identified. Host interactions with *Bacteroides fragilis* have been better deciphered. This pathobiont closely associates to mucosa and produces an anti-inflammatory capsular polysaccharide A (PSA) that induces the differentiation of IL-10 secreting Foxp3 (+) Treg cells [124, 125]. These bacteria thus protect mice from *H. hepaticus*-induced colitis [124]. Enterotoxigenic *B. fragilis* (ETBF) has been recently characterized [126]. This strain secretes a proinflammatory toxin called BFT that stimulates colonic inflammation in predisposed multiple intestinal neoplasia (MIN) mice [127, 128].

An example of pathobiont in humans is *Helicobacter pylori*. This species interacts with the gastric mucosa and is responsible for gastritis, peptic ulcer and gastric adenocarcinoma. Although found in 50 % of the human population, however, only a few percent of this population will develop these pathologies, and the advantageous aspects of colonization by *H. pylori* remain to be demonstrated, even if colonization with this strain seems to decrease the risk of oesophageal carcinomas and asthma [129–131]. Virulent strains of *H. pylori* translocate into gastric epithelial cells the virulence factor CagA that hijacks host signalling pathways involved in inflammation and oncogenesis [132, 133]. However this bacterial protein is not sufficient to induce pathogenesis as asymptomatic carriers exist, suggesting that other factors are required such as host genetic polymorphisms for the IL-1β proinflammatory gene [134]. Other bacteria such as γ-proteobacteria, particularly Enterobacteriaceae, also present proinflammatory characteristics that may turn out to be deleterious for the host in case of host immunological failure or dysbiosis [135].

Conclusion and Perspectives

Compelling evidence shows that regulation of bacterial interactions with the intestinal mucosal surface is a critical stage for the establishment and maintenance of intestinal homeostasis. Discrimination between harmful and harmless bacteria is a matter of survival for the host. PRRs play an essential role in PAMPs and DAMPs recognition, leading to activation of immune defence mechanisms. In a healthy host, this recognition leads to physiological or pathological inflammation, in the case of symbiotic bacteria or pathogenic bacteria respectively. However, the existence of pathobionts indicates the existence of a continuum in between. The recognition of host damage is a more efficient mechanism to detect harmful bacteria rather than the discrimination between symbionts and pathogens.

Many questions remain unanswered concerning host–bacteria interactions. Immunity qualitatively and quantitatively shapes the microbiota. However, despite many metagenomic studies, we still know little about microbiota composition and how mucosa-associated bacterial species are distinct from those that are in the lumen. Virulence factors that are used by pathogens are well described. On the contrary, colonization or symbiosis factors are poorly known. Moreover bacteria–bacteria interactions are even less known and are highly difficult to grasp and to decipher in the context of the intestinal mucosa. These bacteria have co-evolved into a "community behaviour", that could be important to understand, for instance, in the context of mechanisms triggering inflammatory bowel diseases. Interpreting the behaviour and the fate of host–bacteria interactions will provide clues and biomarkers to anticipate and predict the evolution of host–bacteria interactions, and as a consequence the evolution of infectious and inflammatory processes. This ecosystem may also be considered a "gold mine" to identify novel bioactive molecules, from host and microbial origin that will contribute to the development of original therapeutic and preventive approaches. The field of prebiotics, probiotics and postbiotics is likely to undergo a revolution, thanks to the global application of cellular microbiology principles to the study of the symbiosis-to-pathogenesis transition.

References

1. Janeway CA Jr (1989) Approaching the asymptote? Evolution and revolution in immunology. Cold Spring Harb Symp Quant Biol 54(pt 1):1–13, Epub 1989/01/01
2. Kono H, Rock KL (2008) How dying cells alert the immune system to danger. Nat Rev Immunol 8(4):279–289, Epub 2008/03/15
3. Matzinger P (1994) Tolerance, danger, and the extended family. Annu Rev Immunol 12: 991–1045, Epub 1994/01/01
4. Casadevall A, Pirofski LA (2000) Host-pathogen interactions: basic concepts of microbial commensalism, colonization, infection, and disease. Infect Immun 68(12):6511–6518, Epub 2000/11/18
5. Merrell DS, Falkow S (2004) Frontal and stealth attack strategies in microbial pathogenesis. Nature 430(6996):250–256, Epub 2004/07/09
6. Eckburg PB, Bik EM, Bernstein CN, Purdom E, Dethlefsen L, Sargent M et al (2005) Diversity of the human intestinal microbial flora. Science 308(5728):1635–1638, Epub 2005/04/16
7. Qin J, Li R, Raes J, Arumugam M, Burgdorf KS, Manichanh C et al (2010) A human gut microbial gene catalogue established by metagenomic sequencing. Nature 464(7285):59–65, Epub 2010/03/06
8. Tap J, Mondot S, Levenez F, Pelletier E, Caron C, Furet JP et al (2009) Towards the human intestinal microbiota phylogenetic core. Environ Microbiol 11(10):2574–2584, Epub 2009/07/16
9. Backhed F, Ding H, Wang T, Hooper LV, Koh GY, Nagy A et al (2004) The gut microbiota as an environmental factor that regulates fat storage. Proc Natl Acad Sci U S A 101(44): 15718–15723, Epub 2004/10/27
10. Lotz M, Menard S, Hornef M (2007) Innate immune recognition on the intestinal mucosa. Int J Med Microbiol 297(5):379–392, Epub 2007/04/27
11. Nishiguchi MK, Hirsch AM, Devinney R, Vedantam G, Riley MA, Mansky LM (2008) Deciphering evolutionary mechanisms between mutualistic and pathogenic symbioses. Vie Milieu Paris 58(2):87–106, Epub 2008/01/01
12. Hooper LV (2009) Do symbiotic bacteria subvert host immunity? Nat Rev Microbiol 7(5):367–374, Epub 2009/04/17
13. Cerf-Bensussan N, Gaboriau-Routhiau V (2010) The immune system and the gut microbiota: friends or foes? Nat Rev Immunol 10(10):735–744, Epub 2010/09/25
14. Gaboriau-Routhiau V, Rakotobe S, Lecuyer E, Mulder I, Lan A, Bridonneau C et al (2009) The key role of segmented filamentous bacteria in the coordinated maturation of gut helper T cell responses. Immunity 31(4):677–689, Epub 2009/10/17
15. Bach JF (2002) The effect of infections on susceptibility to autoimmune and allergic diseases. N Engl J Med 347(12):911–920, Epub 2002/09/20

16. Blaser MJ, Falkow S (2009) What are the consequences of the disappearing human microbiota? Nat Rev Microbiol 7(12):887–894, Epub 2009/11/10
17. Jones KE, Patel NG, Levy MA, Storeygard A, Balk D, Gittleman JL et al (2008) Global trends in emerging infectious diseases. Nature 451(7181):990–993, Epub 2008/02/22
18. Woolhouse M, Gaunt E (2007) Ecological origins of novel human pathogens. Crit Rev Microbiol 33(4):231–242, Epub 2007/11/23
19. Mogensen TH (2009) Pathogen recognition and inflammatory signaling in innate immune defenses. Clin Microbiol Rev 22(2):240–273, Table of contents. Epub 2009/04/16
20. Vance RE, Isberg RR, Portnoy DA (2009) Patterns of pathogenesis: discrimination of pathogenic and nonpathogenic microbes by the innate immune system. Cell Host Microbe 6(1): 10–21, Epub 2009/07/21
21. Macpherson AJ, Geuking MB, McCoy KD (2005) Immune responses that adapt the intestinal mucosa to commensal intestinal bacteria. Immunology 115(2):153–162, Epub 2005/05/12
22. Van den Abbeele P, Van de Wiele T, Verstraete W, Possemiers S (2011) The host selects mucosal and luminal associations of coevolved gut microorganisms: a novel concept. FEMS Microbiol Rev 35(4):681–704, Epub 2011/03/03
23. May RM, Anderson RM (1983) Epidemiology and genetics in the coevolution of parasites and hosts. Proc R Soc Lond B Biol Sci 219(1216):281–313, Epub 1983/10/22
24. Levin BR, Bergstrom CT (2000) Bacteria are different: observations, interpretations, speculations, and opinions about the mechanisms of adaptive evolution in prokaryotes. Proc Natl Acad Sci U S A 97(13):6981–6985, Epub 2000/06/22
25. Dyall SD, Brown MT, Johnson PJ (2004) Ancient invasions: from endosymbionts to organelles. Science 304(5668):253–257, Epub 2004/04/10
26. Rohmer L, Hocquet D, Miller SI (2011) Are pathogenic bacteria just looking for food? Metabolism and microbial pathogenesis. Trends Microbiol 19(7):341–348, Epub 2011/05/24
27. Stecher B, Macpherson AJ, Hapfelmeier S, Kremer M, Stallmach T, Hardt WD (2005) Comparison of Salmonella enterica serovar Typhimurium colitis in germfree mice and mice pretreated with streptomycin. Infect Immun 73(6):3228–3241, Epub 2005/05/24
28. Stecher B, Robbiani R, Walker AW, Westendorf AM, Barthel M, Kremer M et al (2007) Salmonella enterica serovar typhimurium exploits inflammation to compete with the intestinal microbiota. PLoS Biol 5(10):2177–2189, Epub 2007/09/01
29. Xu J, Bjursell MK, Himrod J, Deng S, Carmichael LK, Chiang HC et al (2003) A genomic view of the human-Bacteroides thetaiotaomicron symbiosis. Science 299(5615):2074–2076, Epub 2003/03/29
30. Hacker J, Blum-Oehler G, Muhldorfer I, Tschape H (1997) Pathogenicity islands of virulent bacteria: structure, function and impact on microbial evolution. Mol Microbiol 23(6): 1089–1097, Epub 1997/03/01
31. Janeway CA Jr, Medzhitov R (2002) Innate immune recognition. Annu Rev Immunol 20:197–216, Epub 2002/02/28
32. Kawai T, Akira S (2010) The role of pattern-recognition receptors in innate immunity: update on Toll-like receptors. Nat Immunol 11(5):373–384, Epub 2010/04/21
33. Elinav E, Strowig T, Henao-Mejia J, Flavell RA (2011) Regulation of the antimicrobial response by NLR proteins. Immunity 34(5):665–679, Epub 2011/05/28
34. Kufer TA, Sansonetti PJ (2011) NLR functions beyond pathogen recognition. Nat Immunol 12(2):121–128, Epub 2011/01/20
35. Chamaillard M, Hashimoto M, Horie Y, Masumoto J, Qiu S, Saab L et al (2003) An essential role for NOD1 in host recognition of bacterial peptidoglycan containing diaminopimelic acid. Nat Immunol 4(7):702–707, Epub 2003/06/11
36. Girardin SE, Boneca IG, Carneiro LA, Antignac A, Jehanno M, Viala J et al (2003) Nod1 detects a unique muropeptide from gram-negative bacterial peptidoglycan. Science 300(5625):1584–1587, Epub 2003/06/07
37. Girardin SE, Boneca IG, Viala J, Chamaillard M, Labigne A, Thomas G et al (2003) Nod2 is a general sensor of peptidoglycan through muramyl dipeptide (MDP) detection. J Biol Chem 278(11):8869–8872, Epub 2003/01/16

38. Hugot JP, Chamaillard M, Zouali H, Lesage S, Cezard JP, Belaiche J et al (2001) Association of NOD2 leucine-rich repeat variants with susceptibility to Crohn’s disease. Nature 411(6837):599–603, Epub 2001/06/01
39. Ogura Y, Bonen DK, Inohara N, Nicolae DL, Chen FF, Ramos R et al (2001) A frameshift mutation in NOD2 associated with susceptibility to Crohn’s disease. Nature 411(6837): 603–606, Epub 2001/06/01
40. Brandl K, Plitas G, Schnabl B, DeMatteo RP, Pamer EG (2007) MyD88-mediated signals induce the bactericidal lectin RegIII gamma and protect mice against intestinal Listeria monocytogenes infection. J Exp Med 204(8):1891–1900, Epub 2007/07/20
41. Vaishnava S, Behrendt CL, Ismail AS, Eckmann L, Hooper LV (2008) Paneth cells directly sense gut commensals and maintain homeostasis at the intestinal host-microbial interface. Proc Natl Acad Sci U S A 105(52):20858–20863, Epub 2008/12/17
42. Rakoff-Nahoum S, Paglino J, Eslami-Varzaneh F, Edberg S, Medzhitov R (2004) Recognition of commensal microflora by toll-like receptors is required for intestinal homeostasis. Cell 118(2):229–241, Epub 2004/07/21
43. Round JL, Lee SM, Li J, Tran G, Jabri B, Chatila TA et al (2011) The Toll-like receptor 2 pathway establishes colonization by a commensal of the human microbiota. Science 332(6032):974–977, Epub 2011/04/23
44. Kobayashi KS, Chamaillard M, Ogura Y, Henegariu O, Inohara N, Nunez G et al (2005) Nod2-dependent regulation of innate and adaptive immunity in the intestinal tract. Science 307(5710):731–734, Epub 2005/02/05
45. Maeda S, Hsu LC, Liu H, Bankston LA, Iimura M, Kagnoff MF et al (2005) Nod2 mutation in Crohn’s disease potentiates NF-kappaB activity and IL-1beta processing. Science 307(5710):734–738, Epub 2005/02/05
46. McGuckin MA, Linden SK, Sutton P, Florin TH (2011) Mucin dynamics and enteric pathogens. Nat Rev Microbiol 9(4):265–278, Epub 2011/03/17
47. Swidsinski A, Weber J, Loening-Baucke V, Hale LP, Lochs H (2005) Spatial organization and composition of the mucosal flora in patients with inflammatory bowel disease. J Clin Microbiol 43(7):3380–3389, Epub 2005/07/08
48. Van der Sluis M, De Koning BA, De Bruijn AC, Velcich A, Meijerink JP, Van Goudoever JB et al (2006) Muc2-deficient mice spontaneously develop colitis, indicating that MUC2 is critical for colonic protection. Gastroenterology 131(1):117–129, Epub 2006/07/13
49. Velcich A, Yang W, Heyer J, Fragale A, Nicholas C, Viani S et al (2002) Colorectal cancer in mice genetically deficient in the mucin Muc2. Science 295(5560):1726–1729, Epub 2002/03/02
50. Mantle M, Thakore E, Hardin J, Gall DG (1989) Effect of Yersinia enterocolitica on intestinal mucin secretion. Am J Physiol 256(2 pt 1):G319–G327, Epub 1989/02/01
51. Nutten S, Sansonetti P, Huet G, Bourdon-Bisiaux C, Meresse B, Colombel JF et al (2002) Epithelial inflammation response induced by Shigella flexneri depends on mucin gene expression. Microbes Infect 4(11):1121–1124, Epub 2002/10/04
52. Bevins CL (2006) Paneth cell defensins: key effector molecules of innate immunity. Biochem Soc Trans 34(pt 2):263–266, Epub 2006/03/21
53. Hornef MW, Putsep K, Karlsson J, Refai E, Andersson M (2004) Increased diversity of intestinal antimicrobial peptides by covalent dimer formation. Nat Immunol 5(8):836–843, Epub 2004/07/06
54. Bates JM, Akerlund J, Mittge E, Guillemin K (2007) Intestinal alkaline phosphatase detoxifies lipopolysaccharide and prevents inflammation in zebrafish in response to the gut microbiota. Cell Host Microbe 2(6):371–382, Epub 2007/12/15
55. Goldberg RF, Austen WG Jr, Zhang X, Munene G, Mostafa G, Biswas S et al (2008) Intestinal alkaline phosphatase is a gut mucosal defense factor maintained by enteral nutrition. Proc Natl Acad Sci U S A 105(9):3551–3556, Epub 2008/02/23
56. Koyama I, Matsunaga T, Harada T, Hokari S, Komoda T (2002) Alkaline phosphatases reduce toxicity of lipopolysaccharides in vivo and in vitro through dephosphorylation. Clin Biochem 35(6):455–461, Epub 2002/11/05

57. Ramasamy S, Nguyen DD, Eston MA, Alam SN, Moss AK, Ebrahimi F et al (2011) Intestinal alkaline phosphatase has beneficial effects in mouse models of chronic colitis. Inflamm Bowel Dis 17(2):532–542, Epub 2010/07/21
58. Shifrin DA Jr, McConnell RE, Nambiar R, Higginbotham JN, Coffey RJ, Tyska MJ (2012) Enterocyte microvillus-derived vesicles detoxify bacterial products and regulate epithelial-microbial interactions. Curr Biol 22(7):627–631, Epub 2012/03/06
59. Macpherson AJ, Uhr T (2004) Induction of protective IgA by intestinal dendritic cells carrying commensal bacteria. Science 303(5664):1662–1665, Epub 2004/03/16
60. Macpherson AJ, Hunziker L, McCoy K, Lamarre A (2001) IgA responses in the intestinal mucosa against pathogenic and non-pathogenic microorganisms. Microbes Infect 3(12): 1021–1035, Epub 2001/10/03
61. Spits H, Di Santo JP (2011) The expanding family of innate lymphoid cells: regulators and effectors of immunity and tissue remodeling. Nat Immunol 12(1):21–27, Epub 2010/11/30
62. Lee YK, Menezes JS, Umesaki Y, Mazmanian SK (2011) Proinflammatory T-cell responses to gut microbiota promote experimental autoimmune encephalomyelitis. Proc Natl Acad Sci U S A 108(Suppl 1):4615–4622, Epub 2010/07/28
63. Chow J, Mazmanian SK (2010) A pathobiont of the microbiota balances host colonization and intestinal inflammation. Cell Host Microbe 7(4):265–276, Epub 2010/04/24
64. Miller MB, Bassler BL (2001) Quorum sensing in bacteria. Annu Rev Microbiol 55: 165–199, Epub 2001/09/07
65. Fujiya M, Musch MW, Nakagawa Y, Hu S, Alverdy J, Kohgo Y et al (2007) The Bacillus subtilis quorum-sensing molecule CSF contributes to intestinal homeostasis via OCTN2, a host cell membrane transporter. Cell Host Microbe 1(4):299–308, Epub 2007/11/17
66. Cossart P (2004) Bacterial invasion: a new strategy to dominate cytoskeleton plasticity. Dev Cell 6(3):314–315, Epub 2004/03/20
67. Hooper LV, Littman DR, Macpherson AJ (2012) Interactions between the microbiota and the immune system. Science 336(6086):1268–1273, Epub 2012/06/08
68. Mueller C, Macpherson AJ (2006) Layers of mutualism with commensal bacteria protect us from intestinal inflammation. Gut 55(2):276–284, Epub 2006/01/13
69. Macpherson AJ, Gatto D, Sainsbury E, Harriman GR, Hengartner H, Zinkernagel RM (2000) A primitive T cell-independent mechanism of intestinal mucosal IgA responses to commensal bacteria. Science 288(5474):2222–2226, Epub 2000/06/24
70. Vaishnava S, Yamamoto M, Severson KM, Ruhn KA, Yu X, Koren O et al (2011) The antibacterial lectin RegIIIgamma promotes the spatial segregation of microbiota and host in the intestine. Science 334(6053):255–258, Epub 2011/10/15
71. Slack E, Hapfelmeier S, Stecher B, Velykoredko Y, Stoel M, Lawson MA et al (2009) Innate and adaptive immunity cooperate flexibly to maintain host-microbiota mutualism. Science 325(5940):617–620, Epub 2009/08/01
72. Lotz M, Gutle D, Walther S, Menard S, Bogdan C, Hornef MW (2006) Postnatal acquisition of endotoxin tolerance in intestinal epithelial cells. J Exp Med 203(4):973–984, Epub 2006/04/12
73. Rogler G, Brand K, Vogl D, Page S, Hofmeister R, Andus T et al (1998) Nuclear factor kappaB is activated in macrophages and epithelial cells of inflamed intestinal mucosa. Gastroenterology 115(2):357–369, Epub 1998/07/25
74. Smith PD, Ochsenbauer-Jambor C, Smythies LE (2005) Intestinal macrophages: unique effector cells of the innate immune system. Immunol Rev 206:149–159, Epub 2005/07/29
75. Ogura Y, Lala S, Xin W, Smith E, Dowds TA, Chen FF et al (2003) Expression of NOD2 in Paneth cells: a possible link to Crohn's ileitis. Gut 52(11):1591–1597, Epub 2003/10/23
76. Ortega-Cava CF, Ishihara S, Rumi MA, Kawashima K, Ishimura N, Kazumori H et al (2003) Strategic compartmentalization of Toll-like receptor 4 in the mouse gut. J Immunol 170(8): 3977–3985, Epub 2003/04/12
77. Viala J, Chaput C, Boneca IG, Cardona A, Girardin SE, Moran AP et al (2004) Nod1 responds to peptidoglycan delivered by the Helicobacter pylori cag pathogenicity island. Nat Immunol 5(11):1166–1174, Epub 2004/10/19

78. Gewirtz AT, Navas TA, Lyons S, Godowski PJ, Madara JL (2001) Cutting edge: bacterial flagellin activates basolaterally expressed TLR5 to induce epithelial proinflammatory gene expression. J Immunol 167(4):1882–1885, Epub 2001/08/08
79. Reed KA, Hobert ME, Kolenda CE, Sands KA, Rathman M, O'Connor M et al (2002) The Salmonella typhimurium flagellar basal body protein FliE is required for flagellin production and to induce a proinflammatory response in epithelial cells. J Biol Chem 277(15):13346–13353, Epub 2002/02/01
80. Burns K, Clatworthy J, Martin L, Martinon F, Plumpton C, Maschera B et al (2000) Tollip, a new component of the IL-1RI pathway, links IRAK to the IL-1 receptor. Nat Cell Biol 2(6):346–351, Epub 2000/06/15
81. Zhang G, Ghosh S (2002) Negative regulation of toll-like receptor-mediated signaling by Tollip. J Biol Chem 277(9):7059–7065, Epub 2001/12/26
82. Lu M, Zhang M, Takashima A, Weiss J, Apicella MA, Li XH et al (2005) Lipopolysaccharide deacylation by an endogenous lipase controls innate antibody responses to Gram-negative bacteria. Nat Immunol 6(10):989–994, Epub 2005/09/13
83. Munford RS (2005) Detoxifying endotoxin: time, place and person. J Endotoxin Res 11(2):69–84, Epub 2005/06/14
84. Alhawi M, Stewart J, Erridge C, Patrick S, Poxton IR (2009) Bacteroides fragilis signals through Toll-like receptor (TLR) 2 and not through TLR4. J Med Microbiol 58(pt 8): 1015–1022, Epub 2009/06/17
85. Weintraub A, Zahringer U, Wollenweber HW, Seydel U, Rietschel ET (1989) Structural characterization of the lipid A component of Bacteroides fragilis strain NCTC 9343 lipopolysaccharide. Eur J Biochem 183(2):425–431, Epub 1989/08/01
86. Kelly D, Campbell JI, King TP, Grant G, Jansson EA, Coutts AG et al (2004) Commensal anaerobic gut bacteria attenuate inflammation by regulating nuclear-cytoplasmic shuttling of PPAR-gamma and RelA. Nat Immunol 5(1):104–112, Epub 2003/12/24
87. Tien MT, Girardin SE, Regnault B, Le Bourhis L, Dillies MA, Coppee JY et al (2006) Anti-inflammatory effect of Lactobacillus casei on Shigella-infected human intestinal epithelial cells. J Immunol 176(2):1228–1237, Epub 2006/01/06
88. Piccinini AM, Midwood KS (2010) DAMPening inflammation by modulating TLR signalling. Mediators Inflamm 2010:1–21. doi: 10.1155/2010/672395, Epub 2010/08/14
89. Pedra JH, Cassel SL, Sutterwala FS (2009) Sensing pathogens and danger signals by the inflammasome. Curr Opin Immunol 21(1):10–16, Epub 2009/02/19
90. Mariathasan S, Weiss DS, Newton K, McBride J, O'Rourke K, Roose-Girma M et al (2006) Cryopyrin activates the inflammasome in response to toxins and ATP. Nature 440(7081): 228–232, Epub 2006/01/13
91. Pelegrin P, Surprenant A (2006) Pannexin-1 mediates large pore formation and interleukin-1beta release by the ATP-gated P2X7 receptor. EMBO J 25(21):5071–5082, Epub 2006/10/13
92. Skoberne M, Beignon AS, Bhardwaj N (2004) Danger signals: a time and space continuum. Trends Mol Med 10(6):251–257, Epub 2004/06/05
93. Lamkanfi M, Dixit VM (2011) Modulation of inflammasome pathways by bacterial and viral pathogens. J Immunol 187(2):597–602, Epub 2011/07/08
94. Santaolalla R, Abreu MT (2012) Innate immunity in the small intestine. Curr Opin Gastroenterol 28(2):124–129, Epub 2012/01/14
95. Franchi L, Munoz-Planillo R, Nunez G (2012) Sensing and reacting to microbes through the inflammasomes. Nat Immunol 13(4):325–332, Epub 2012/03/21
96. Nakanishi K, Yoshimoto T, Tsutsui H, Okamura H (2001) Interleukin-18 is a unique cytokine that stimulates both Th1 and Th2 responses depending on its cytokine milieu. Cytokine Growth Factor Rev 12(1):53–72, Epub 2001/04/20
97. Sutton CE, Lalor SJ, Sweeney CM, Brereton CF, Lavelle EC, Mills KH (2009) Interleukin-1 and IL-23 induce innate IL-17 production from gammadelta T cells, amplifying Th17 responses and autoimmunity. Immunity 31(2):331–341, Epub 2009/08/18
98. Franchi L, Eigenbrod T, Munoz-Planillo R, Nunez G (2009) The inflammasome: a caspase-1-activation platform that regulates immune responses and disease pathogenesis. Nat Immunol 10(3):241–247, Epub 2009/02/18

99. Lamkanfi M, Walle LV, Kanneganti TD (2011) Deregulated inflammasome signaling in disease. Immunol Rev 243(1):163–173, Epub 2011/09/03
100. Iwai H, Kim M, Yoshikawa Y, Ashida H, Ogawa M, Fujita Y et al (2007) A bacterial effector targets Mad2L2, an APC inhibitor, to modulate host cell cycling. Cell 130(4):611–623, Epub 2007/08/28
101. Kim M, Ogawa M, Fujita Y, Yoshikawa Y, Nagai T, Koyama T et al (2009) Bacteria hijack integrin-linked kinase to stabilize focal adhesions and block cell detachment. Nature 459(7246):578–582, Epub 2009/06/03
102. Boyle EC, Brown NF, Finlay BB (2006) Salmonella enterica serovar Typhimurium effectors SopB, SopE, SopE2 and SipA disrupt tight junction structure and function. Cell Microbiol 8(12):1946–1957, Epub 2006/07/28
103. Bruno VM, Hannemann S, Lara-Tejero M, Flavell RA, Kleinstein SH, Galan JE (2009) Salmonella typhimurium type III secretion effectors stimulate innate immune responses in cultured epithelial cells. PLoS Pathog 5(8):e1000538, Epub 2009/08/08
104. Muller AJ, Hoffmann C, Galle M, Van Den Broeke A, Heikenwalder M, Falter L et al (2009) The S. typhimurium effector SopE induces caspase-1 activation in stromal cells to initiate gut inflammation. Cell Host Microbe 6(2):125–136
105. Bartoleschi C, Pardini MC, Scaringi C, Martino MC, Pazzani C, Bernardini ML (2002) Selection of Shigella flexneri candidate virulence genes specifically induced in bacteria resident in host cell cytoplasm. Cell Microbiol 4(9):613–626, Epub 2002/10/23
106. Lenz LL, Mohammadi S, Geissler A, Portnoy DA (2003) SecA2-dependent secretion of autolytic enzymes promotes Listeria monocytogenes pathogenesis. Proc Natl Acad Sci U S A 100(21):12432–12437, Epub 2003/10/07
107. Ham H, Sreelatha A, Orth K (2011) Manipulation of host membranes by bacterial effectors. Nat Rev Microbiol 9(9):635–646, Epub 2011/07/19
108. Viboud GI, Bliska JB (2005) Yersinia outer proteins: role in modulation of host cell signaling responses and pathogenesis. Annu Rev Microbiol 59:69–89, Epub 2005/04/26
109. Neish AS, Naumann M (2011) Microbial-induced immunomodulation by targeting the NF-kappaB system. Trends Microbiol 19(12):596–605, Epub 2011/10/01
110. Newman RM, Salunkhe P, Godzik A, Reed JC (2006) Identification and characterization of a novel bacterial virulence factor that shares homology with mammalian Toll/interleukin-1 receptor family proteins. Infect Immun 74(1):594–601, Epub 2005/12/22
111. Krachler AM, Woolery AR, Orth K (2011) Manipulation of kinase signaling by bacterial pathogens. J Cell Biol 195(7):1083–1092, Epub 2011/11/30
112. Arbibe L, Kim DW, Batsche E, Pedron T, Mateescu B, Muchardt C et al (2007) An injected bacterial effector targets chromatin access for transcription factor NF-kappaB to alter transcription of host genes involved in immune responses. Nat Immunol 8(1):47–56, Epub 2006/12/13
113. Schotte P, Denecker G, Van Den Broeke A, Vandenabeele P, Cornelis GR, Beyaert R (2004) Targeting Rac1 by the Yersinia effector protein YopE inhibits caspase-1-mediated maturation and release of interleukin-1beta. J Biol Chem 279(24):25134–25142, Epub 2004/04/03
114. Brzuszkiewicz E, Bruggemann H, Liesegang H, Emmerth M, Olschlager T, Nagy G et al (2006) How to become a uropathogen: comparative genomic analysis of extraintestinal pathogenic Escherichia coli strains. Proc Natl Acad Sci U S A 103(34):12879–12884, Epub 2006/08/17
115. Round JL, Mazmanian SK (2009) The gut microbiota shapes intestinal immune responses during health and disease. Nat Rev Immunol 9(5):313–323, Epub 2009/04/04
116. Talham GL, Jiang HQ, Bos NA, Cebra JJ (1999) Segmented filamentous bacteria are potent stimuli of a physiologically normal state of the murine gut mucosal immune system. Infect Immun 67(4):1992–2000, Epub 1999/03/20
117. Ivanov II, Atarashi K, Manel N, Brodie EL, Shima T, Karaoz U et al (2009) Induction of intestinal Th17 cells by segmented filamentous bacteria. Cell 139(3):485–498, Epub 2009/10/20

118. Littman DR, Pamer EG (2011) Role of the commensal microbiota in normal and pathogenic host immune responses. Cell Host Microbe 10(4):311–323, Epub 2011/10/25
119. Stepankova R, Powrie F, Kofronova O, Kozakova H, Hudcovic T, Hrncir T et al (2007) Segmented filamentous bacteria in a defined bacterial cocktail induce intestinal inflammation in SCID mice reconstituted with CD45RBhigh CD4+ T cells. Inflamm Bowel Dis 13(10):1202–1211, Epub 2007/07/04
120. Wu HJ, Ivanov II, Darce J, Hattori K, Shima T, Umesaki Y et al (2010) Gut-residing segmented filamentous bacteria drive autoimmune arthritis via T helper 17 cells. Immunity 32(6):815–827, Epub 2010/07/14
121. Cahill RJ, Foltz CJ, Fox JG, Dangler CA, Powrie F, Schauer DB (1997) Inflammatory bowel disease: an immunity-mediated condition triggered by bacterial infection with Helicobacter hepaticus. Infect Immun 65(8):3126–3131, Epub 1997/08/01
122. Erdman SE, Poutahidis T, Tomczak M, Rogers AB, Cormier K, Plank B et al (2003) CD4+ CD25+ regulatory T lymphocytes inhibit microbially induced colon cancer in Rag2-deficient mice. Am J Pathol 162(2):691–702, Epub 2003/01/28
123. Kullberg MC, Ward JM, Gorelick PL, Caspar P, Hieny S, Cheever A et al (1998) Helicobacter hepaticus triggers colitis in specific-pathogen-free interleukin-10 (IL-10)-deficient mice through an IL-12- and gamma interferon-dependent mechanism. Infect Immun 66(11): 5157–5166, Epub 1998/10/24
124. Mazmanian SK, Round JL, Kasper DL (2008) A microbial symbiosis factor prevents intestinal inflammatory disease. Nature 453(7195):620–625, Epub 2008/05/30
125. Round JL, Mazmanian SK (2010) Inducible Foxp3+ regulatory T-cell development by a commensal bacterium of the intestinal microbiota. Proc Natl Acad Sci U S A 107(27): 12204–12209, Epub 2010/06/23
126. Sears CL (2009) Enterotoxigenic Bacteroides fragilis: a rogue among symbiotes. Clin Microbiol Rev 22(2):349–369, Table of contents. Epub 2009/04/16
127. Housseau F, Sears CL (2010) Enterotoxigenic Bacteroides fragilis (ETBF)-mediated colitis in Min (Apc+/-) mice: a human commensal-based murine model of colon carcinogenesis. Cell Cycle 9(1):3–5, Epub 2009/12/17
128. Wu S, Rhee KJ, Albesiano E, Rabizadeh S, Wu X, Yen HR et al (2009) A human colonic commensal promotes colon tumorigenesis via activation of T helper type 17 T cell responses. Nat Med 15(9):1016–1022, Epub 2009/08/25
129. Chow WH, Blaser MJ, Blot WJ, Gammon MD, Vaughan TL, Risch HA et al (1998) An inverse relation between cagA+strains of Helicobacter pylori infection and risk of esophageal and gastric cardia adenocarcinoma. Cancer Res 58(4):588–590, Epub 1998/03/04
130. Dorer MS, Talarico S, Salama NR (2009) Helicobacter pylori's unconventional role in health and disease. PLoS Pathog 5(10):e1000544, Epub 2009/10/27
131. Sonnenberg A, Lash RH, Genta RM (2010) A national study of Helicobacter pylori infection in gastric biopsy specimens. Gastroenterology 139(6):1894–1901.e2, quiz e12. Epub 2010/08/24
132. Odenbreit S, Puls J, Sedlmaier B, Gerland E, Fischer W, Haas R (2000) Translocation of Helicobacter pylori CagA into gastric epithelial cells by type IV secretion. Science 287(5457):1497–1500, Epub 2000/02/26
133. Polk DB, Peek RM Jr (2010) Helicobacter pylori: gastric cancer and beyond. Nat Rev Cancer 10(6):403–414, Epub 2010/05/25
134. El-Omar EM, Carrington M, Chow WH, McColl KE, Bream JH, Young HA et al (2000) Interleukin-1 polymorphisms associated with increased risk of gastric cancer. Nature 404(6776):398–402, Epub 2000/04/04
135. Mukhopadhya I, Hansen R, El-Omar EM, Hold GL (2012) IBD-what role do Proteobacteria play? Nat Rev Gastroenterol Hepatol 9(4):219–230, Epub 2012/02/22

Chapter 15
Cytokines in Inflammatory Bowel Disease

Matthias A. Engel and Markus F. Neurath

Abstract Erroneous communication between the innate and adaptive immune systems through cytokines results in exaggerated or attenuated immune response. It is not known whether the pathologic immune response in inflammatory bowel disease has its origin in a dysbalance of pro- and anti-inflammatory cytokine release or whether it is secondary in subsequence of a defective intestinal barrier or the destructive power of aggressive microbiota in the gut lumen.

Many cytokines have been found upregulated in patients with inflammatory bowel diseases in correlation with disease activity. A central role seem to play cytokines that coordinate the T helper cell response. Although big scientific efforts have been made until today, only TNF blockers reached the clinical routine and many anti-cytokine strategies were only effective in rodent models of colitis. This chapter gives an overview about relevant pathomechanisms in mucosal immunology of the gut and focuses on the key cytokines that have been identified as targets for novel therapeutic strategies in human IBD.

M.A. Engel, M.D. • M.F. Neurath, M.D., Ph.D. (✉)
Division of Gastroenterology, First Department of Medicine,
Universitätsklinikum Erlangen, Friedrich-Alexander-Universität Erlangen-Nurnberg,
Ulmenweg 18, 91054 Erlangen, Germany
e-mail: Matthias.Engel@uk-erlangen.de; Markus.Neurath@uk-erlangen.de

M. D'Amato and J.D. Rioux (eds.), *Molecular Genetics of Inflammatory Bowel Disease*,
DOI 10.1007/978-1-4614-8256-7_15,

Introduction

Inflammatory bowel diseases (IBD) comprise ulcerative colitis (UC) and Crohn's disease (CD). Both are chronic, relapsing inflammatory conditions of the gut. UC is usually limited to the colon but CD can affect the whole GI tract discontinuously. While UC affects the mucosa/submucosa only, CD may involve the whole bowel wall leading to fistula formation. The pathogenesis of both UC and CD is not completely understood. The aberrant immune response is thought to result from a pathologic interaction between the microbial flora of the intestinal lumen and the mucosal immune system in genetically predisposed individuals. There are three potential levels of disease origin that do all result in an exaggerated immune response: an abnormal composition of the microbial flora, disintegration of the intestinal epithelium, or pathologic interaction between the different immune cell subtypes of the mucosal immune system per se may all ultimately result in colitis.

Immune cells communicate through peptides termed cytokines. An unknown initial pathogenetic insult in IBD pathogenesis leads to the perturbation of the first line of mucosal defense comprising the innate immune cell system. The consecutive release of increased amounts of proinflammatory cytokines ultimately leads to an overwhelming inflammatory, destructive process within the gut. An individual cytokine profile driven by Th1/Th17 or Th2 cells leads to the manifestation of the two clinically separate disease entities, Crohn's disease and ulcerative colitis, respectively.

Understanding the characteristic cytokine profiles that are involved in IBD pathogenesis is the key to novel therapeutic approaches. Elucidating these pathways gives the opportunity to selectively dampen the inflammatory reaction by either blockage of certain proinflammatory cytokines or induction of anti-inflammatory cytokine pathways.

Pleiotropic Cytokines with Central Roles in Immune Regulation and IBD

The first line of mucosal immunity comprises dendritic cells and macrophages that secrete characteristic cytokines to first induce a local inflammatory response. These cells have also antigen-presenting capabilities (antigen-presenting cells, APCs) and thus involve the adaptive immune system in the second line. TNF-α is one of the central proinflammatory cytokines. It is secreted by cells of the innate immune system (macrophages, monocytes, neutrophils, NK-cells) following their stimulation by bacterial lipopolysaccharides. In addition, Th1 and Th2 cells produce TNF-α. TNF-α exists in two forms: membrane-bound TNF (mTNF) and a soluble form (sTNF) which is generated by proteolytic cleavage via the TNF-α-converting enzyme. The two receptors, tumor necrosis factor receptor 1 (TNFR1) (p55) and TNFR2 (p75), mediate its biological effects [1, 2]. Numerous studies have shown increased TNF-α and TNF-receptor I and II levels in the intestinal mucosa and plasma in patients with IBD and its correlation with clinical and laboratory indices [3–8]. Anti-TNF-antibody treatment is meanwhile an established therapy in steroid-refractory CD patients, in fistulizing CD and active UC [9]. However, the molecular mechanism of its therapeutic effect is not yet completely understood. A central mechanism seems to be the induction of lamina propria T-cell apoptosis mediated by the proapoptotic proteins Bax and Bak [10, 11]. In line with these results, infliximab induced apoptosis in a T-cell line. An interesting observation and therapeutic feature was that apoptosis was only induced in activated (CD3/CD28 stimulated) but not unstimulated T cells. Supporting data was derived from another TNF-antibody (adalimumab) that also induced apoptosis in peripheral blood monocytes [12]. However, in contrast to this paradigm, another TNF-antibody, certolizumab [13], which was clinically effective did not mediate increased levels of apoptosis in peripheral blood monocytes or lymphocytes [14]. Further studies are needed to clarify this discrepancy. Interestingly, Atreya et al. recently found that the clinically effective anti-TNF antibodies infliximab, adalimumab, and certolizumab pegol all induced T-cell apoptosis by blocking the interaction between mTNF on APCs and TNFR2 on T cells [15]. Thus, these three antibodies appear to target T-cell apoptosis via a specific molecular pathway.

The introduction of anti-TNF therapies in 1998 had a major impact on IBD management and prognosis [16]. Although often very effective TNF blockers bear the risk for severe infections and some patients have to cope with loss of response and development of antidrug antibodies. In some patients, the loss of response is associated with low serum trough levels [17] as a consequence of increased clearance. Optimizing serum levels by repeated measurements and customized application of the drugs may increase response rates. The therapeutic strategies to battle antidrug antibodies include scheduled use of anti-TNF agents [18] and concomitant use of immunosuppressive drugs [19, 20] or corticosteroids [21]. Another option for patients that present with low trough levels in the presence of antidrug antibodies is switching the anti-TNF agent. However, a deeper understanding of therapeutic mechanisms of TNF blockage is necessary to understand why a relevant proportion of IBD patients are nonresponders to this type of therapy.

Alternative strategies to inhibit TNF signaling are currently tested in clinical studies. For example, the small molecule ATN-103 is a single heavy-chain variable domain [22], the smallest known antigen-binding antibody fragment, and demonstrates higher stability than conventional antibodies. For the about 30–40 % of CD patients that are primary nonresponders to classical anti-TNF agents, another promising strategy is the immunization of patients against TNF [23]. However, since such immunization generates a humoral response, a boost is required every 3–4 months. Another approach is the oral administration of Lactococcus lactis secreting an anti-TNF Fab. However, large clinical trials will be necessary to demonstrate the efficacy of this anti-TNF agent.

In addition to TNF, some reports suggest a functional relevance of the cytokine IL-1. There are two functionally almost equivalent forms of IL1, IL1-α, and IL1-β (both 17 kDa) that are encoded by two different genes. IL1-β is the predominant form in humans [IL1-α in mice]. There is an inverse correlation between IBD activity and the interleukin 1 receptor a (IL-1Ra)/IL-1 ratio, suggesting a critical role of the IL-1 cytokine system in contribution to chronic gut inflammation [24]. The main source of IL-1 in the colonic mucosa of IBD patients are stimulated colonic macrophages that induce interleukin IL-1-converting enzyme (ICE) to produce large amounts of mature IL1-β [25]. Proinflammatory biological effects of IL-1 include the increased expression of adhesion molecules on vascular endothelium and chemokines [26, 27]. IL1-β and TNF-α induce the synthesis of matrix metalloproteinases (MMPs) that exert devastating effects by causing tissue degradation and ulceration [28]. Both IL1-β and TNF-α are produced in excess in IBD. The majority of these IL-1/TNF-induced inflammatory pathways are mediated by activation of a central proinflammatory transcription factor denoted as NF-kappaB. NF-kappaB inhibitors showed therapeutic benefit in mice with experimental colitis which led to the development of NF-kappaB blockers for IBD treatment [29]. But large clinical trials are still missing. However, NF-kappaB function depends on the cell type. While it augments proinflammatory cytokine production by immune cells, it is essential for maintaining epithelial integrity of intestinal epithelial cells [30]. Thus, selective targeting rather than general inhibition of NF-kappaB activity in the gut is desirable in IBD.

IL-6 is another central cytokine in immunity per se, and increased levels are correlated with clinical and histological severity of CD and UC [31, 32]. Intestinal macrophages and CD4+ T cells are the main producers of IL-6 [33]. Interestingly, elevated serum IL-6 levels after steroid-induced remission were highly predictive for relapse [34, 35], ascribing this cytokine an important role as a serological biomarker in patients with IBD. IL-6 exerts its biological effects not only through binding to its membrane-bound receptor but also through a soluble IL-6 receptor (sIL-6R) and can thus stimulate numerous cells that express only the signal-transducing subunit gp130, a process referred to as IL-6 trans-signaling. The soluble IL-6 receptor is mainly produced by macrophages [36]. A recombinant sgp130 protein fused to the Fc region of human IgG1, which selectively blocked sIL-6R [37], had the same efficacy as an IL-6R antibody in experimental colitis models which emphasizes the biological importance of trans-signaling. Steroid-responsive IBD patients showed a decrease in serum sIL-6R concentrations under therapy [38]. Upstream, LPS-induced activation of macrophages induces a cytokine named macrophage-migration inhibitory factor (MIF) which again induces IL-6. Experimental murine colitis was successfully treated with monoclonal antibodies against MIF [39]. Downstream, IL-6 signals through signal transducer and activator of transcription-3 (STAT3) and STAT3 phosphorylation were found to be increased in $CD4^+$ T cells in both CD and UC [33, 40, 41]. This IL-6/STAT3 pathway links the innate to the adaptive immune system. STAT-3 induces the antiapoptotic factors Bcl-2 and Bcl-xL thereby mediating T-cell resistance against apoptosis [33, 42], and consecutive accumulation of T cells results in chronic colitis in humans [33, 43]. This hypothesis was elegantly proven by the use of a monoclonal antibody against the IL-6 receptor which led to abolishment of murine colitis by inhibiting the recruitment of leukocytes and increasing T-cell apoptosis [35]. In a proof of concept study, lamina propria T cells from patients with IBD were treated with anti-IL-6R agents, which led to enhanced in vitro apoptosis [33]. Targeting total STAT3 activity directly may not be beneficial as STAT3 not only conveys proinflammatory but also anti-inflammatory effects, however [44]. The humanized antihuman IL-6R IgG1 antibody tocilizumab (Roche/Chugai Pharmaceutical Co. Ltd., Tokyo, Japan) that binds both membrane bound and soluble forms of the human IL-6R was tested in a first small placebo-controlled CD study [45]. The results were promising with clinical response rates of 80 % compared to 31 % in the placebo group.

It has to be emphasized that the named cytokines TNF-α, IL-1β, and IL-6 are associated with both forms of human IBD as they exert multiple downstream effects [46, 47]. These cytokines activate the NF-kappaB and mitogen-activated protein kinases (MAPK) pathways, and both have tremendous effects in proinflammatory cascades. Moreover, IL-6 and IL-1β were also shown to act upstream to induce Th17 responses [48, 49]. Recently, TL1A (tumor necrosis factor-like ligand) was linked to IBD. This cytokine is produced by both antigen-presenting and T cells. It activates the DR3 receptor (which belongs to the family of TNF receptors) that is mainly found on T cells. TLA1 is induced in APCs by TLR ligands on one hand and FcR cross-linking on the other. T cells secrete TLA1 upon TCR stimulation [50]. This cytokine has relevant biological effects in colitis. In the DSS model TLA1 injection increased colonic Th1 and Th17 responses in mice [51], and TLA1 antibodies or DR3 blockers ameliorated TNBS and DSS colitis [52, 53]. TL1A inhibits the induction of adaptive Foxp3+ regulatory T cells but induces the expansion of existing ones [54]. Mice having a TL1A transgene expressed in T cells show increased baseline T-cell and B-cell activation via TCR stimulation which ultimately leads to a unique spontaneous gut inflammation predominantly in the terminal ileum resembling Crohn's disease [52]. The fact that this type of inflammation was driven by IL-13 synthesis once again shuffles the concept of polarized Th cell responses in IBD, since oxazolone colitis which resembles human UC in some aspects is also driven by IL-13 (see below). In human studies elevated levels of TLA1 were found in both forms of IBD suggesting TLA1 to function as a biological enhancer rather than a direct proinflammatory cytokine driving Th cell differentiation [50, 51, 55, 56].

Cytokines and the Adaptive Immune System in IBD

The second line of immune defense comprises the adaptive immune system, and there is consensus about the important role of CD4+ T cells in the pathogenesis of IBD. Intestinal lamina propria T cells are unique in a way that they are hyporesponsive to T-cell receptor stimulation [57]. Hence, mucosal T lymphocytes are critically dependent on costimulation driven by luminal antigens and mediated by the innate immune system which is capable of producing cytokines. In addition to differences in localization in the GI tract, CD and UC differ in the characteristic cytokine profile driven by specific subsets of T helper cells (see Fig. 15.1). CD is associated with a Th1/Th17 T-cell-mediated response. IL-12 is the major cytokine of Th1-mediated

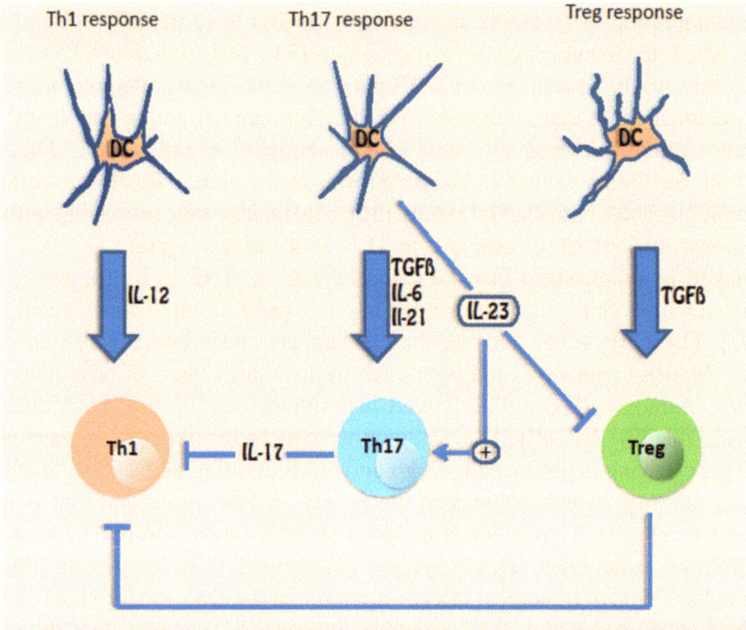

Fig. 15.1 Complex interplay between the Th1, Th17, and Treg lineages of T helper cells. Th17 cells appear to have a dual role at different stages of the inflammatory reaction. They may act both permissive and inhibitory of the Th1 response. IL23 is critically involved in these interactions

immune responses. IL-12 is a heterodimeric 70 kDa glycoprotein (IL12-p70) consisting of a 40 kDa subunit (IL12-p40) and a 35 kDa subunit (IL12-p35) linked by disulfide bonds that are essential for the biological activity of IL-12. IL-12 was originally identified to play a pivotal role in TNBS colitis, a rodent model of colitis that resembles CD. Treatment with a monoclonal antibody against IL-12p40 abrogated TNBS colitis in mice [58]. It was suggested that the prevention of Th1 T-cell activation and the induction of Fas-mediated T-cell apoptosis were responsible for the anti-inflammatory net effect [58–61]. Increased levels of IL-12 were also found in the lamina propria of patients with CD [59]. However, it is not yet fully understood whether other members of the IL-12 family play a more dominant role in intestinal inflammation. The IL-12 cytokine family comprises IL-12 p35/p40, IL23 p19/p40, IL-27 EBI3 (Epstein–Barr virus-induced gene 3)/p28 and IL-35 p35/EBI3 [62], and several other heterodimeric proteins. Recent data suggested that IL23 rather than IL-12 drives experimental colonic inflammation [63], independently of T-cell activation. IL23 thus may directly activate cells of the innate mucosal immune system to produce proinflammatory cytokines. Clinical studies utilizing a human anti-IL-12p40 antibody for the treatment of CD showed some efficacy in CD patients resistant to TNF-alpha blockade [64, 65]. The results of a Phase III study with ustekinumab (Stelara®, Centocor, PA, USA) are currently awaited.

The fact that anti-IL-12p40 antibody treatment is effective in CD does not necessarily identify the Th1 response as the master T helper cell population in this disease. Both IL-12 and IL23 are heterodimers that consist of one p40 chain [66]. The function of IL23 is to interact with differentiated Th17 cells, which leads to stabilization of differentiation and expansion of proinflammatory Th17 cells [49, 67–69]. If IL23 is lacking Th17 cells produce the anti-inflammatory cytokine IL-10 and therefore become weak inducers of inflammation [70]. As a matter of fact, the major cytokines in CD arise from Th1 and Th17 CD4+ T-cell differentiation and consist of interferon-γ and *IL-22/IL-17*. Several studies underline the importance of the Th17 response in the mouse model of transfer colitis. *Rag* (recombination-activating gene)-deficient mice lacking IL-23p19 did not develop colitis upon transfer of naïve T cells, whereas *Rag* KO deficient in IL-12p35 did [63, 71].

Another experimental therapeutic approach utilized the same principle of blocking the IL-12/IL23 pathways by selective inhibition of the translocation of c-Rel, a member of the NF-kappaB family of transcription factors, into the nucleus, thus blocking both p40 and p35 promoter activity [72]. The compound apilimod mesylate (Synta Pharmaceuticals, MA, USA) was recently tested in a randomized, double-blind, placebo-controlled trial in 220 active CD patients. Apilimod mesylate was found to be safe; however, the clinical response rate was even lower than in the placebo group. It is hard to say whether this contradicts the principle of therapeutic efficacy of p40 block in CD. Alternatively, local drug concentrations of this orally administered drug might have been too low to efficiently block p40 gene expression. Further clinical trials with different dosages will be required to finally evaluate the potential of this drug.

The effects of the major Th17 cytokine IL-17 in colitis are contradictory and were critically depending on the mouse model used. *Il17a* knockout (KO) mice demonstrated increased susceptibility to dextran sulfate sodium [73] colitis [74]. Contradictory data were shown in the adoptive transfer model, where both protective and neutral roles of IL-17A were described [75, 76].

However, the transfer of IL-17R-deficient T cells resulted in a comparable degree of colitis than wild-type T cells did [75]. Interestingly, *Il17ra KO* mice or wild-type mice treated with an IL-17RA IgG1 fusion protein were protected against trinitrobenzenesulfonic acid (TNBS)-induced colitis [77]. Another relevant cytokine might be IL-17F [78]. A recent report showed that *Il-17f* KO mice were largely resistant against DSS colitis [74]. *Il23*-deficient mice show greatly reduced levels of other proinflammatory cytokines [TNF-α and IFN-γ] [79]. IL23 enhanced intestinal T-cell proliferation and lamina propria levels of IL-17 and IFN-γ, supposedly derived from a special subset of TH17 cells that also produce IFN-γ. Compelling evidence for the role of the TH17 T helper cell subset was generated from a modified model of transfer colitis where SCID mice were transferred previously polarized Th1 or Th17 that mainly produce IFN-γ and IL-17, respectively. Surprisingly, only Th17 cells but not Th1 could induce transfer colitis. Furthermore, in Th17-induced transfer colitis, anti-IL-23p19 treatment showed both prophylactic and therapeutic efficacy. It has to be mentioned that the TH17 cells that produced primarily IL-17 before T-cell transfer demonstrated also high levels of IFN-γ production during transfer colitis. Disappointingly, targeting IFN-γ with fontolizumab (HuZAF™), a humanized IgG1 antibody against recombinant human IFN-γ (Protein Design Labs), was ineffective in the treatment of CD in an i.v./s.c. regimen [80, 81]. However, the achieved reduction of CRP levels in the patients' serum raises hope for potentially reasonable effects of this compound when using it in a combined approach. Combination with TNF blockers seems promising due to the fact that in vitro neutralization of IFN-γ did not reduce TNF production by monocytes, and this sustained TNF production may be responsible for the therapeutic failure of IFN-γ blocking strategies.

There is interesting data on the plasticity of Th17 cells. Adequately stimulated, e.g., with IL-12 or IL-23, they can convert into the Th1 lineage [79, 82, 83]. The "class switch" from IL-17A to IFN-γ production occurs in a TGF-β1-free environment [84, 85]. IL-12, IL-23, and Smad7 (which abrogates TGF-beta1 activity) are highly upregulated in the gut of IBD patients which makes it conceivable that IFN-γ-producing T cells are derived at least in part from classical IL-17A-expressing Th17 cells [58]. Per se Th17 cells can produce IL-17A, IL-17F, IL-21, IL-22, IL-26, and IL-9 [89–92], but the fact that not all of the Th17 cells produce the complete arsenal of cytokines suggests that the "biological truth" is even more complex and that different subsets of Th17 cells may exist. In summary, both Th1 and Th17 cytokines seem to promote inflammation in the gut, and compounds interfering with Th1 and Th17 cell activity could be useful future therapeutics for IBD. Therefore, another promising target may be IL-21. This cytokine stimulates both Th1 and Th17 immunity [93, 94]. Increased levels of IL-21 were detected in IBD patients and murine models of inflammatory bowel disease. The production of Th17-related cytokines was found reduced when mice were treated with a neutralizing IL-21R fusion protein that attenuated DSS colitis. In accord, DSS and TNBS colitis were abrogated in *Il-21* KO mice, and this protection was accompanied by a distinct reduction of IL-17A and F levels [95]. Many other biological features of IL-21 ultimately promote gut inflammation. These effects include inhibition of differentiation of regulatory T cells (Tregs) and at the same time increasing the resistance of CD4+ T cells towards Treg-mediated immune suppression [95]. Furthermore, IL-21 increases levels of tissue-degrading proteases by stimulation of stromal cells [96]. Via the IL-21-receptor in human intestinal epithelial cells, IL-21 stimulates production and secretion of T-cell chemoattractant macrophage inflammatory protein-3a [97].

To date there is no unequivocal evidence for a predominant role of Th17 cytokines in experimental colitis. Transfer colitis in RAG1-deficient mice also deficient of IL-23p19 and Th17 responses only develops in the absence of IL-10 or TGF-β suggesting that these cytokines effectively suppress proinflammatory Th17 responses [98]. Moreover, transfer colitis develops in *Rag1/Il23* KO mice that also lack Foxp3+ regulatory T cells, although the absence of IL-23p19 hampers an adequate Th17 response [79]. In addition, there is evidence that IL23 directly suppresses Treg development, and thus the biological effect of IL23 blockage may be rather to facilitate regulatory T-cell responses than proinflammatory TH17-mediated effects. However, the recent discovery that decreased susceptibility to both forms of

IBD was associated with a single-nucleotide polymorphism in the *IL23R* gene raised new interest in the role of Th17 responses in IBD. Although there is consensus that Th17 responses may play some role in the inflammatory process of Crohn's disease, they are overrided by a quantitatively stronger Th1 response. Unfortunately, two different neutralizing antibodies against the IL-23/p40 subunit were recently tested in clinical trials in CD with relatively disappointing outcomes [64, 68]. Moreover, this year the first randomized, double-blind, placebo-controlled clinical trial with an anti-IL-17A monoclonal antibody [99] for moderate to severe Crohn's disease was published. Disappointingly, this therapeutic regimen was ineffective with even higher rates of adverse events compared to placebo [99]. However, inactivation of the Th17 master transcription factor RORgammaT prevented T-cell transfer colitis effectively suggesting that other approaches targeting Th17 cells may be more promising than anti-IL-17A antibody therapy [75].

However, recent studies even gave rise to the idea of an anti-inflammatory role of Th17 cells. They produced large amounts of IL-22 which was shown to ameliorate experimental colitis. Furthermore *Il22* KO mice showed DSS colitis aggravation compared to controls [100]. In addition to its direct immunological effects, IL-22 conveys intestinal epithelial integrity by induction of STAT3 [101]. IL-22 promotes colitis protection through stimulation of epithelial cell growth, goblet cell restitution, as well as mucus and antimicrobial production [100, 102].

Another cytokine of the IL-12 family, IL-27, has also been shown to have important effects on T-cell function. It activated the transcription factors STAT1 and STAT3 in naive CD4+ T cells. Whereas STAT1 phosphorylation by IL-27 induced the expression of the Th1 master transcription factor, T box protein expressed in T cells (T-bet), STAT3 activation was pivotal for T-cell proliferation. Recently, the role of IL-27 in a T-cell-mediated model of colitis (transfer colitis) was studied in IL-27 receptor α (Il27ra)-deficient mice [103]. These mice demonstrated attenuated colitis and reduced weight loss. This suggested proinflammatory role of IL-27 seems to be mediated by a restrained development of regulatory T cells that are known to play an anti-inflammatory role in vivo. Additionally, T effector cell proliferation was reduced in IL-27ra-deficient mice, which was accompanied by reduced levels of IFN-γ. Thus, previously described distinct effects of IFN-γ on epithelial cell function may also contribute to the effects observed in IL-27ra-deficient mice [104].

Elevated IFN-γ levels induced tight junction internalization which resulted in a disruption of the intestinal epithelial barrier [105]. It was also shown to inhibit epithelial cell migration [106] and to impair epithelial wound healing via endocytosis of β1-integrin and gap junction communication via connexin 43 [107, 108]. Intriguingly, IFN-γ also enables intestinal epithelial cells to function as nonprofessional APCs by induction of the expression of major histocompatibility complex [109] class II molecules [110, 111]. IL-27 influences intestinal epithelial cell proliferation and restitution also by direct activation of ERK, p38 MAPKs, as well as Akt, STAT1, STAT3, and STAT6. IL-27 was furthermore ascribed a pivotal role in the induction of antibacterial and anti-inflammatory proteins such as indoleamine 2, 3-dioxygenase (IDO1) and DMTB1 [112]. However, there is a vital debate about whether IL-27 acts pro- or anti-inflammatory in vivo. In a recent study, IL-27 inhibited chemically induced murine colitis by TNBS (2, 4, 6-trinitrobenzene sulfonic acid). Supposedly, this effect was mediated by inhibition of Th17 cell differentiation, a subpopulation of proinflammatory T helper cells [113].

The recently identified cytokine IL-35 (p35/EBI3) was suggested to control regulatory T-cell function since p35- and EBI3-deficient T cells failed to cure experimental colitis. Since EBI3 is also a part of the IL-27 heterodimer, *Ebi3*-deficient mice and *Il27*-deficient mice with spontaneous or T-cell transfer-induced colitis were compared. EBI3-deficient mice showed worse colitis compared to *Il27-deficient* mice or wild types suggesting a superior role of IL-35 over IL-27 in colon protection. Cytokine levels characteristic for the TH1 and Th17 profile were elevated in the intestinal mucosa of *Ebi3* knockout mice. Recombinant IL-35 significantly inhibited various experimental colitis models which were accompanied by reduced levels of Th1 and Th17 cytokines [114].

The discovery of the oxazolone model of colitis confirmed the concept of two distinct pathways of T helper cell-mediated forms of colitis (see Fig. 15.2). Oxazolone colitis resembled human UC in some aspects, since it led to a superficial mucosal inflammation associated with micro ulcerations of the epithelium and showed a proximo-distal gradient of colitis aggravation. Initial cytokine analyses demonstrated increased amounts of IL-4 and TGF-β but not IFN-γ—a cytokine pattern that mirrors Th2 conditions. Anti-IL-4 treatment prevented the disease; however, intriguingly, anti-IL-12p40 exacerbated oxazolone colitis. A more chronic variant of the oxazolone model where mice were presensitized with oxazolone revealed that IL-13 produced by natural killer T cells (NKT cells) was the driving cytokine [115]. In line with this observation, *Cd1*-deficient mice and Jα281 KO mice that show deficits in antigen presentation to NKT cells or NKT cell function per se were protected from oxazolone colitis [27].

However, human UC is characterized by a typical T-cell cytokine response, since it is associated with increased IL-13 and IL-5 (but not IL-4) production in the absence of an increased IFN-γ response. Flow cytometric studies revealed that UC rather than CD patients have high levels of lamina propria CD4+ T cells producing IL-13 [116–118]. B cells transfected with CD1d preferentially activate NKT cells. When these cells were co-incubated with lamina propria T cells from UC patients, they produced high levels of IL-13. However, these lamina propria NKT cells were atypical as they were lacking invariant T-cell receptors [119]. Until today clinical evidence is lacking that IL-13 plays a role in the pathogenesis of UC or that blockage of this cytokine is of clinical relevance. A recent small clinical trial, however, demonstrated that clinical response and remission in patients treated with IFN-β-1a was accompanied by reduced IL-13 levels in serum and lamina propria mononuclear cells [120]. Currently, a trial is running to evaluate clinical efficacy of an IL-13 antibody in CD patients with perianal fistulae [115].

IL-2 is a promiscuous cytokine which is mainly produced by activated T cells and promotes the differentiation of CD4+ T cells into Th1 and Th2 effector subsets, is critical for Treg development, and at the same time inhibits Th17 differentiation [121]. Its receptor (IL-2Ra, CD25) levels are found to be increased in UC [122] and

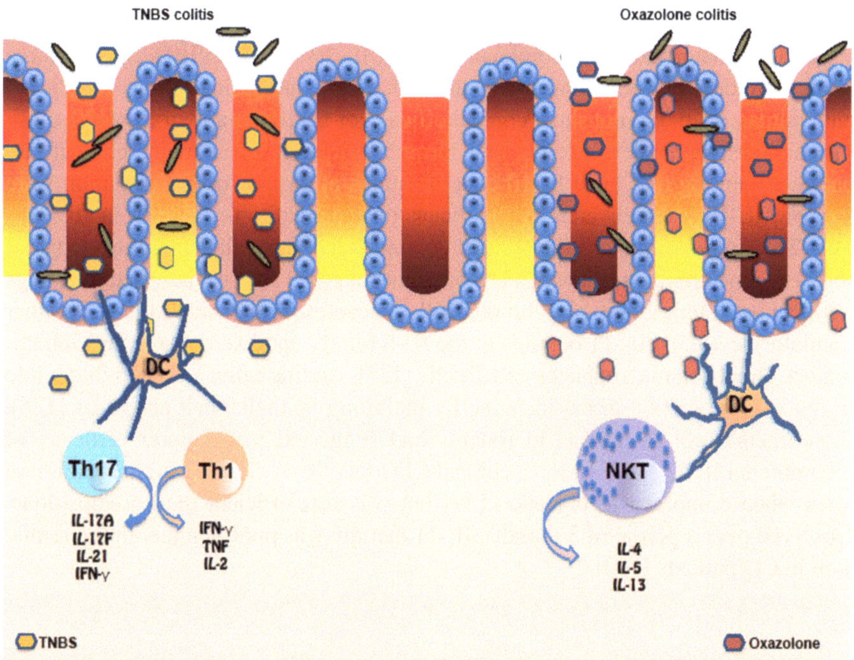

Fig. 15.2 The TNBS and oxazolone models are due to an initial toxic insult (coadministration of ethanol) leading to barrier disruption and translocation of bacterial antigens from the commensal microflora. TNBS and oxazolone act as haptens and bind to unknown antigens of the intestinal epithelium and/or microbiota. The flood of antigen exposure leads to an exaggerated immune response that is characterized by a mixed Th1/Th17 phenotype in TNBS colitis. In contrast, oxazolone colitis is associated with a Th2-like cytokine pattern

CD patients [123]. The fundamental role of IL-2 in immunity is underlined by the observation that the lacking of IL-2 or CD25 causes severe autoimmune diseases. Monoclonal antibodies neutralizing IL-2R (CD25) (daclizumab [Roche] and basiliximab [Novartis]) were tested in UC (but not yet CD) patients. Both compounds initially showed promising response rates in open-label studies [124], which were not confirmed in large randomized, double-blind, placebo-controlled trials [125–127].

Targeting Regulatory Cytokines

Instead of blocking proinflammatory cytokines, another approach to restore intestinal homeostasis is to support the function of the regulatory cytokines such as TGF-β, IL-11, or IL-10. IL-10 has pleiotropic inhibitory effects on various immune cells such as macrophages, T, and B cells but also promotes the development of Tregs [128].

Importantly, IL-10-deficient mice spontaneously develop colitis [129]; thus, it seemed mandatory to test recombinant human IL-10 in human IBD. Several large double-blinded, placebo-controlled studies that used increasing doses of recombinant human IL-10 did not show clinical efficacy of IL-10 therapy in CD [130–132]. Instead, dose-dependent anemia and thrombocytopenia were major clinical setbacks. Assuming that local IL-10 concentrations at the site of inflammation were too low, IL-10 was delivered by the means of genetically modified L. lactis (LL-Thy12) in a subsequent study [133]. Again, the first small non-controlled trial showed some clinical response in UC patients but could not be confirmed by a large multicenter double-blind, placebo-controlled, dose-escalation study [134]. Another regulatory cytokine IL-11 belongs to the IL-6 family and exerts many anti-inflammatory effects on macrophages and T cells [135]. Additionally, it plays a direct role in mucosal protection in the intestine by inhibiting epithelial cell apoptosis [136]. Experimental colitis models in rodents had suggested a therapeutic efficacy of recombinant IL-11 [137, 138]. In clinical CD trials, IL-11 induced higher remission rates when compared to placebo [139] but was less efficient than prednisolone. Observed over a period of 3 months, IL-11 therapy was poor in maintaining remission in CD patients [140].

Conclusive Remarks

To date, anti-TNF strategies remain the most effective biological therapies for IBD. It remains unclear whether study designs, unavailability of compounds at the site of inflammation, or the heterogeneity of patient populations is responsible for the poor efficacy of alternative cytokine therapies. In the future, identification of individual cytokine patterns by analysis of mucosal biopsies in combination with blood tests before initiating a treatment regimen may lead to successful personalized therapies. The failure of many approaches targeting only a single cytokine makes it reasonable to thoroughly investigate combined anti-cytokine therapies in detailed large placebo-controlled trials.

References

1. Eissner G, Kolch W, Scheurich P (2004) Ligands working as receptors: reverse signaling by members of the TNF superfamily enhance the plasticity of the immune system. Cytokine Growth Factor Rev 15(5):353–366, Epub 2004/09/29
2. Wajant H, Pfizenmaier K, Scheurich P (2003) Tumor necrosis factor signaling. Cell Death Differ 10(1):45–65
3. Breese EJ, Michie CA, Nicholls SW, Murch SH, Williams CB, Domizio P et al (1994) Tumor necrosis factor alpha-producing cells in the intestinal mucosa of children with inflammatory bowel disease. Gastroenterology 106(6):1455–1466

4. Murch SH, Braegger CP, Walker-Smith JA, MacDonald TT (1993) Location of tumour necrosis factor alpha by immunohistochemistry in chronic inflammatory bowel disease. Gut 34(12):1705–1709
5. Neurath MF, Meyer zum Buschenfelde KH (1996) Protective and pathogenic roles of cytokines in inflammatory bowel diseases. J Investig Med 44(9):516–521
6. Reimund JM, Wittersheim C, Dumont S, Muller CD, Baumann R, Poindron P et al (1996) Mucosal inflammatory cytokine production by intestinal biopsies in patients with ulcerative colitis and Crohn's disease. J Clin Immunol 16(3):144–150
7. Sartor RB (1994) Cytokines in intestinal inflammation: pathophysiological and clinical considerations. Gastroenterology 106(2):533–539
8. MacDonald TT, Hutchings P, Choy MY, Murch S, Cooke A (1990) Tumour necrosis factor-alpha and interferon-gamma production measured at the single cell level in normal and inflamed human intestine. Clin Exp Immunol 81(2):301–305, Epub 1990/08/01
9. Present DH, Rutgeerts P, Targan S, Hanauer SB, Mayer L, van Hogezand RA et al (1999) Infliximab for the treatment of fistulas in patients with Crohn's disease. N Engl J Med 340(18):1398–1405
10. Lugering A, Schmidt M, Lugering N, Pauels HG, Domschke W, Kucharzik T (2001) Infliximab induces apoptosis in monocytes from patients with chronic active Crohn's disease by using a caspase-dependent pathway. Gastroenterology 121(5):1145–1157
11. ten Hove T, van Montfrans C, Peppelenbosch MP, van Deventer SJ (2002) Infliximab treatment induces apoptosis of lamina propria T lymphocytes in Crohn's disease. Gut 50(2):206–211
12. Shen C, Assche GV, Colpaert S, Maerten P, Geboes K, Rutgeerts P et al (2005) Adalimumab induces apoptosis of human monocytes: a comparative study with infliximab and etanercept. Aliment Pharmacol Ther 21(3):251–258
13. Sandborn WJ, Feagan BG, Stoinov S, Honiball PJ, Rutgeerts P, Mason D et al (2007) Certolizumab pegol for the treatment of Crohn's disease. N Engl J Med 357(3):228–238, Epub 2007/07/20
14. Nesbitt A, Fossati G, Bergin M, Stephens P, Stephens S, Foulkes R et al (2007) Mechanism of action of certolizumab pegol (CDP870): in vitro comparison with other anti-tumor necrosis factor alpha agents. Inflamm Bowel Dis 13(11):1323–1332
15. Atreya R, Zimmer M, Bartsch B, Waldner MJ, Atreya I, Neumann H et al (2011) Antibodies against tumor necrosis factor (TNF) induce T-cell apoptosis in patients with inflammatory bowel diseases via TNF receptor 2 and intestinal CD14(+) macrophages. Gastroenterology 141(6):2026–2038
16. Rutgeerts P, Van Assche G, Vermeire S (2006) Review article: Infliximab therapy for inflammatory bowel disease—seven years on. Aliment Pharmacol Ther 23(4):451–463
17. Afif W, Loftus EV Jr, Faubion WA, Kane SV, Bruining DH, Hanson KA et al (2010) Clinical utility of measuring infliximab and human anti-chimeric antibody concentrations in patients with inflammatory bowel disease. Am J Gastroenterol 105(5):1133–1139
18. Hanauer SB, Wagner CL, Bala M, Mayer L, Travers S, Diamond RH et al (2004) Incidence and importance of antibody responses to infliximab after maintenance or episodic treatment in Crohn's disease. Clin Gastroenterol Hepatol 2(7):542–553, Epub 2004/06/30
19. Baert F, Noman M, Vermeire S, Van Assche G, D'Haens G, Carbonez A et al (2003) Influence of immunogenicity on the long-term efficacy of infliximab in Crohn's disease. N Engl J Med 348(7):601–608
20. Maser EA, Villela R, Silverberg MS, Greenberg GR (2006) Association of trough serum infliximab to clinical outcome after scheduled maintenance treatment for Crohn's disease. Clin Gastroenterol Hepatol 4(10):1248–1254
21. Farrell RJ, Alsahli M, Jeen YT, Falchuk KR, Peppercorn MA, Michetti P (2003) Intravenous hydrocortisone premedication reduces antibodies to infliximab in Crohn's disease: a randomized controlled trial. Gastroenterology 124(4):917–924

22. Van Bockstaele F, Holz JB, Revets H (2009) The development of nanobodies for therapeutic applications. Curr Opin Investig Drugs 10(11):1212–1224
23. Semerano L, Assier E, Delavallee L, Boissier MC (2011) Kinoid of human tumor necrosis factor-alpha for rheumatoid arthritis. Expert Opin Biol Ther 11(4):545–550, Epub 2011/03/10
24. Cappello M, Keshav S, Prince C, Jewell DP, Gordon S (1992) Detection of mRNAs for macrophage products in inflammatory bowel disease by in situ hybridisation. Gut 33(9):1214–1219, Epub 1992/09/01
25. Mayer L. Inflammatory Bowel Disease, 5th ed., Kirsner JB ed., Philadelphia: W. B. Saunders Company 2000;280–296
26. Fantini MC, Monteleone G, Macdonald TT (2007) New players in the cytokine orchestra of inflammatory bowel disease. Inflamm Bowel Dis 13(11):1419–1423
27. Strober W, Fuss IJ (2011) Proinflammatory cytokines in the pathogenesis of inflammatory bowel diseases. Gastroenterology 140(6):1756–1767
28. Pender SL, MacDonald TT (2004) Matrix metalloproteinases and the gut—new roles for old enzymes. Curr Opin Pharmacol 4(6):546–550
29. Atreya I, Atreya R, Neurath MF (2008) NF-kappaB in inflammatory bowel disease. J Intern Med 263(6):591–596
30. Nenci A, Becker C, Wullaert A, Gareus R, van Loo G, Danese S et al (2007) Epithelial NEMO links innate immunity to chronic intestinal inflammation. Nature 446(7135):557–561, Epub 2007/03/16
31. Gross V, Andus T, Caesar I, Roth M, Scholmerich J (1992) Evidence for continuous stimulation of interleukin-6 production in Crohn's disease. Gastroenterology 102(2):514–519
32. Hyams JS, Fitzgerald JE, Treem WR, Wyzga N, Kreutzer DL (1993) Relationship of functional and antigenic interleukin 6 to disease activity in inflammatory bowel disease. Gastroenterology 104(5):1285–1292
33. Atreya R, Mudter J, Finotto S, Mullberg J, Jostock T, Wirtz S et al (2000) Blockade of interleukin 6 trans signaling suppresses T-cell resistance against apoptosis in chronic intestinal inflammation: evidence in Crohn disease and experimental colitis in vivo. Nat Med 6(5):583–588
34. Louis E, Belaiche J, van Kemseke C, Franchimont D, de Groote D, Gueenen V et al (1997) A high serum concentration of interleukin-6 is predictive of relapse in quiescent Crohn's disease. Eur J Gastroenterol Hepatol 9(10):939–944
35. Van Kemseke C, Belaiche J, Louis E (2000) Frequently relapsing Crohn's disease is characterized by persistent elevation in interleukin-6 and soluble interleukin-2 receptor serum levels during remission. Int J Colorectal Dis 15(4):206–210
36. Hosokawa T, Kusugami K, Ina K, Ando T, Shinoda M, Imada A et al (1999) Interleukin-6 and soluble interleukin-6 receptor in the colonic mucosa of inflammatory bowel disease. J Gastroenterol Hepatol 14(10):987–996, Epub 1999/10/26
37. Jostock T, Mullberg J, Ozbek S, Atreya R, Blinn G, Voltz N et al (2001) Soluble gp130 is the natural inhibitor of soluble interleukin-6 receptor transsignaling responses. Eur J Biochem 268(1):160–167
38. Mitsuyama K, Toyonaga A, Sasaki E, Ishida O, Ikeda H, Tsuruta O et al (1995) Soluble interleukin-6 receptors in inflammatory bowel disease: relation to circulating interleukin-6. Gut 36(1):45–49
39. de Jong YP, Abadia-Molina AC, Satoskar AR, Clarke K, Rietdijk ST, Faubion WA et al (2001) Development of chronic colitis is dependent on the cytokine MIF. Nat Immunol 2(11):1061–1066
40. Holub MC, Mako E, Devay T, Dank M, Szalai C, Fenyvesi A et al (1998) Increased interleukin-6 levels, interleukin-6 receptor and gp130 expression in peripheral lymphocytes of patients with inflammatory bowel disease. Scand J Gastroenterol Suppl 228:47–50
41. Ina K, Itoh J, Fukushima K, Kusugami K, Yamaguchi T, Kyokane K et al (1999) Resistance of Crohn's disease T cells to multiple apoptotic signals is associated with a Bcl-2/Bax mucosal imbalance. J Immunol 163(2):1081–1090

42. Ishiguro Y (1999) Mucosal proinflammatory cytokine production correlates with endoscopic activity of ulcerative colitis. J Gastroenterol 34(1):66–74, Epub 1999/04/16
43. Iimura M, Nakamura T, Shinozaki S, Iizuka B, Inoue Y, Suzuki S et al (2000) Bax is down-regulated in inflamed colonic mucosa of ulcerative colitis. Gut 47(2):228–235
44. Mitsuyama K, Sata M, Rose-John S (2006) Interleukin-6 trans-signaling in inflammatory bowel disease. Cytokine Growth Factor Rev 17(6):451–461
45. Ito H, Takazoe M, Fukuda Y, Hibi T, Kusugami K, Andoh A et al (2004) A pilot randomized trial of a human anti-interleukin-6 receptor monoclonal antibody in active Crohn's disease. Gastroenterology 126(4):989–996, discussion 47
46. Strober W, Fuss I, Mannon P (2007) The fundamental basis of inflammatory bowel disease. J Clin Invest 117(3):514–521
47. Strober W, Fuss IJ, Blumberg RS (2002) The immunology of mucosal models of inflammation. Annu Rev Immunol 20:495–549
48. Meng G, Zhang F, Fuss I, Kitani A, Strober W (2009) A mutation in the Nlrp3 gene causing inflammasome hyperactivation potentiates Th17 cell-dominant immune responses. Immunity 30(6):860–874
49. Veldhoen M, Hocking RJ, Atkins CJ, Locksley RM, Stockinger B (2006) TGFbeta in the context of an inflammatory cytokine milieu supports de novo differentiation of IL-17-producing T cells. Immunity 24(2):179–189
50. Prehn JL, Thomas LS, Landers CJ, Yu QT, Michelsen KS, Targan SR (2007) The T cell costimulator TL1A is induced by FcgammaR signaling in human monocytes and dendritic cells. J Immunol 178(7):4033–4038
51. Kamada N, Hisamatsu T, Honda H, Kobayashi T, Chinen H, Takayama T et al (2010) TL1A produced by lamina propria macrophages induces Th1 and Th17 immune responses in cooperation with IL-23 in patients with Crohn's disease. Inflamm Bowel Dis 16(4):568–575
52. Meylan F, Song YJ, Fuss I, Villarreal S, Kahle E, Malm IJ et al (2011) The TNF-family cytokine TL1A drives IL-13-dependent small intestinal inflammation. Mucosal Immunol 4(2):172–185
53. Takedatsu H, Michelsen KS, Wei B, Landers CJ, Thomas LS, Dhall D et al (2008) TL1A (TNFSF15) regulates the development of chronic colitis by modulating both T-helper 1 and T-helper 17 activation. Gastroenterology 135(2):552–567
54. Schreiber TH, Wolf D, Tsai MS, Chirinos J, Deyev VV, Gonzalez L et al (2010) Therapeutic Treg expansion in mice by TNFRSF25 prevents allergic lung inflammation. J Clin Invest 120(10):3629–3640
55. Bamias G, Martin C III, Marini M, Hoang S, Mishina M, Ross WG et al (2003) Expression, localization, and functional activity of TL1A, a novel Th1-polarizing cytokine in inflammatory bowel disease. J Immunol 171(9):4868–4874
56. Prehn JL, Mehdizadeh S, Landers CJ, Luo X, Cha SC, Wei P et al (2004) Potential role for TL1A, the new TNF-family member and potent costimulator of IFN-gamma, in mucosal inflammation. Clin Immunol 112(1):66–77
57. Duchmann R, Lochs H, Kruis W (1999) Morbus Crohn, Colitis ulcerosa. Wenn Bakterien die Darmwand attackieren.... [Crohn disease, ulcerative colitis. When bacteria attack the intestinal wall....]. MMW Fortschritte der Medizin 141(51–52):48–51. Epub 2000/08/19
58. Neurath MF, Fuss I, Kelsall BL, Stuber E, Strober W (1995) Antibodies to interleukin 12 abrogate established experimental colitis in mice. J Exp Med 182(5):1281–1290
59. Davidson NJ, Hudak SA, Lesley RE, Menon S, Leach MW, Rennick DM (1998) IL-12, but not IFN-gamma, plays a major role in sustaining the chronic phase of colitis in IL-10-deficient mice. J Immunol 161(6):3143–3149
60. Fuss IJ, Marth T, Neurath MF, Pearlstein GR, Jain A, Strober W (1999) Anti-interleukin 12 treatment regulates apoptosis of Th1 T cells in experimental colitis in mice. Gastroenterology 117(5):1078–1088
61. Neurath MF, Finotto S, Fuss I, Boirivant M, Galle PR, Strober W (2001) Regulation of T-cell apoptosis in inflammatory bowel disease: to die or not to die, that is the mucosal question. Trends Immunol 22(1):21–26

62. Goriely S, Neurath MF, Goldman M (2008) How microorganisms tip the balance between interleukin-12 family members. Nat Rev Immunol 8(1):81–86
63. Uhlig HH, McKenzie BS, Hue S, Thompson C, Joyce-Shaikh B, Stepankova R et al (2006) Differential activity of IL-12 and IL-23 in mucosal and systemic innate immune pathology. Immunity 25(2):309–318
64. Mannon PJ, Fuss IJ, Mayer L, Elson CO, Sandborn WJ, Present D et al (2004) Anti-interleukin-12 antibody for active Crohn's disease. N Engl J Med 351(20):2069–2079
65. Sandborn WJ, Feagan BG, Fedorak RN, Scherl E, Fleisher MR, Katz S et al (2008) A randomized trial of Ustekinumab, a human interleukin-12/23 monoclonal antibody, in patients with moderate-to-severe Crohn's disease. Gastroenterology 135(4):1130–1141
66. Oppmann B, Lesley R, Blom B, Timans JC, Xu Y, Hunte B et al (2000) Novel p19 protein engages IL-12p40 to form a cytokine, IL-23, with biological activities similar as well as distinct from IL-12. Immunity 13(5):715–725
67. Bettelli E, Oukka M, Kuchroo VK (2007) T(H)-17 cells in the circle of immunity and autoimmunity. Nat Immunol 8(4):345–350
68. Ivanov II, Zhou L, Littman DR (2007) Transcriptional regulation of Th17 cell differentiation. Semin Immunol 19(6):409–417
69. McGeachy MJ, Chen Y, Tato CM, Laurence A, Joyce-Shaikh B, Blumenschein WM et al (2009) The interleukin 23 receptor is essential for the terminal differentiation of interleukin 17-producing effector T helper cells in vivo. Nat Immunol 10(3):314–324
70. McGeachy MJ, Bak-Jensen KS, Chen Y, Tato CM, Blumenschein W, McClanahan T et al (2007) TGF-beta and IL-6 drive the production of IL-17 and IL-10 by T cells and restrain T(H)-17 cell-mediated pathology. Nat Immunol 8(12):1390–1397
71. Hue S, Ahern P, Buonocore S, Kullberg MC, Cua DJ, McKenzie BS et al (2006) Interleukin-23 drives innate and T cell-mediated intestinal inflammation. J Exp Med 203(11):2473–2483
71. Burakoff R, Barish CF, Riff D, Pruitt R, Chey WY, Farraye FA et al (2006) A phase 1/2A trial of STA 5326, an oral interleukin-12/23 inhibitor, in patients with active moderate to severe Crohn's disease. Inflamm Bowel Dis 12(7):558–565
73. Dideberg V, Kristjansdottir G, Milani L, Libioulle C, Sigurdsson S, Louis E et al (2007) An insertion-deletion polymorphism in the interferon regulatory Factor 5 (IRF5) gene confers risk of inflammatory bowel diseases. Hum Mol Genet 16(24):3008–3016, Epub 2007/09/21
74. Yang XO, Chang SH, Park H, Nurieva R, Shah B, Acero L et al (2008) Regulation of inflammatory responses by IL-17F. J Exp Med 205(5):1063–1075
75. Leppkes M, Becker C, Ivanov II, Hirth S, Wirtz S, Neufert C et al (2009) RORgamma-expressing Th17 cells induce murine chronic intestinal inflammation via redundant effects of IL-17A and IL-17F. Gastroenterology 136(1):257–267
76. O'Connor W Jr, Kamanaka M, Booth CJ, Town T, Nakae S, Iwakura Y et al (2009) A protective function for interleukin 17A in T cell-mediated intestinal inflammation. Nat Immunol 10(6):603–609
77. Zhang Z, Zheng M, Bindas J, Schwarzenberger P, Kolls JK (2006) Critical role of IL-17 receptor signaling in acute TNBS-induced colitis. Inflamm Bowel Dis 12(5):382–388
78. Moseley TA, Haudenschild DR, Rose L, Reddi AH (2003) Interleukin-17 family and IL-17 receptors. Cytokine Growth Factor Rev 14(2):155–174
79. Ahern PP, Schiering C, Buonocore S, McGeachy MJ, Cua DJ, Maloy KJ et al (2010) Interleukin-23 drives intestinal inflammation through direct activity on T cells. Immunity 33(2):279–288, Epub 2010/08/25
80. Reinisch W, de Villiers W, Bene L, Simon L, Racz I, Katz S et al (2010) Fontolizumab in moderate to severe Crohn's disease: a phase 2, randomized, double-blind, placebo-controlled, multiple-dose study. Inflamm Bowel Dis 16(2):233–242
81. Reinisch W, Hommes DW, Van Assche G, Colombel JF, Gendre JP, Oldenburg B et al (2006) A dose escalating, placebo controlled, double blind, single dose and multidose, safety and tolerability study of fontolizumab, a humanised anti-interferon gamma antibody, in patients with moderate to severe Crohn's disease. Gut 55(8):1138–1144

82. Annunziato F, Cosmi L, Santarlasci V, Maggi L, Liotta F, Mazzinghi B et al (2007) Phenotypic and functional features of human Th17 cells. J Exp Med 204(8):1849–1861
83. Boniface K, Blumenschein WM, Brovont-Porth K, McGeachy MJ, Basham B, Desai B et al (2010) Human Th17 cells comprise heterogeneous subsets including IFN-gamma-producing cells with distinct properties from the Th1 lineage. J Immunol 185(1):679–687
84. Lee YK, Mukasa R, Hatton RD, Weaver CT (2009) Developmental plasticity of Th17 and Treg cells. Curr Opin Immunol 21(3):274–280
85. Volpe E, Servant N, Zollinger R, Bogiatzi SI, Hupe P, Barillot E et al (2008) A critical function for transforming growth factor-beta, interleukin 23 and proinflammatory cytokines in driving and modulating human T(H)-17 responses. Nat Immunol 9(6):650–657
86. Monteleone G, Biancone L, Marasco R, Morrone G, Marasco O, Luzza F et al (1997) Interleukin 12 is expressed and actively released by Crohn's disease intestinal lamina propria mononuclear cells. Gastroenterology 112(4):1169–1178
87. Monteleone G, Kumberova A, Croft NM, McKenzie C, Steer HW, MacDonald TT (2001) Blocking Smad7 restores TGF-beta1 signaling in chronic inflammatory bowel disease. J Clin Invest 108(4):601–609
88. Schmidt C, Giese T, Ludwig B, Mueller-Molaian I, Marth T, Zeuzem S et al (2005) Expression of interleukin-12-related cytokine transcripts in inflammatory bowel disease: elevated interleukin-23p19 and interleukin-27p28 in Crohn's disease but not in ulcerative colitis. Inflamm Bowel Dis 11(1):16–23
89. Beriou G, Bradshaw EM, Lozano E, Costantino CM, Hastings WD, Orban T et al (2010) TGF-beta induces IL-9 production from human Th17 cells. J Immunol 185(1):46–54
90. Bettelli E, Korn T, Oukka M, Kuchroo VK (2008) Induction and effector functions of T(H)17 cells. Nature 453(7198):1051–1057
91. Korn T, Bettelli E, Oukka M, Kuchroo VK (2009) IL-17 and Th17 Cells. Annu Rev Immunol 27:485–517
92. Stephens GL, Swerdlow B, Benjamin E, Coyle AJ, Humbles A, Kolbeck R et al (2011) IL-9 is a Th17-derived cytokine that limits pathogenic activity in organ-specific autoimmune disease. Eur J Immunol 41(4):952–962
93. Fina D, Sarra M, Fantini MC, Rizzo A, Caruso R, Caprioli F et al (2008) Regulation of gut inflammation and th17 cell response by interleukin-21. Gastroenterology 134(4):1038–1048
94. Monteleone G, Monteleone I, Fina D, Vavassori P, Del Vecchio BG, Caruso R et al (2005) Interleukin-21 enhances T-helper cell type I signaling and interferon-gamma production in Crohn's disease. Gastroenterology 128(3):687–694
95. Peluso I, Fantini MC, Fina D, Caruso R, Boirivant M, MacDonald TT et al (2007) IL-21 counteracts the regulatory T cell-mediated suppression of human CD4+ T lymphocytes. J Immunol 178(2):732–739
96. Monteleone G, Caruso R, Fina D, Peluso I, Gioia V, Stolfi C et al (2006) Control of matrix metalloproteinase production in human intestinal fibroblasts by interleukin 21. Gut 55(12):1774–1780
97. Caruso R, Fina D, Peluso I, Stolfi C, Fantini MC, Gioia V et al (2007) A functional role for interleukin-21 in promoting the synthesis of the T-cell chemoattractant, MIP-3alpha, by gut epithelial cells. Gastroenterology 132(1):166–175
98. Izcue A, Hue S, Buonocore S, Arancibia-Carcamo CV, Ahern PP, Iwakura Y et al (2008) Interleukin-23 restrains regulatory T cell activity to drive T cell-dependent colitis. Immunity 28(4):559–570
99. Hueber W, Sands BE, Lewitzky S, Vandemeulebroecke M, Reinisch W, Higgins PD et al (2012) Secukinumab, a human anti-IL-17A monoclonal antibody, for moderate to severe Crohn's disease: unexpected results of a randomised, double-blind placebo-controlled trial. Gut 61(12):1693–1700
100. Sugimoto K, Ogawa A, Mizoguchi E, Shimomura Y, Andoh A, Bhan AK et al (2008) IL-22 ameliorates intestinal inflammation in a mouse model of ulcerative colitis. J Clin Invest 118(2):534–544

101. Pickert G, Neufert C, Leppkes M, Zheng Y, Wittkopf N, Warntjen M et al (2009) STAT3 links IL-22 signaling in intestinal epithelial cells to mucosal wound healing. J Exp Med 206(7):1465–1472
102. Brand S, Beigel F, Olszak T, Zitzmann K, Eichhorst ST, Otte JM et al (2006) IL-22 is increased in active Crohn's disease and promotes proinflammatory gene expression and intestinal epithelial cell migration. Am J Physiol Gastrointest Liver Physiol 290(4):G827–G838
103. Cox JH, Kljavin NM, Ramamoorthi N, Diehl L, Batten M, Ghilardi N (2011) IL-27 promotes T cell-dependent colitis through multiple mechanisms. J Exp Med 208(1):115–123
104. Wirtz S, Tubbe I, Galle PR, Schild HJ, Birkenbach M, Blumberg RS et al (2006) Protection from lethal septic peritonitis by neutralizing the biological function of interleukin 27. J Exp Med 203(8):1875–1881
105. Bruewer M, Luegering A, Kucharzik T, Parkos CA, Madara JL, Hopkins AM et al (2003) Proinflammatory cytokines disrupt epithelial barrier function by apoptosis-independent mechanisms. J Immunol 171(11):6164–6172
106. Leaphart CL, Dai S, Gribar SC, Richardson W, Ozolek J, Shi XH et al (2008) Interferon-gamma inhibits enterocyte migration by reversibly displacing connexin43 from lipid rafts. Am J Physiol Gastrointest Liver Physiol 295(3):G559–G569
107. Leaphart CL, Qureshi F, Cetin S, Li J, Dubowski T, Baty C et al (2007) Interferon-gamma inhibits intestinal restitution by preventing gap junction communication between enterocytes. Gastroenterology 132(7):2395–2411
108. Tong Q, Vassilieva EV, Ivanov AI, Wang Z, Brown GT, Parkos CA et al (2005) Interferon-gamma inhibits T84 epithelial cell migration by redirecting transcytosis of beta1 integrin from the migrating leading edge. J Immunol 175(6):4030–4038
109. The_MHC_Consortium (1999) Complete sequence and gene map of a human major histocompatibility complex. The MHC sequencing consortium. Nature 401(6756):921–923, Epub 1999/11/30
110. Colgan SP, Parkos CA, Matthews JB, D'Andrea L, Awtrey CS, Lichtman AH et al (1994) Interferon-gamma induces a cell surface phenotype switch on T84 intestinal epithelial cells. Am J Physiol 267(2 Pt 1):C402–C410
111. Ruemmele FM, Gurbindo C, Mansour AM, Marchand R, Levy E, Seidman EG (1998) Effects of interferon gamma on growth, apoptosis, and MHC class II expression of immature rat intestinal crypt (IEC-6) cells. J Cell Physiol 176(1):120–126
112. Diegelmann J, Olszak T, Goke B, Blumberg RS, Brand S (2012) A novel role for interleukin-27 (IL-27) as mediator of intestinal epithelial barrier protection mediated via differential signal transducer and activator of transcription (STAT) protein signaling and induction of antibacterial and anti-inflammatory proteins. J Biol Chem 287(1):286–298
113. Sasaoka T, Ito M, Yamashita J, Nakajima K, Tanaka I, Narita M et al (2011) Treatment with IL-27 attenuates experimental colitis through the suppression of the development of IL-17-producing T helper cells. Am J Physiol Gastrointest Liver Physiol 300(4):G568–G576
114. Wirtz S, Billmeier U, McHedlidze T, Blumberg RS, Neurath MF (2011) Interleukin-35 mediates mucosal immune responses that protect against T-cell-dependent colitis. Gastroenterology 141(5):1875–1886
115. Jacobo EM, Dario FM, Kathleen SH (2000) Behavioral and psychological signs and symptoms of dementia: a practicing psychiatrist's viewpoint. Dialogues Clin Neurosci 2(2): 139–155, Epub 2000/06/01
116. Fuss IJ, Becker C, Yang Z, Groden C, Hornung RL, Heller F et al (2006) Both IL-12p70 and IL-23 are synthesized during active Crohn's disease and are down-regulated by treatment with anti-IL-12 p40 monoclonal antibody. Inflamm Bowel Dis 12(1):9–15
117. Hart AL, Al-Hassi HO, Rigby RJ, Bell SJ, Emmanuel AV, Knight SC et al (2005) Characteristics of intestinal dendritic cells in inflammatory bowel diseases. Gastroenterology 129(1):50–65
118. Matsuoka K, Inoue N, Sato T, Okamoto S, Hisamatsu T, Kishi Y et al (2004) T-bet upregulation and subsequent interleukin 12 stimulation are essential for induction of Th1 mediated immunopathology in Crohn's disease. Gut 53(9):1303–1308

119. Fuss IJ, Heller F, Boirivant M, Leon F, Yoshida M, Fichtner-Feigl S et al (2004) Nonclassical CD1d-restricted NK T cells that produce IL-13 characterize an atypical Th2 response in ulcerative colitis. J Clin Invest 113(10):1490–1497
120. Mannon PJ, Hornung RL, Yang Z, Yi C, Groden C, Friend J et al (2011) Suppression of inflammation in ulcerative colitis by interferon-beta-1a is accompanied by inhibition of IL-13 production. Gut 60(4):449–455
121. Hoyer KK, Dooms H, Barron L, Abbas AK (2008) Interleukin-2 in the development and control of inflammatory disease. Immunol Rev 226:19–28
122. Balzano A, Bove A, Leonardi E, Bevilacqua N, Grande G, Ascierto PA et al (1997) The soluble interleukin-2 receptor as an indicator of clinical evolution in patients with ulcerative colitis. Eur J Gastroenterol Hepatol 9(2):173–177
123. Nielsen OH, Ciardelli T, Wu Z, Langholz E, Kirman I (1995) Circulating soluble interleukin-2 receptor alpha and beta chain in inflammatory bowel disease. Am J Gastroenterol 90(8):1301–1306
124. Van Assche G, Dalle I, Noman M, Aerden I, Swijsen C, Asnong K et al (2003) A pilot study on the use of the humanized anti-interleukin-2 receptor antibody daclizumab in active ulcerative colitis. Am J Gastroenterol 98(2):369–376
125. Creed TJ, Norman MR, Probert CS, Harvey RF, Shaw IS, Smithson J et al (2003) Basiliximab (anti-CD25) in combination with steroids may be an effective new treatment for steroid-resistant ulcerative colitis. Aliment Pharmacol Ther 18(1):65–75
126. Creed TJ, Probert CS, Norman MN, Moorghen M, Shepherd NA, Hearing SD et al (2006) Basiliximab for the treatment of steroid-resistant ulcerative colitis: further experience in moderate and severe disease. Aliment Pharmacol Ther 23(10):1435–1442
127. Van Assche G, Sandborn WJ, Feagan BG, Salzberg BA, Silvers D, Monroe PS et al (2006) Daclizumab, a humanised monoclonal antibody to the interleukin 2 receptor (CD25), for the treatment of moderately to severely active ulcerative colitis: a randomised, double blind, placebo controlled, dose ranging trial. Gut 55(11):1568–1574
128. Saraiva M, O'Garra A (2010) The regulation of IL-10 production by immune cells. Nat Rev Immunol 10(3):170–181
129. Kuhn R, Lohler J, Rennick D, Rajewsky K, Muller W (1993) Interleukin-10-deficient mice develop chronic enterocolitis. Cell 75(2):263–274
130. Colombel JF, Rutgeerts P, Malchow H, Jacyna M, Nielsen OH, Rask-Madsen J et al (2001) Interleukin 10 (Tenovil) in the prevention of postoperative recurrence of Crohn's disease. Gut 49(1):42–46
131. Fedorak RN, Gangl A, Elson CO, Rutgeerts P, Schreiber S, Wild G et al (2000) Recombinant human interleukin 10 in the treatment of patients with mild to moderately active Crohn's disease. The Interleukin 10 Inflammatory Bowel Disease Cooperative Study Group. Gastroenterology 119(6):1473–1482
132. Schreiber S, Fedorak RN, Nielsen OH, Wild G, Williams CN, Nikolaus S et al (2000) Safety and efficacy of recombinant human interleukin 10 in chronic active Crohn's disease. Crohn's Disease IL-10 Cooperative Study Group. Gastroenterology 119(6):1461–1472
133. Braat H, Rottiers P, Hommes DW, Huyghebaert N, Remaut E, Remon JP et al (2006) A phase I trial with transgenic bacteria expressing interleukin-10 in Crohn's disease. Clin Gastroenterol Hepatol 4(6):754–759, Epub 2006/05/24
134. Vermeire S, Rutgeerts P, D'haens G et al (2010) A Phase 2a Randomized placebo-controlled double-blind multi-center dose escalation study to evaluate the safety, tolerability, pharmacodynamics and efficacy of AG011 in patients with moderately active ulcerative colitis. Digestive Disease Week. New Orleans, LA, USA
135. Trepicchio WL, Wang L, Bozza M, Dorner AJ (1997) IL-11 regulates macrophage effector function through the inhibition of nuclear factor-kappaB. J Immunol 159(11):5661–5670
136. Kiessling S, Muller-Newen G, Leeb SN, Hausmann M, Rath HC, Strater J et al (2004) Functional expression of the interleukin-11 receptor alpha-chain and evidence of antiapoptotic effects in human colonic epithelial cells. J Biol Chem 279(11):10304–10315

137. Peterson RL, Wang L, Albert L, Keith JC Jr, Dorner AJ (1998) Molecular effects of recombinant human interleukin-11 in the HLA-B27 rat model of inflammatory bowel disease. Lab Invest 78(12):1503–1512
138. Qiu BS, Pfeiffer CJ, Keith JC Jr (1996) Protection by recombinant human interleukin-11 against experimental TNB-induced colitis in rats. Dig Dis Sci 41(8):1625–1630
139. Sands BE, Winston BD, Salzberg B, Safdi M, Barish C, Wruble L et al (2002) Randomized, controlled trial of recombinant human interleukin-11 in patients with active Crohn's disease. Aliment Pharmacol Ther 16(3):399–406
140. Herrlinger KR, Witthoeft T, Raedler A, Bokemeyer B, Krummenerl T, Schulzke JD et al (2006) Randomized, double blind controlled trial of subcutaneous recombinant human interleukin-11 versus prednisolone in active Crohn's disease. Am J Gastroenterol 101(4): 793–797

Part IV
Concluding Remarks and Future Perspectives

Chapter 16
Towards Personalized Therapy in Inflammatory Bowel Disease

Dermot P.B. McGovern

Abstract There has been very significant progress to characterizing the IBD genome with more than 160 IBD loci now identified. Now, the real challenge for researchers is the bidirectional translation of these findings in order to understand the functional and clinical consequences of this genetic variation. The influence of other factors including, but not limited to, epigenetic phenomena and environmental exposures, including the microbiome, cannot be ignored. The realistic hope is that, in the near future, patients will benefit from these advances as multi-omic modalities facilitate the path to a more personalized approach to disease management.

Introduction

The recent pace of discovery in our understanding of the genetic etiology of the inflammatory bowel diseases has been staggering. However, like many fields in scientific research, these advances have posed as many questions as they have answered. The rapid pace of "gene" discovery has meant that understanding the functional sequelae of disease-associated genetic variants and the sub-phenotype and disease behavior associations with known IBD loci are lagging far behind! It is a sobering thought for the research community that more than 10 years after the discovery of *NOD2/CARD15* as the first IBD gene, there remains considerable debate regarding the functional consequences of the three common CD-associated *NOD2* variants. The fact that *NOD2* was not an immediately targetable therapeutic target was a significant disappointment and "reality check" for the field. Nevertheless,

D.P.B. McGovern, M.D., Ph.D., F.R.C.P. (✉)
Cedars-Sinai Medical Center, 8730 Alden Drive, Los Angeles, CA 90048, USA
e-mail: Dermot.McGovern@cshs.org

M. D'Amato and J.D. Rioux (eds.), *Molecular Genetics of Inflammatory Bowel Disease*,
DOI 10.1007/978-1-4614-8256-7_16, © Springer Science+Business Media New York 2013

the huge investments in projects such as the Human Genome Project, the HapMap project, and the 1,000 Genomes Project mandate some significant progress towards improving outcomes for patients in genetically complex diseases.

Very Recent Advances

The recurrent observations of shared immune disease-associated loci from both GWAS and also more targeted studies prompted the international community to collaborate in the development of the immunochip (Illumina™, San Diego, CA, USA). Research groups, studying type 1 diabetes, ankylosing spondylitis, celiac disease, multiple sclerosis, psoriasis, autoimmune thyroid disease, IgA deficiency, primary biliary cirrhosis, systemic lupus erythematosis (SLE), rheumatoid arthritis as well as IBD, "pooled" both their collective experience and association data (including associated loci that had not reached the accepted criteria for genome-wide significance in their individual scans) in the design of the immunochip. The immunochip consists of approximately 200,000 SNPs in total containing not only index SNPs from immune disease-associated loci but also multiple polymorphisms around these loci that will greatly facilitate future fine-mapping studies. The immunochip also contains a dense SNP map across the HLA, which may be very informative particularly in UC and Crohn's colitis. The benefit of international collaboration has been clearly demonstrated. The very large number of these chips likely to be sold enabled Illumina to retail the chip at approximately $40 thereby allowing many thousands of samples to be genotyped in a very cost-effective manner. The IIBDGC adopted a strategy of performing a meta-analysis of all available IBD GWAS (no combined IBD GWAS of all available studies had been published previously) together with newly genotyped immunochip cases. In total, over 38,000 cases and over 37,000 controls were included in the study [1]. This included 14,763 CD cases, 10,920 UC cases, and nearly 16,000 controls. In summary a total of 71 new IBD loci (meeting the association criteria of $p > 5.0 \times 10^{-8}$) were identified, bringing to 163 the total number of IBD loci. Interestingly, the majority of these loci, 110 in total, were associated with both CD and UC, while 30 were CD-specific and 23 UC-specific. The *HLA* region was very UC-specific with little signal seen from CD. In contrast both *NOD2* and *PTPN22* were very significantly associated with CD, and in fact the CD-risk alleles at these two loci conferred protection from UC. While many loci were shared between UC and CD, it is clear that some have a much greater effect in one condition than the other (e.g., *IL23R*—CD > UC), whereas other loci such as *IL12B* or *CARD9* appear to have equal effects in both UC and CD.

Over 70 % of the identified IBD loci shared association with other genetically complex conditions with particular enrichment seen for ankylosing spondylitis (13-fold enrichment), psoriasis (14-fold enrichment), and type 1 diabetes (tenfold enrichment). Other interesting overlaps included associations seen with known leprosy and primary immunodeficiency syndromes (including Mendelian susceptibility to mycobacterial disease (46-fold enrichment)).

The identification of susceptibility loci is, of course, just the first process in identifying causative variants. The investigators in the immunochip paper recognized this and so performed in silico analyses in order to highlight and help prioritize genes at the identified loci. Nevertheless the real task of identifying the functional consequences of these genetic variants has yet to be realized.

Missing Heritability

It has been calculated that the 160 loci identified in the recently published paper account for 13.6 % of CD and 7.5 % of UC heritability. This is likely to be a very significant underestimate as this study only included the "lead" tagging SNP for each locus. This assertion is supported by previously published GWAS meta-analysis that utilized only a single SNP for tagging the *NOD2* locus. In this study the *NOD2* contribution to CD was estimated at only 0.8 %, whereas other studies, which have included the three common CD-associated *NOD2* variants, have estimated the contribution to be of the order of 5 % [2]. This suggests that fine mapping of these loci will greatly increase the estimated heritability for these loci. In addition to refining the true signal at loci, these studies will likely discover additional associated variants at many of the loci. In parallel to these studies, additional variants may be identified through sequencing strategies. Sequencing of known IBD-associated loci has already revealed additional disease-associated variants at loci such as *NOD2* and *CARD9* [3].

Additional genetic strategies will also be required including epigenetic approaches. Epigenetic phenomena have been, to date, largely understudied in IBD. One study performed by a group from Edinburgh identified more than 1,000 differentially methylated sites in patients with ileal CD [4]. Altered methylation in pathways, involved in immune responses and defense response to bacteria as well as dendritic cell activity, and the TH17 pathway were implicated in this study. Other studies have assessed the role of epigenetics at single loci such as interferon-γ (INFγ) and determined the effect of altered methylation patterns on gene expression [5]. Additional studies have found no real or limited associations with altered methylation [6]. It is clear that this area of research in IBD genetics is in its infancy and that further studies are warranted in larger and well-characterized cohorts.

It has been understood for some time that environmental factors also contribute to IBD pathogenesis and future studies should assess the role of gene–environment interactions in disease susceptibility and natural history. Such studies should concentrate on well-established environmental factors including smoking and NSAIDs. Recent advances in our understanding of the microbiome and its role in the development of IBD have prompted researchers to study the role of host genetic profile on the microbiome. One group examined the role of CD-associated *fucosyltransferase 2* (*FUT2*) variants and suggested that *FUT2* status explained substantial differences in community composition, diversity, and structure, indicating that *FUT2* is an important genetic factor influencing host-microbial diversity [7]. Studies of genetically manipulated mice have suggested that researchers should not solely concentrate on the bacterial component of the microbiome but should also be considering other aspects including the fungiome and viriome. The potential role of the viriome and its interaction with host genetic profile in IBD pathogenesis was demonstrated by the elegant studies in the hypomorphic *atg16l1* mouse that developed a Crohn's-like phenotype, but only in the presence of both commensal flora and murine norovirus [8]. Similarly *dectin-1* (*clec7a*) knockout mice develop a more severe colitis that is responsive to antifungal therapy [9]. The intriguing finding in humans, from the same study, was that *DECTIN-1* variants were associated with a more severe form of colitis requiring colectomy.

Future "Genetic" Directions

It is clear that while there has been significant progress in identifying IBD susceptibility loci, there is much work to be done, and the examples above suggest a number of important indicators of where the field needs to move including:

- Fine-mapping and sequencing approaches in order to discover both common and rare functional variants
- Clustering or networking of genetic variants for studying function in order to understand the functional complexities in a more realistic "physiological" environment
- An emphasis on non-Caucasian populations for both novel discovery and trans-ethnic mapping of susceptibility loci that transcend ethnic boundaries
- The study of genetic variants in the context of the environment around them including their role in shaping the microbiome
- The inclusion of nonclassical genetic associations with both disease susceptibility and natural history including studying the epigenome, the role of microRNAs, and copy number variation among others

The Future and Personalized Medicine

The IBDs are potentially a perfect vehicle for the development of personalized medicine given the clinically heterogenous nature of both ulcerative colitis and Crohn's disease. In addition, the recent introduction of the biological therapies, with anti-TNF at the vanguard, means that clinicians now have access to disease-modifying therapies. Furthermore Eric Green and Mark Guyer in their Nature Perspectives article "Charting a course for genomic medicine from base pairs to bedside" identified a number of criteria that needed to be met in order to bridge the bench-bedside divide including [10]:

1. Defining the genetic components of disease.
2. An understanding of the role of the human microbiome in health and disease.

3. Making genomics-based diagnostics/prognostics routine.
4. Add practical systems for clinical genomics informatics.

It can be argued that researchers in IBD are making significant advances in all of these fields. Points 1 and 2 have been discussed in this chapter and in more detail elsewhere in this book. IBD clinicians have been used to incorporating clinical and other parameters for prognostication for many years as clinical and social factors including extensive disease, smoking, and the presence of extra-intestinal manifestations are associated with more severe disease in Crohn's disease. Increasingly, clinicians are using other biomarkers including serum CRP and IBD-associated serology to complement the known clinical risk factors. Clinicians have also developed some experience in using genetic information in the clinic setting as the majority of gastroenterologists will check a thiopurine methyltransferase (TPMT) genotype or activity prior to introducing a thiopurine therapy in order to avoid bone marrow toxicity in the approximately 10 % of individuals who carry a *TPMT* polymorphism. It will not be a huge leap for clinicians to integrate genomic data into these algorithms as long as these are available in a readily digestible form. This is the fourth challenge issued by Green and Guyer in their plea for "practical systems for clinical genomics informatics." Clinicians will need easy ways to enter and visualize individual patients' data with meaningful outputs that can be readily applied to the individual patient in their clinic. A number of groups are working on developing these tools using both genetic and integrated multi-omic datasets.

It is easy to see how, in the near future, clinicians may utilize these data in their assessment of a recently diagnosed or possible case of IBD. These include the following.

Diagnosis

IBD can present in similar ways to many conditions including colorectal cancer and so it is extremely unlikely that genomics alone would be utilized as a diagnostic test. In addition the low pretest probability of these conditions makes genetic testing of low yield for IBD. This concern is supported by one study that utilized a number of commercially available direct to consumer genetic "tests" for IBD with contradictory "diagnoses" in identical patients [11]. Where genomics may show some utility as a diagnostic aid, is in distinguishing between UC and colonic CD in the cases that are currently termed as IBD unspecified. The ability to accurately diagnose CD or UC in these cases would be of great benefit to clinicians as different therapeutic and particularly surgical options are optimal for the different diagnoses. While many genes are shared between UC and CD, it is clear that they may have differential effect sizes on these conditions and it is possible that composite scores based on genetic variants and other parameters may be of clinical utility in this setting.

Prognosis

It would be a major clinical advance to have validated biomarkers that predict which patients were at risk of severe disease and therefore those most likely to benefit from earlier intervention with more aggressive therapy. This would allow clinicians to target therapies with the potential for severe adverse events to individuals where the risk-benefit is appropriate. A number of studies have shown genetic associations with more severe disease including the relationship between the rare HLA class II allele, *HLA DRB1*0103*, and extensive UC. This allele is found in approximately 2 % of the Caucasian population, but this increases up to 11 % of those with a pancolitis and up to 16 % in patients requiring colectomy [12]. Another locus that has been associated with a more severe phenotype of both extensive and severe UC and more complicated CD is the multidrug resistance gene (*ABCB1/MDR1*) and its product, P-glycoprotein 170. This gene is a relevant candidate gene for both disease susceptibility and natural history given the *mdr1* knockout mouse is susceptible to colitis and that drugs used to treat IBD including ciclosporin and corticosteroids are substrates for the P-glycoprotein 170. Ho et al. identified an *MDR1* haplotype strongly associated with UC and particularly extensive UC and, using conditional analyses, was able to demonstrate at least two signals from this locus including a well-documented C345T polymorphism known to alter drug pharmacokinetics [13].

Learning from the experience of using whole-genome approaches to identify associated loci, one study utilized this approach in order to identify SNPs associated with medically refractory UC (MRUC) [14]. In the study by Haritunians et al., over 300 cases of UC requiring colectomy for disease resistant to medical therapy were compared firstly with over 500 cases with "mild" UC cases not requiring colectomy. Demographic analyses demonstrated associations between MRUC and both extensive disease and a positive family history of UC. In addition, a number of SNPs were associated with the need for surgery although none reached a level consistent with genome-wide significance. Nevertheless, the authors combined the "risk" SNPs in a scoring system that was strongly associated with the need for surgery. Individuals with UC who fell into the highest group had an 84 % chance of surgery by 5 years compared to the lowest risk group who had a 0 % chance of going to surgery. This model clearly requires independent validation but demonstrates the potential clinical utility of combining a number of markers of modest effect. This study also highlights the potential benefit of utilizing an unbiased genome-wide approach to searching for disease behavior genes. Furthermore, the experience of the subsequent power of combining cohorts when searching for susceptibility genes suggests that researchers should, in the future, perform similar studies on sub-phenotypes but in appropriately powered cohorts. A major challenge for researchers is ensuring uniformity across centers in sub-phenotypic definitions.

Similar studies have been performed in CD including a meta-analysis confirming the relationship between *NOD2* and stricturing disease [15]. A study from the Leuven group identified an association between an SNP located adjacent to *IL12B* and stricturing disease and the need for surgery in CD, and a Dutch study demonstrated that cumulative numbers of CD-associated variants were associated with more younger onset, more severe disease, and the need for surgery [16, 17]. What is clear is that future studies should be performed on large, well-characterized, and preferably, prospective cohorts. Prospective cohorts such as the Crohn's and Colitis Foundation of America sponsored RISK cohort have been developed primarily to answer these questions. In addition, it is likely that severity-associated variants are likely to be independent of disease-associated variants and that researchers should adopt unbiased whole-genome approaches to these questions.

Researchers should not limit themselves to genetic variant-based analyses alone. A group from the UK identified a gene expression profiling of CD8+ T cells that is associated with more severe disease in both CD and UC [18]. Interestingly, this approach identified many genes that overlapped between CD and UC severity as well as between the IBDs and SLE. In addition the authors estimated that their gene expression profile was superior to ASCA in predicting time to medication escalation in CD. Taken together, these data suggest that a number of "-omic" profiles including serological, genetic, gene expression, and microbiome may be profiled in a newly diagnosed case of IBD in order to prognosticate as to the likelihood of that individual having a more severe disease course if left untreated. These models will inform both clinicians and patients about a more appropriate risk-benefit approach to induction and maintenance of remission.

Response to Therapy

One of the great hopes for clinicians with the advances of genetic discovery was in the field of pharmacogenetics and the potential ability, a priori, to predict risk of adverse events and likelihood of response to therapy thereby avoiding unnecessary toxicity in patients not likely to respond to treatment. These expectations have, to date, unfortunately not been realized. There have been a number of studies that have identified polymorphisms associated with response/nonresponse to agents such as anti-TNF therapies, but the majority of these have been performed in relatively small and retrospective cohorts. One study in CD identified clinical factors together with an apoptotic pharmacogenetic index (API) that could assign patients into a low, medium, and high response to first infusion of CD although the authors concede that further studies were warranted [19]. UC individuals homozygous for IBD-associated *IL23R* variants were more likely to respond to infliximab than individuals homozygous for IBD risk-decreasing variants (74.1 % vs. 34.6 %; $p=0.001$) [20]. The Leuven group utilized preclinical colonic mucosal gene expression and tested for association in UC infliximab responders and nonresponders [21]. Five genes involved in adaptive immune responses were the most differentially regulated (osteoprotegerin, stanniocalcin-1, prostaglandin-endoperoxide synthase 2, interleukin 13 receptor alpha, and interleukin 11), and a combination of these expression profiles "predicted" nonresponse to infliximab with a 95 and 96 % sensitivity. While these studies suggest that genomic modalities may be useful in discriminating subsets of disease and individuals that differentially respond to interventions, there have not been validated studies in prospective cohorts facilitating the transfer of these "tests" to the clinic. This is in part due to the pharmaceutical industries wariness of including these translational type analyses in prospective clinical trials, arguably the best vehicle for investigating these issues. Recent philosophical and policy changes from the FDA and other regulatory agencies have encouraged the pharmaceutical industry to incorporate parallel translational studies. This encouragement, together with the realization that the presence of biomarkers that help distinguish populations likely to respond to their particular agent may give them a competitive edge on their competitors (in addition to being good for patients!), is likely to significantly advance the field.

Another strategy of potential therapeutic utility is through an understanding of which pathways are important in driving any given individuals disease. Given the clinical and genetic heterogeneity of the IBDs, it will come as no surprise that no one therapy is effective in all patients. An understanding of the underlying pathogenesis of disease in a given individual may highlight therapeutic approaches that are optimal for that patient. A possible example of this involves the *IL10* pathway. *IL10* variation is associated with both CD and UC and these associations compliment the animal model findings such as the *il10* knockout mice that are more sensitive to colitis [1]. IL10, being anti-inflammatory cytokine, is a good therapeutic target for IBD although results to date have been disappointing [22, 23]. However, it may be that this strategy may only show benefit in individuals where the IL10 pathway appears to be important in disease pathogenesis. An extreme example of this approach and its possible efficacy has been shown to be of benefit in an infant with severe IBD caused by an *IL10R2* variant who responded well to bone marrow transplantation after failing conventional medical strategies [24]. There are a number targetable pathways identified by genetic associations with IBD that may benefit from this approach and now that the pharmaceutical industry is overcoming its reluctance to perform these studies, real advances may be imminent. In addition an understanding of these pathways may greatly reduce the number of individuals needed for clinical trials as informed clinical trial design with mechanism of action studies in targeted individuals may become a viable first alternative to large-scale randomized controlled trials. This would, likely, lead to a very significant reduction in both the time and cost of the development of urgently needed new therapies in the future.

Summary and Future Directions

The pace of discovery has been truly amazing in identifying susceptibility genes for IBD with over 160 independent loci now identified. Sub-phenotype analyses and an understanding of the genetic contribution to response to therapy and natural history are in their infancy, but researchers studying in this field would do well to heed the lessons of the benefits of large-scale international collaboration seen in the disease susceptibility studies. The spectacular success seen in identifying susceptibility

IBD loci is only the start of a process that will ultimately, hopefully, lead to a much better understanding of the underlying causes of these conditions. There are likely additional loci which may, in part, be identified from the identification of more homogenous groups based on clinical and para-clinical or intermediate phenotypes such as disease extent, behavior, and response to therapy. Again, it is likely that large-scale and multicenter studies will be necessary in order to achieve the necessary power for these approaches. Other areas for development include the need for non-Caucasian-based studies including Hispanic, African ancestry populations and additional Asian studies in order to generate "IBD genomes" in these populations. Furthermore, other genetic modalities including the microbiome and epigenetic phenomena have, in preliminary studies, demonstrated the ability to distinguish subgroups of IBD and need to be developed further [5, 9, 25].

Clinicians can foresee the time when a newly diagnosed patient or suspected new onset IBD case comes to see their physician who, in conjunction with a history, physical exam, and radiological work up, will request an "IBD genome." This will help clarify the diagnosis (CD vs. UC) and identify which pathogenic subgroup or pathways are important in the patient as well as revealing information about prognosis, natural history, development of complications, and response to therapy. Clinicians may well have access to additional modalities including microbiome and gene expression profiles that, in addition to the other parameters, are combined into models that can be adapted and easily communicated to both clinician and patient. The recent study "Personal omics profiling reveals dynamic molecular and medical phenotypes" [26] demonstrated the utility of repeated multi-omic analyses over 1 year in a single individual. The development of cheaper, near-patient, and validated multi-omic markers that monitor disease activity would be a further enhancement of the molecular monitoring of these debilitating diseases, perhaps allowing clinicians to anticipate flares of disease, for example.

The great hope, and expectation, is that the huge investment that has generated these genetic advances will lead to significant enhancement of clinical practice—there is a robust platform from which to start, but the major challenges for bridging the gap from the "bench to the bed" are yet to be overcome and this remains the major challenge for researchers in the future.

References

1. Jostins L, Ripke S, Weersma RK, Duerr RH, McGovern DP, Hui KY et al (2012) Host-microbe interactions have shaped the genetic architecture of inflammatory bowel disease. Nature 491(7422):119–124, Epub 2012/11/07
2. Franke A, McGovern DP, Barrett JC, Wang K, Radford-Smith GL, Ahmad T et al (2010) Genome-wide meta-analysis increases to 71 the number of confirmed Crohn's disease susceptibility loci. Nat Genet 42(12):1118–1125, Epub 2010/11/26
3. Rivas MA, Beaudoin M, Gardet A, Stevens C, Sharma Y, Zhang CK et al (2011) Deep resequencing of GWAS loci identifies independent rare variants associated with inflammatory bowel disease. Nat Genet 43(11):1066–1073, Epub 2011/10/11

4. Nimmo ER, Prendergast JG, Aldhous MC, Kennedy NA, Henderson P, Drummond HE et al (2012) Genome-wide methylation profiling in Crohn's disease identifies altered epigenetic regulation of key host defense mechanisms including the Th17 pathway. Inflamm Bowel Dis 18(5):889–899, Epub 2011/10/25
5. Gonsky R, Deem RL, Landers CJ, Derkowski CA, Berel D, McGovern DP et al (2011) Distinct IFNG methylation in a subset of ulcerative colitis patients based on reactivity to microbial antigens. Inflamm Bowel Dis 17(1):171–178, Epub 2010/09/18
6. Harris RA, Nagy-Szakal D, Pedersen N, Opekun A, Bronsky J, Munkholm P et al (2012) Genome-wide peripheral blood leukocyte DNA methylation microarrays identified a single association with inflammatory bowel diseases. Inflamm Bowel Dis 18(12):2334–2341, Epub 2012/04/03
7. Rausch P, Rehman A, Kunzel S, Hasler R, Ott SJ, Schreiber S et al (2011) Colonic mucosa-associated microbiota is influenced by an interaction of Crohn disease and FUT2 (secretor) genotype. Proc Natl Acad Sci U S A 108(47):19030–19035, Epub 2011/11/10
8. Cadwell K, Liu JY, Brown SL, Miyoshi H, Loh J, Lennerz JK et al (2008) A key role for autophagy and the autophagy gene Atg16l1 in mouse and human intestinal Paneth cells. Nature 456(7219):259–263, Epub 2008/10/14
9. Iliev ID, Funari VA, Taylor KD, Nguyen Q, Reyes CN, Strom SP et al (2012) Interactions between commensal fungi and the C-type lectin receptor Dectin-1 influence colitis. Science 336(6086):1314–1317, Epub 2012/06/08
10. Green ED, Guyer MS (2011) Charting a course for genomic medicine from base pairs to bedside. Nature 470(7333):204–213, Epub 2011/02/11
11. Ng PC, Murray SS, Levy S, Venter JC (2009) An agenda for personalized medicine. Nature 461(7265):724–726, Epub 2009/10/09
12. Ahmad T, Armuzzi A, Neville M, Bunce M, Ling KL, Welsh KI et al (2003) The contribution of human leucocyte antigen complex genes to disease phenotype in ulcerative colitis. Tissue Antigens 62(6):527–535, Epub 2003/11/18
13. Ho GT, Soranzo N, Nimmo ER, Tenesa A, Goldstein DB, Satsangi J (2006) ABCB1/MDR1 gene determines susceptibility and phenotype in ulcerative colitis: discrimination of critical variants using a gene-wide haplotype tagging approach. Hum Mol Genet 15(5):797–805, Epub 2006/01/26
14. Haritunians T, Taylor KD, Targan SR, Dubinsky M, Ippoliti A, Kwon S et al (2010) Genetic predictors of medically refractory ulcerative colitis. Inflamm Bowel Dis 16(11):1830–1840, Epub 2010/09/18
15. Economou M, Trikalinos TA, Loizou KT, Tsianos EV, Ioannidis JP (2004) Differential effects of NOD2 variants on Crohn's disease risk and phenotype in diverse populations: a metaanalysis. Am J Gastroenterol 99(12):2393–2404, Epub 2004/12/02
16. Weersma RK, Stokkers PC, van Bodegraven AA, van Hogezand RA, Verspaget HW, de Jong DJ et al (2009) Molecular prediction of disease risk and severity in a large Dutch Crohn's disease cohort. Gut 58(3):388–395, Epub 2008/10/01
17. Henckaerts L, Van Steen K, Verstreken I, Cleynen I, Franke A, Schreiber S et al (2009) Genetic risk profiling and prediction of disease course in Crohn's disease patients. Clin Gastroenterol Hepatol 7(9):972–980.e2, Epub 2009/05/09
18. Lee JC, Lyons PA, McKinney EF, Sowerby JM, Carr EJ, Bredin F et al (2011) Gene expression profiling of CD8+ T cells predicts prognosis in patients with Crohn disease and ulcerative colitis. J Clin Invest 121(10):4170–4179, Epub 2011/09/29
19. Hlavaty T, Ferrante M, Henckaerts L, Pierik M, Rutgeerts P, Vermeire S (2007) Predictive model for the outcome of infliximab therapy in Crohn's disease based on apoptotic pharmacogenetic index and clinical predictors. Inflamm Bowel Dis 13(4):372–379, Epub 2007/01/09
20. Jurgens M, Laubender RP, Hartl F, Weidinger M, Seiderer J, Wagner J et al (2010) Disease activity, ANCA, and IL23R genotype status determine early response to infliximab in patients with ulcerative colitis. Am J Gastroenterol 105(8):1811–1819, Epub 2010/03/04

21. Arijs I, Li K, Toedter G, Quintens R, Van Lommel L, Van Steen K et al (2009) Mucosal gene signatures to predict response to infliximab in patients with ulcerative colitis. Gut 58(12): 1612–1619, Epub 2009/08/25
22. Fedorak RN, Gangl A, Elson CO, Rutgeerts P, Schreiber S, Wild G et al (2000) Recombinant human interleukin 10 in the treatment of patients with mild to moderately active Crohn's disease. The Interleukin 10 Inflammatory Bowel Disease Cooperative Study Group. Gastroenterology 119(6):1473–1482
23. Schreiber S, Fedorak RN, Nielsen OH, Wild G, Williams CN, Nikolaus S et al (2000) Safety and efficacy of recombinant human interleukin 10 in chronic active Crohn's disease. Crohn's Disease IL-10 Cooperative Study Group. Gastroenterology 119(6):1461–1472
24. Glocker EO, Kotlarz D, Boztug K, Gertz EM, Schaffer AA, Noyan F et al (2009) Inflammatory bowel disease and mutations affecting the interleukin-10 receptor. N Engl J Med 361(21): 2033–2045, Epub 2009/11/06
25. Michail S, Durbin M, Turner D, Griffiths AM, Mack DR, Hyams J et al (2012) Alterations in the gut microbiome of children with severe ulcerative colitis. Inflamm Bowel Dis 18(10): 1799–1808, Epub 2011/12/16
26. Chen R, Mias GI, Li-Pook-Than J, Jiang L, Lam HY, Miriami E et al (2012) Personal omics profiling reveals dynamic molecular and medical phenotypes. Cell 148(6):1293–1307, Epub 2012/03/20

Index

M. D'Amato and J.D. Rioux (eds.), *Molecular Genetics of Inflammatory Bowel Disease*, DOI 10.1007/978-1-4614-8256-7,

MIX
Papier aus verantwortungsvollen Quellen
Paper from responsible sources
FSC® C105338

If you have any concerns about our products,
you can contact us on
ProductSafety@springernature.com

In case Publisher is established outside the EU,
the EU authorized representative is:
Springer Nature Customer Service Center GmbH
Europaplatz 3, 69115 Heidelberg, Germany

Printed by Libri Plureos GmbH
in Hamburg, Germany